Computer Math
Problem Solving
for Information Technology

Charles Marchant Reeder

Prentice
Hall

Upper Saddle River, New Jersey 07458

Library of Congress Cataloging-in-Publication Data

Reeder, Charles Marchant
    Computer math : problem solving for information technology / Charles Marchant Reeder.
      p. cm.
    Includes index.
    ISBN 0-13-061319-3
    1. Mathematics. 2. Computer science—Mathematics. I. Title.
QA39.3.R442002
510—dc21
                              2001034577

Editor-in-Chief: Stephen Helba
Executive Assistant: Nancy Kesterson
Executive Acquisitions Editor: Frank Mortimer, Jr.
Editorial Assistant: Barbara Rosenberg
Managing Editor: Mary Carnis
Production Management: WordCrafters Editorial Services, Inc.
Production Editor: Linda Zuk
Production Liaison: Adele M. Kupchik
Director of Manufacturing and Production: Bruce Johnson
Manufacturing Buyer: Cathleen Petersen
Creative Director: Cheryl Asherman
Senior Design Coordinator: Miguel Ortiz
Formatting: Clarinda Co.
Electronic Art Creation: Creative Communications for the 21st Century
Marketing Manager: Tim Peyton
Marketing Assistant: Melissa Orsborn
Composition: Clarinda Co.
Printer/Binder: Banta, Harrisonburg, VA
Copyeditor: Patsy Fortney
Proofreader: Maria McColligan
Interior Design: Linda Zuk
Cover Illustration: Quint Buchholz, *Albert Einstein*, PicturePress, Munich, Germany
Cover Printer: Banta, Harrisonburg, VA

Intel Pentium®, is a Registered trademark of the Intel Corporation.
Quotation on page 291 from Dorothy Fields, "Pick Yourself Up." Words and music by Jerome Kern, Dorothy Fields, © copyright 1936 Universal-Polygram International Publishing Inc., ASCAP. International copyright secured. All rights reserved.

Pearson Education LTD, *London*
Pearson Education Australia PTY, Limited, *Sydney*
Pearson Education Singapore, Pte. Ltd.
Pearson Education North Asia Ltd., *Hong Kong*
Pearson Education Canada, Ltd., *Toronto*
Pearson Educacion de Mexico, S.A. de C.V.
Pearson Education—Japan
Pearson Education Malaysia, Pte. Ltd.

10 9 8 7 6 5 4 3 2 1
ISBN 0-13-061319-3

In memory of my father, Charles Charretton Reeder,
who shared with me his love for problem solving.

Contents

Chapter 2 ■ Exponents 47

Chapter 3 ■ Number Systems 83

Chapter 4 ■ Unit Analysis   115

Chapter 5 ■ A Little Algebra   163

Chapter 6 ■ Graphing   191

Chapter 7 ■ Computer Programming Concepts   223

Chapter 8 ■ Computer Logic   261

Chapter 9 ■ Structured Program Design   291

Appendix A ■ Arithmetic Review   327

Preface

This book was written to fill a need for a quarter-long course presenting the quantitative and algorithmic tools appropriate for students pursuing a two-year degree in computer information systems. The book was designed for the typical CIS student at Edmonds Community College in Lynnwood, WA. The profile of this student is typical of information technology students enrolled at many other colleges.

The Typical Student

The typical student for whom this book was designed

- has only two years of funding available to become employable;

- has very fragile math skills upon entry, usually at the prealgebra level;

- cannot afford the extra year of study needed to follow traditional college mathematics;

- does not plan to transfer to a university;

- finds that traditional college math courses do not meet his or her professional needs;

- must take college-level math to meet degree guidelines; and

- must take math to gain entry into programming classes.

This book was written especially for a course designed to better serve this population of students who are currently working at or above the prealgebra level.

Topics Covered

Problem solving is central to the everyday needs of information technology professionals. Thus, the book begins with an easy-to-use problem-solving methodology. The remaining chapters present a variety of tools, both mathematical and algorithmic, that can be used to support that methodology.

The topics of mathematics that have been selected are the most essential to students of information technology. They include the following:

- Exponents and number systems

- Unit analysis

- Beginning algebra and graphing

Topics related to algorithms are included to prepare the student for a first course in computer programming. The final three chapters in the book introduce computer programming concepts and include the following topics:

- Introduction to a simple programming language

- Building and testing simple programs

- Computer logic

- Algorithm design tools (IPO charts, pseudocode, flowcharts, and structure charts)

- Modular design

Computer topics covered in this book are especially important for students preparing to study a so-called "visual" language such as Visual Basic. Because a significant portion of a course in such a language must focus on the visual interface, much of the coding is relegated to small, event-driven procedures. Therefore, coverage of traditional programming concepts are often cut short. Students who are introduced to traditional programming concepts in this book are at an advantage when continuing on to study a visual language.

Features of the Book

Each chapter begins with a section entitled "How to Use This Chapter," which puts the material about to be presented into the overall context of the course. This is followed by an introductory problem designed for students to tackle in small groups. Here students are allowed to "play" with some of the concepts of the chapter before they encounter all the conceptual details. Students should approach these preliminary problems with an open mind, even though they have not yet been "taught" how to work the problem. The "play" experience is designed to make learning the formal concepts easier.

Throughout each chapter, example problems and practice problems are presented. Wherever appropriate, these problems are related to information technology. In addition, sidebars are included showing how the current topic can be applied to computers. Practice problems are divided into three levels of difficulty, allowing problems to be assigned that are most appropriate to the specific skills of the students in the class. Each chapter concludes with a section in which the topics introduced in the previous sections of the chapter are applied to solving problems.

I hope this text will serve students and instructors alike by providing a course on problem solving designed especially for the needs of information technology students at the associate degree level. I welcome your comments.

Marc Reeder
Seattle, Washington
mreeder@qwest.net

Acknowledgments

It took the help of many people to develop an idea into a textbook. First I want to thank Judy Forth, who used my preliminary manuscript as a text for her course in Computer Math and Problem Solving at Edmonds Community College. Her feedback and that of her students was most valuable. The ability to field test the text greatly enriched it. Judy was also involved in proofreading the final manuscript. Her comments were essential in getting the kinks out of the manuscript and making this a much more effective text than it might have been.

Next, special thanks to all those who reviewed the preliminary manuscript. Their comments and suggestions were appreciated and have certainly made this a better text: Ron Davidson, Highline Community College; Hilda Halliday, Skagit Valley College; Robert Koehler, SUNY College-Buffalo; and Metha Schuler, Santa Rosa Junior College.

Next, I want to thank the people involved in the production of the book. Elizabeth Sugg was the editor at Prentice Hall who saw merit in my manuscript. She shared my vision that this unusual approach to teaching math and problem solving to students of information technology would work beyond the boundaries of my college. Without Elizabeth, the project would have never begun. Frank Mortimer was the editor who managed the project. Working with Frank was an absolute joy. His personal touch made the project fun. His professional insight made the project click. Linda Zuk was the production editor who made the manuscript into a textbook and amazed me as I saw the transformation take place. Patsy Fortney was the copyeditor who dotted my *i*'s and crossed my *t*'s and made sure everything was readable and consistent. These were just the people who I worked with directly. There were many others that I only heard about indirectly. Of these I would like to thank Miguel Ortiz, who designed the awesome cover; Adele Kupchik, the production liaison who kept everything moving; and Tim Peyton, the marketing manager who is even now helping to get the word out about the text.

Thanks to all the other friends, family, and colleagues who have contributed more specific comments and encouragement: Charles Ardary, David Chalif, Richard Davis, Jim Frances, Bill Friend, Sabrina Friend, Russ Herman, David Holdzkom, Roslyn Holdzkom, Julie Jackson, Jeremy Kostner, Emily Reeder, John Reeder, and Jerry Rosenberg.

Finally, and most importantly, to my wife Candace, who put food beside the keyboard and gave me the essential encouragement to see this project through, I couldn't have done it without you.

Tips for the Student

Math vs. Computer Notation

Because this book is a bridge between mathematics and information technology, students must learn to cope with the notations employed in each field. On the job you will see both types of notation, so get comfortable using both now. Before you begin this course, look over the summary of notation given in the **table in Appendix A, Section A-2,** for details.

Here are the most important examples, showing several equivalent ways to express the same thing:

Operation	*Math Notation*	*Computer Notation*
Division	$\dfrac{(2+3)}{4}$ or $(2+3) \div 4$ or $(2+3)/4$	$(2+3)/4$
Multiplication	$3xy$	$3 * x * y$
Parentheses	$5[2+(3+1)/2]$	$5 * (2 + (3+1)/2)$
Powers	3^2	$3 \char`\^ 2$ or $3 * 3$

Computer notation is used exclusively in the chapters on computers (Chapters 7–9). In the other chapters of the book, you will encounter primarily the math notation. But in some cases the computer notation is used there as well. So, it is best get comfortable using both notations. Be sure to consult the table in Appendix A, Section A-2, for details.

The Successful Student

This book is designed for the student whose math skills are a little rusty. If you are the typical student described in the Preface, it is likely that your student skills are also a little rusty. Here is a list of tips I always share with my students on the first day of class. If you follow them, your chances in succeeding in this course will be greatly improved.

1. **Attend every class.** Students who miss class are most at risk of failure. Never miss a class unless it is a true emergency. If such an emergency does come up, get someone to take notes and collect handouts for you. Also, notify your instructor beforehand, just as you would notify your employer if you are going to miss a day of work.

2. **Prepare for class.** Always read the material *before* it is covered in class. Come to class ready to participate in discussions and to ask questions about points you did not understand. Expect to spend at least two hours at home for each hour spent in class.

3. **Do all assignments.** Always turn in assignments on time. Sloppy work makes a poor impression, so present your work in a clean and easy-to-read format. Put your name at the top of every page. Staple loose pages together. Show the step-by-step solution of problems, never just the answer.

4. **Speak up in class.** Ask questions and participate in class discussions. There are no "dumb" questions. Often when you don't understand something, there are other students who also don't understand. However, don't dominate the class time.

5. **Talk to the instructor.** Either before or after class, chat with the instructor. Visit the instructor during office hours. Asking questions lets your instructor get to know you.

6. **Be attentive.** Pay attention to what is happening in class. Never use class time to work on outside projects. Always appear to be focused and involved. Instructors always notice which students appear bored.

7. **Join a study group.** Discussing and checking assignments and studying for exams with a group of your fellow students is an excellent way to master the material. But never turn in someone else's work.

8. **Get help.** Extra help is usually available for those who need it. A visit to your instructor's office during office hours may be enough. In other cases a tutor might be necessary. Don't allow yourself to get too far behind before taking action.

Tips for the Instructor

Material to Skip

As this book evolved, it came to contain more material than could be covered in a quarter-long course of about 55 class hours. To correct this, some of the original material has been delegated to appendixes that, while no longer part of the main body of the course, may still be of interest to some students.

Even with the appendix material out of the mainstream of the course, instructors will find it a challenge to cover all of the material in Chapters 1 through 9. Each instructor should decide what material is best omitted from the course to fit the needs of the students in the time available.

Following are some topics that have been omitted from the course at Edmonds Community College without compromising the integrity of the course:

- Introductory problems from Chapter 2, Chapter 5, and Chapter 6

- Topics in the sections called "Problem Solving with . . .": Memory Addressing in Computers in Chapter 3 and Boolean Variables and Flags in Chapter 8

- Parts of Chapter 6 entitled Spreadsheet Charts and Solving Two Linear Equations.

Notation

At the start of the course, instructors should make it clear to their students that two different systems of notation are used in this text. For example, computer notation uses an asterisk for multiplication and a slash for division. Math notation uses adjacent symbols to indicate multiplication. For division and fractions, you will see either a slash or a fraction bar. The computer chapters use computer notation exclusively. However, the math chapters use both.

Help your students become comfortable switching between these two notations, as they will have to deal with both when on the job. **See Appendix A, Section A-2, for more details on notation.**

Problem Solutions

An instructor supplement includes full solutions to all of the introductory problems and all of the practice problems. A student supplement includes full solutions to the even-numbered problems whose answers appear at the end of each chapter.

Problem Solving

Minds are like parachutes. They only function when they are open.

—James Dewar

1-1 ■ Introduction

How to Use This Chapter

The first chapter in this book introduces a method for planning and carrying out the solution of problems. This method will prove useful in attacking a wide variety of problems ranging from mathematical story problems to real-life problems you will encounter on the job in an information technology career.

The method will be demonstrated using only the simple tools of basic reasoning and arithmetic. So, at the start the types of problems you can solve will be limited by a lack of more advanced tools. In the chapters that follow, many tools useful to the problem solver will be introduced. These new tools, both mathematical tools and tools associated with information technology, will expand your problem-solving horizons.

For now, it is essential that you fully grasp the problem-solving method. To do this, you must formally write down all aspects of planning the solution of a problem. You will be given a framework to help you do this. Although this is time consuming and awkward, it is an important step in learning to use the method. Once the steps involved in planning the solution of a problem are understood, this method can be applied informally by doing the bulk of the planning mentally and writing down only the essential parts of the solution.

Practice! Practice! Practice!

The story goes that a visitor stops a native New Yorker on a Manhattan street and asks, "How do you get to Carnegie Hall?" The New Yorker quickly responds, "Practice! Practice! Practice!"

No one will dispute that practice is an essential part of mastering any of the performing arts. Students, however, sometimes fail to grasp the importance of practice in mastering problem solving and its supporting mathematical tools. In fact, students who have failed to answer an exam question correctly will often exclaim, "But I understand perfectly everything we did in class."

In the computer field, while a program must be correctly designed and documented, design and documentation are not enough. The program must also run correctly on the computer. A single comma out of place among thousands of lines of instructions can cause a program to malfunction.

In music, mathematics, and computer programming, understanding the underlying theory is a first step. However, understanding the theory is not enough. To get the correct result, you must be able to apply that theory successfully. That requires practice.

 The best way to learn the techniques of problem solving is to solve a lot of problems.

Background

The information technology workplace is full of interesting and challenging problems. Professionals working in the field are expected to have a knack for approaching the solution to problems, both big and small, in a systematic manner and without trepidation. This textbook will help you develop the problem-solving skills that are useful and necessary in an information technology career.

To be a successful problem solver, you must have a variety of tools at your disposal. Like a master carpenter, you need your own toolbox. But instead of screwdrivers and chisels, the information technology toolbox contains tools such as algorithms, truth tables, flowcharts, and mathematics. (Increasingly, these tools are being automated in software packages.) Being able to use these tools is essential, but having dozens of tools at your disposal is of little use unless you also know how to select the best tool for the job at hand.

A variety of tools will be introduced in later chapters. Consider these tools a starter kit for your information technology toolbox. In courses that follow, additional tools will be made available to you. The more tools you have in your toolbox and the more skillful you become in using them, the more successful you will be at tackling problems of greater complexity and variety.

Before you can put your tools to work, like the carpenter, you need a plan. This chapter will develop a method for planning and carrying out the solution to a variety of problems.

Each chapter will begin with an introductory problem (see the section that follows). Before the details of a particular topic are presented, you will be given a chance to "play" with some of the upcoming issues without being encumbered by the concepts that will be presented later. This will make the material presented in the remainder of the chapter more interesting.

Chapter Objectives

In this chapter you will learn:

- How to use a formal method for planning the solution of problems

- The importance of having a variety of problem-solving tools at your disposal

- The importance of diagrams in solving problems

- The usefulness of breaking down a difficult problem into several easier problems

- The importance of checking your solution

Introductory Problem

This chapter's introductory problem requires a team of two persons.

String Handcuffs

- Each person needs to make a set of "handcuffs" from a piece of string about 3 feet long. Tie a slipknot at each end of the string. (If you don't know how, ask your instructor.)

- The first person puts on the handcuffs by placing a slipknot around each wrist and drawing it snug but not tight. Notice the handcuffs complete an unbroken "ring" that includes the shoulders and both arms.

- The next person puts on the handcuffs by putting one hand in a slipknot and passing the other slipknot through the partner's "ring" before putting before putting it on the remaining wrist. Again, pull both slipknots close over the wrist, but not tight.

Each team member is connected to the other by an unbroken "ring" formed by arm, body, arm, and string. The ring of one person is interlocked with that of the other. It's somewhat like those brass rings that magicians so mysteriously snap apart.

To solve the problem you must separate the two "rings" so you and your partner are no longer connected. Do this without breaking the string or removing any slipknot from around any wrist. Random trial and error is a good way to begin. But at some point it will be helpful to carefully think about your situation and try to examine things systematically. You may break the rules to gain insight. But for your final solution you must follow the rules.

From time to time, as you try to solve the problem, your instructor may give you hints.

After You Have Solved the Problem

Write a detailed set of instructions telling how to solve the problem. Then have another team carry out your instructions. Correct and refine your instructions if some parts are incorrect or unclear.

Answer the following questions:

- What insight let you discover the solution?

- What was slightly misleading about the wording of the problem?

After Your Written Solution Has Been Tested

At the end, your instructor will present you with a formal solution to compare with your own.

To be a successful problem solver, you have to have a plan, a way to go about finding a solution to the problem that confronts you. Many such plans can be found in the literature of problem solving. This section will present two problem-solving plans, one by Polya and another by Rubinstein.

Polya's Four Phases

George Polya was a professor of mathematics at Stanford University when his book *How to Solve It* was first published in 1945. The book is still in print today, a testament to its usefulness. While Polya's book was intended primarily for solving mathematical problems, many of the ideas he presented are transferable to all kinds of other problems.

Polya presents four phases involved in the solution of a problem:

1. Understand the problem.

2. Devise a plan to solve it.

3. Carry out the plan.

4. Look back at the solution.

The four phases seem very simple, almost obvious. As each of the phases is discussed, you will notice that they have a decidedly mathematical flavor. Don't let this deter you. You can learn a lot from Polya's phases. In the next section these phases will be recast into a problem-solving method that is more appropriate for information technology. You will find that both methods are very similar.

The following outline is reorganized from a chart in *How to Solve It,* by Polya (Princeton Press, 1973).

Understanding the Problem

First, you have to understand the problem.

What is the unknown?

What are the data?

What is the condition?

Is it possible to satisfy the condition?

Is the condition sufficient to determine the unknown, or is it insufficient, or redundant, or contradictory?

Draw a figure.

Introduce suitable notation.

Separate the various parts of the condition.

Can you write them down?

Devising a Plan

Second, find the condition between the data and the unknown. You may be obliged to consider auxiliary problems if an immediate connection cannot be found. You should eventually obtain a plan of the solution.

Have you seen it before?
Or have you seen the same problem in a slightly different form?

Do you know a related problem?
> Do you know a theorem that can be useful?

Look at the unknown!
> Try to think of a familiar problem having the same or a similar unknown.

If you have a related problem that has been solved before:
> Can you use it?
> Can you use its method?
> Can you use some auxiliary element in order to make its use possible?

Can you restate the problem?
> Can you restate it differently?

Go back to definitions.

If you cannot solve the proposed problem, first try to solve some related problem.
> Can you imagine a more accessible, related problem?
> A more general problem?
> An analogous problem?
> Can you solve a part of the problem?
> Keep only a part of the condition, drop the other part.
>> To what extent is the unknown now determined?
>> How can it vary?
> Can you derive something useful from the data?
> Can you think of other data appropriate to determine the unknown?
> Can you change the unknown or the data, or both if necessary, so the new unknown and the new data are nearer to each other?
> Did you use all the data?
> Did you use the whole condition?
> Have you taken into account all essential notions involved in the problem?

Carrying Out the Plan

Third, carry out the plan.

Carry out your plan of the solution.

Check each step.
> Can you see clearly that the step is correct?
> Can you prove that it is correct?

Looking Back

Fourth, examine the solution obtained.

Can you check the result?

Can you check the argument?

Can you derive the result differently?

Can you see it at a glance?

Can you use the result or the method for some other problem?

Rubinstein's Anatomy of a Problem

In his book *Tools for Thinking and Problem Solving* (Prentice Hall, 1986), Moshe Rubinstein presents an approach designed to work with a wide variety of problems including "real-life" problems. While Polya's approach was developed primarily for students of mathematics, Rubinstein's approach is much more general since it applies to professional problems, social problems, and political problems, as well as mathematical problems.

Rubinstein thinks that a problem exists in the mind of the problem solver when the following elements are present:

- A perceived present or initial state

- A perceived desired goal or end state

- Perceived obstacles that prevent bridging the gap between the present and goal state

The initial state represents what you are given (Polya's data and conditions). The goal state represents what you are trying to find (Polya's unknown). The gap between these two states must be bridged by a process (or processes) that transforms the initial state into the goal state (Polya's plan).

Notice that each of Rubinstein's three elements is prefaced by the word *perceived*. This means that the problem must be analyzed within the context of the problem solver. Problems are not solved in a vacuum. A problem in one situation may be a very different problem in another situation, or no problem at all.

Rubinstein's anatomy of a problem is presented in Figure 1.1.

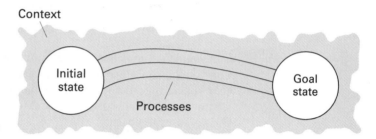

Figure 1.1

Practice Problems

For the following problems, merely identify the data, condition, and unknown (Polya's components). Every problem need not have a condition, but if it does, it will often be preceded by the word *if*. For now, there is no need to develop a plan or carry out the solution. You will have an opportunity to do that later.

► Example:

A computer and monitor together cost $900. If the monitor costs half the cost of the computer, what is the cost of the computer?

Data:	A computer and monitor together cost $900.
Condition:	The monitor costs half the cost of the computer.
Unknown:	The cost of the computer.

Introductory Problems

1-2.1 A computer-generated report includes inventory information for 200 items. The information for each item is given on a separate line. If 50 lines will fit on a page, how many pages of printer paper will it take to print the report?

1-2.2 A box contains 30 cookies. If each cookie weighs $\frac{1}{2}$ ounce, what is the weight of the cookies?

1-2.3 A box contains 30 cookies. If the empty box weighs 2 ounces and each cookie weighs $\frac{1}{2}$ ounce, what is the weight of the box full of cookies?

Intermediate Problems

1-2.4 A box of cookies contained 30 cookies when purchased. The empty box weighs 2 ounces, and each cookie weighs $\frac{1}{2}$ ounce. If one third of the cookies have been eaten, what is the combined weight of the box and the remaining cookies?

1-2.5   A computer-generated report includes inventory information for 220 items. The information for each item is given on a separate line. If 50 lines will fit on one page, how many pages will it take to print the report?

More Difficult Problems

1-2.6   A report generated by a computer includes information for 1000 customers. There are two types of customers, regular and preferred. The information for each regular customer takes one line on the report. The information for each preferred customer takes two lines on the report. Each page can contain up to 50 lines. If the number of regular customers is three times the number of preferred customers, how many pages will the report require?

1-2.7   Students in the CIS department can receive internship credit for work experience in a position related to information technology. At the beginning of the term, each student with an approved internship position enrolls for a specified number of credits. At the end of the term each student submits a time sheet showing the number of hours worked. Credit is awarded at the rate of one credit for each 30 hours worked. Credits are rounded to the nearest half credit. (For example, values in the range 1.75 and up to but not including 2.25 will round to 2.0. Values in the range 2.25 and up to but not including 2.75 will round to 2.5.) The credits awarded can never exceed the credits enrolled, regardless of the number of hours worked. Compute the number of credits awarded to each student from the following enrollment data:

Name	Credits Enrolled	Hours Worked
Adams, Jane	1	30
Beck, Carl	2	15
Jones, Michael	2	90
Smith, Mary	2	50
Walker, William	5	75

Brainstorming Problems

Try solving these problems with careful reasoning. Your instructor has the solutions.

1-2.8   A bear goes for a walk. He walks 3 miles south, then 3 miles east, and finally 3 miles north. At the end of this walk the bear has arrived back where he started. What color is the bear? white

1-2.9   You are driving a bus on the Main Street line. At the start of the run down Main Street the bus is empty. At First Avenue three passengers get on. At Second Avenue two more passengers get on. At Fifth Avenue four passengers get on and two get off. At Seventh Avenue one passenger gets off and two get on. How old is the bus driver?

1-2.10   A 10-pound solid iron ball is dropped from 5 feet above the surface of a lake that is 50 feet deep. The time for the ball to disappear is measured with a stopwatch. This measurement is made under two different conditions as follows:

Condition 1:   It is Noon on a bright, sunny day with the wind from the north at 4 miles per hour. The temperature of the lake at the surface is 40 degrees Fahrenheit. The lake is crystal clear with visibility downward 25 feet.

Condition 2:   It is 2:00 PM on a cloudy day with the wind from the south at 10 miles per hour. The temperature of the lake at the surface is 20 degrees Fahrenheit. Visibility downward is less than a few inches below the surface.

Under which condition will the ball take the longest time to disappear? Explain.

1-2.11   Three lamps (labeled 1, 2, and 3) are sitting on a table in a room. In an adjacent room out of sight of the lamps are three switches (labeled A, B, and C). Each switch controls one of the lamps. The wiring is out of sight and there is no way to see which switch is connected to which lamp. What is the fewest number of trips from the switch room to the lamp room that will allow you to match each switch with its lamp?

1-2.12 Four men want to cross a bridge at night so they can proceed on to the town on the other side. All begin on the side opposite the town. They need a flashlight to cross the bridge, and they have only one flashlight among them. Also, a maximum of only two men can cross the bridge at one time, and whoever crosses must share the flashlight. They can cross two at a time but someone must return the flashlight to the starting side so the next two can cross. Each man walks at a different speed and any pair crossing at the same time must walk at the slower man's pace. Here are their walking rates:

> Man #1 takes 1 minute to cross the bridge.
>
> Man #2 takes 2 minutes to cross the bridge.
>
> Man #3 takes 5 minutes to cross the bridge.
>
> Man #4 takes 10 minutes to cross the bridge.

Can you find a strategy that will get all four men together on the other side of the bridge (under the above constraints) in only 17 minutes?

1-2.13 Ten stacks of gold coins have ten coins each. Nine of the stacks have coins made of pure gold, and one stack has coins made of an alloy that looks exactly like gold. Each pure gold coin weighs 10 grams and each alloy coin weighs 9 grams. You have an electronic scale that can display the weight of any pile of coins with 99.99 percent accuracy. What is the fewest number of weighings needed to determine which stack contains the alloy coins?

Developing the Knack

Some people have a knack for solving technical problems. When asked how they do it, they often reply, "I don't know. I just 'see' the way to the solution." The rest of us have to work hard at learning to solve problems. At first the work is frustrating because mistakes in reasoning and calculation are frequent. To develop a knack for problem solving, you will need the following two things:

- A step-by-step method of attacking the problem
- The determination to overcome mistakes and to push on to a solution

This text will provide the method. You must provide your own determination.

Use It or Lose It

 A bumper sticker seen on a car: "Reality: Use It or Lose It."

Each of us has a conceptual comfort zone, a set of concepts with which we are comfortable. As students of information technology, you are moving through a period in which you are continually expected to expand your comfort zone to embrace a variety of new ideas and new ways of thinking. You must learn to solve unfamiliar problems. But unlike many other fields of study, this field is evolving so rapidly that the techniques and tools you are now learning will soon be outmoded. New techniques and tools will replace them. Lifelong learning is the norm. You must either learn to deal with an ever-expanding comfort zone, or find yourself replaced by someone else who *can* deal with it.

The tools you use will change very rapidly, especially the software tools. The underlying concepts change much more slowly. Developing excellent problem-solving skills is good insurance against obsolescence. Your bumper sticker might read: "Problem Solving: Use It or Lose It."

Choosing an Approach

Some problems are more easily solved using a graphic solution, while others yield to algebra. Computer programming problems involve algorithms. Deciding which approach to employ involves selecting the appropriate tool from the many at your disposal. Once you have selected a tool, you must then decide the best way to use it for the given situation.

Applying the old adage, "Always use the best tool for the job," requires experience. Having seen a tool employed to solve a similar problem will help you decide how to employ that tool in the current situation. However, don't be too rigid. Seeing a new application for an old tool is one of the most exciting aspects of problem solving.

Avoiding Conceptual Boxes

Perhaps the most difficult blocking issue to be overcome is that of conceptual boxes. Whether a novice or a professional problem solver, we are all boxed in by our prior experience. It's ironic that prior experience can both help and hinder problem solving. On the one hand, an essential part of developing and extending one's problem-solving abilities is practice. On the other hand, concepts once learned are hard to abandon to new ways of thinking. It's hard to teach old dogs new tricks.

Here is a classic puzzle that demonstrates the issue:

In the following figure, draw exactly four straight lines to connect all nine dots. Once started, you may not lift your pencil from the paper until finished.

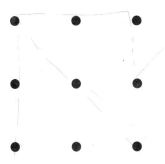

The usual approach, by trial and error, involves drawing all manner of lines within the square formed by the dots. But all of these trials will fail to solve the problem. It is only by breaking out of the conceptual box that a solution will be found. In this case the conceptual box is also the physical box that contains the dots.

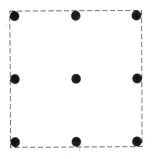

Try your hand at solving the puzzle. The solution is on the next page.

The solution is easily discovered once you start looking for it outside the box.

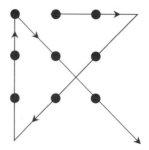

Such conceptual boxes are all around us and are invisible until someone discovers a breakthrough. For the beginning problem solver this presents an especially great challenge. Numerous personal breakthroughs are required along the path of learning. There is no easy solution. Unless you are willing to wrestle with the problem in the first place, no breakthrough will be forthcoming. There is no substitute for a rigorous workout in the mental gymnasium.

Problem-Solving Analogy

Here is an analogy that demonstrates the problem of overcoming conceptual boxes. This problem-solving situation, while greatly oversimplified, shows you how determination to look beyond the obvious is needed to solve many problems.

The Situation
Your problem-solving teacher holds up a manila folder and presents the class with the following challenge:

> Inside this folder is a piece of $8\frac{1}{2}$-×-11-inch paper. I challenge you to figure out the color of this piece of paper.

The teacher lays the folder down on the desk and walks out of the room.

Problem-Solving Approaches
Each student wrestles with the problem. Here are a few examples of how different students approach the problem.

> FRED: "There is no way I can know the color of the piece of paper. This problem is unsolvable. I give up."

> SALLY: "The most common color for $8\frac{1}{2}$-×-11-inch paper is white. Therefore, it is most likely white. But it might be some other color. The teacher occasionally passes out assignments on yellow paper and uses pink paper for quizzes. Therefore, it might be either yellow or pink. But, $8\frac{1}{2}$-×-11-inch paper comes in many other colors. How can I tell which paper is in the folder? I'm stuck."

> JANE: "There is no way to know for sure what color the piece of paper is unless I look inside the folder. The teacher didn't forbid us to look, so I will go and see for myself."

So, as Fred and Sally (and the rest of the class) sit in a muddle of confusion, Jane walks up to the teacher's desk and opens the folder to see a single piece of white paper. Jane has solved the problem.

Analyzing the Solution
While the solution may seem obvious and the problem trivial, it is a perfect analogy of the way many students approach problem solving. If no solution presents itself immediately, many students give up.

Jane realized that reasoning alone couldn't solve the problem. It was necessary to look inside the folder. The rest of the class incorrectly assumed that opening the folder was prohibited and gave up too quickly. The solution was only a few feet away, but they weren't in a position to see it.

To be a successful problem solver, you must be determined to pursue a problem to its solution. If one approach doesn't work, look for another. Don't allow yourself to be boxed into an area from which it is impossible to "see" the solution. The paper problem was solved only when a student moved to a position where the solution was in view. In this analogy the viewpoint (looking down at the open folder) was physical. In most other problems the viewpoint is conceptual. You must put yourself into a mental position to "see" the solution.

> **!** The moral of this story is, don't give up too quickly. Keep shifting your mental viewpoint until the solution comes out of the shadows and into view.

Take a Walk

There is a Latin saying: "Solvatur ambulando," which translates as "It is solved by walking."

Timing is everything when solving difficult problems. Your mind has to be ready to make the leap toward the solution. If you are tired and unfocused, it's time to take a break. Often the best way to relax and get your mind back on track is to take a walk.

There are many variations on this. They include taking a shower, taking a bath, driving, running, or getting some other exercise. Often, in the relaxed time before you fall asleep, a key piece of the solution of a problem will pop into your mind. Have a notepad by your bed and write your thoughts down clearly so you can use them in the morning.

IPO stands for Input-Process-Output. It is a traditional way of examining and understanding processes that occur throughout the field of information technology. To better understand a process, you must first clearly identify the input that feeds into it and the output that comes out of it. When you understand the input and output, you can determine the step-by-step process needed to convert the input into output.

A specific use of IPO is the IPO chart, a technique used in the design of computer programs. (See the sidebar on page 13.) This section will expand the IPO chart into a *method* for general problem solving. Hopefully, it will become the basis for helping you develop a knack for solving problems.

The IPO method identifies the important steps in the solution of any problem. First, we will rearrange input, process, and output into an order that is convenient to the problem solver. Since the output depends on the input, the output should be identified first. The appropriate input can only be identified after you know the output. Identifying the process needed to convert the input into the output comes next. So, the first three sections of the IPO method identify the output, the input, and the process, in that order.

Next, we need to add a section that applies the process to derive the output from the input. We will call this the solution section. Finally, we need a section to check that the solution is correct. This gives the following five steps in the solution of any problem:

1. Output: The result you are trying to find

2. Input: The information that you are given

3. Process: The information and steps needed to convert input into output

4. Solution: Applying the process to give the desired output

5. Check: Confirming that the solution is correct

The process section of the IPO method is the tricky part. Make the process simpler by breaking it into several smaller steps. Do this by asking the following questions:

What **notation** will be useful in the solution of the problem?
   Define the terms and notation to be used.

What **additional information** is needed?
   Is it known?
   Can you look it up in a reference?
   Can it be calculated?
   What equations are needed?
   In the equations, what is known and what is unknown?
   Are the physical units consistent throughout?
   Does the notation match that given in the Notation section?

Would a **diagram, table, or chart** be useful?
   Use a diagram, table, or chart to organize information more concisely.
   Diagrams range from simple line figures to elaborate drawings.
   Tables and charts are a good way to organize data.

What is the best **approach** to the solution of the problem?
   Many problems are extensions of other problems you have previously seen.
   Can other problems you have previously solved be applied to this problem?
   Outline the steps needed to convert the input into the output.

This section will present an approach to the solution of problems called the IPO method. This method has been adapted from a technique for designing computer programs called the IPO chart. We will examine the IPO chart in detail in Chapter 9. For now it may be useful to be introduced to the IPO chart so you can better appreciate the origins of the IPO method.

The IPO chart is a simple chart used to identify the three most basic parts of any computer program: input, process, and output. Ordered from the point of view of the computer:

- The program receives some input.
- The input is processed.
- The result is presented as output.

However, from the point of view of the program designer, things need to be rearranged:

- First, the designer must define the desired **output.**
- Next, the designer must determine what **input** is needed to generate that output.
- Finally, the designer must determine the **process** needed to convert input into output.

Check the result:

After input, process, and output have all been specified, it is necessary to confirm that the processing does in fact convert the given input into the desired output.

Here is a sample IPO chart:

Problem: Enter two numbers from the keyboard and display their sum on the screen.

Input	Process	Output
number1 number2	Input the two numbers from the keyboard Compute sum = number1 + number2 Display sum on screen	sum

Check using sample data:

	Value
Enter the value 5 from the keyboard and store it as number1.	5
Enter the value 7 from the keyboard and store it as number2.	7
Add the value stored as number1 to the value stored as number2.	12
Store the result of the addition as sum.	12
Display the value stored as sum on the screen.	12

The IPO Worksheet

By putting all of this together, we arrive at the following worksheet that will prove useful in attacking many types of problems.

Problem Statement	State the problem in clearly written sentences. Before you start the solution, you must understand the problem.	
Output	State what you are trying to find.	
Input	State what information you are given.	
Process	Notation (if needed)	Define the "shorthand" notation that will be used to represent various quantities in your solution.
	Additional Information	Identify what information is lacking in the input statement that is essential to obtaining the solution, for example, equations, constant values, conversion factors, etc.
	Diagram (if needed)	Use a graphic presentation to restate the problem, for example, diagrams, figures, charts, graphs, etc., that clarify the problem.
	Approach	List the specific steps needed to convert the input into the output. If a step is not clear, break it down into several substeps that are clear. Continue doing this until all steps are completely understood. • If equations are needed, write them out here using the notation above. • In this section you only say *how* you are going to solve the problem. It is not until the following section (solution) that you actually carry out this approach to arrive at the solution.
Solution	Carry out the steps in the approach section to yield the solution. All values must be accompanied by their appropriate units. (Only at this point will you do any calculations.)	
Check	Before you can have confidence in your results, the solution must be checked. When possible, use an alternative solution.	

 The IPO worksheet includes all the essential information you need to solve a problem. Although the IPO method will not tell you how to solve a problem, it will help you plan the solution.

Yes, it is tedious. But understanding each step is essential. Once you have mastered it, you can dispense with the formal worksheet and do the steps mentally.

Applying the IPO Method

The IPO method provides a framework for solving problems. It lays out the essential steps needed to solve a problem by forcing answers to the following questions:

What are you trying to find?

What facts have you been given?

What notation will you use in your solution?

What additional background information is needed?

Will a diagram or chart help clarify the problem?

What approach will you use to solve the problem?

How can the approach be applied to yield the desired solution?

Is the solution you have found correct?

Of course, this method will not automatically solve the problem for you. Knowing what approach to use comes only from the experience of solving many problems. But knowing the answers to these questions is an essential part of the solution of any problem.

The IPO worksheet provides the space to formally answer these questions. Different problems will require different amounts of space. Some problems will have sections with little or no information while other problems will require several pages for just one section. As you are learning to use the IPO method, make a worksheet like this one, carefully writing the necessary information for each section.

Yes, It Is Tedious!

The IPO worksheet is tedious! But it is essential to become familiar with *all* of the steps presented in the worksheet. At the beginning, applying the IPO worksheet formally is a helpful learning tool. However, once you have mastered it, you can apply the method more informally. For many problems, just working through the IPO worksheet in your mind will be enough.

 Once mastered, you can apply the IPO method as a mental checklist.

Learning by Example

On the pages that follow, the IPO method will be formally applied to solve an example problem. Each of the components of the IPO method will be discussed in detail. The information is presented on opposite pages. The left-hand page presents the complete solution using the IPO worksheet with one component shaded gray. The right-hand page presents the discussion about that component.

► Example:

A group of five friends are having dinner, and someone proposes a toast. Following the toast, glasses are clinked together, two at a time. What is the total number of clinks?

Problem Statement

Problem Statement	A group of five friends are having dinner, and someone proposes a toast. Following the toast, glasses are clinked together, two at a time. What is the total number of clinks?
Output	Total number of clinks among the group
Input	Group of five persons clink glasses two at a time

Process	Notation	In the diagram below: • Each person is represented by a dot. • Each clink is represented by a line that connects two dots.
	Additional Information	None needed
	Diagram	This problem is a natural for a graphic solution. Draw five dots equally spaced around a circle. Connect the dots to form a pentagon. Also, connect the dots to form a five-sided star.
	Approach	Interpret the diagram as follows: Each dot represents one person. Each connecting line represents one clink. Determine the number of clinks by counting the lines: Count the lines that form the pentagon. Count the lines that form the star. Add the two counts to get the total clinks.

Solution	There are 5 lines forming the pentagon. There are 5 lines forming the star. The total number of lines is 5 + 5 = 10. Since each line represents one clink, **total clinks = 10.**
Check	Use a similar problem to check your graphic solution: Form a group of five persons and have each person shake hands with every other person. Each of the five persons will shake hands four times. This gives a total of 20. But 20 is twice the number of unique handshakes, since each handshake is between two persons. Thus, the number of unique handshakes is 10. Since each handshake represents a clink, the result checks.

Problem Statement Section

The problem statement section of the IPO worksheet states the problem clearly and in full.

 On the opposite page, an IPO worksheet is presented for the example problem. Notice that the problem statement section is shaded in gray. On subsequent pairs of pages, each of the other sections of the IPO worksheet is explained. The example IPO worksheet is repeated for each section on the left side with a different section shaded in gray. The IPO method for that section is described on the right side.

Written Problems

If the problem is presented in written form, read it carefully. Then reread it, several times if necessary, until you completely understand what you need to find (output) and what information you have been given (input).

Verbal Problems

If the problem is presented in verbal form, write it down. Then follow the instructions for written problems.

Problems Not Fully Formulated

The hardest type of problem to solve is one that is not fully formulated. Before you can solve a problem, you must know what it is you are trying to find (output) and be sure that the given information is sufficient to derive that output.

Start by writing a draft of the problem statement. Attempt to solve the problem as drafted. If no solution is forthcoming, or if the solution is not what you wanted, perhaps the problem statement was not correctly formulated. In this case, revise your draft problem statement and start again.

Application to Information Technology

Problems that are not fully formulated are common in information technology. Before something can be fixed, you must understand what it is you are trying to fix.

Here is an example of this type of problem:

Produce a report to help managers improve production of a new line of equipment.

To solve this problem, many questions need to be answered. What will the report look like? What information should it contain? Who should receive the report? How should they use it to improve production?

The systems analyst is trained to solve this type of problem. Before starting on the solution, the analyst must ensure that the problem has been defined correctly so the most useful information can be included in the report.

Output

Problem Statement	A group of five friends are having dinner, and someone proposes a toast. Following the toast, glasses are clinked together, two at a time. What is the total number of clinks?	
Output	Total number of clinks among the group	
Input	Group of five persons clink glasses two at a time	
Process	Notation	In the diagram below: • Each person is represented by a dot. • Each clink is represented by a line that connects two dots.
	Additional Information	None needed
	Diagram	This problem is a natural for a graphic solution. Draw five dots equally spaced around a circle. Connect the dots to form a pentagon. Also, connect the dots to form a five-sided star.
	Approach	Interpret the diagram as follows: Each dot represents one person. Each connecting line represents one clink. Determine the number of clinks by counting the lines: Count the lines that form the pentagon. Count the lines that form the star. Add the two counts to get the total clinks.
Solution	There are 5 lines forming the pentagon. There are 5 lines forming the star. The total number of lines is $5 + 5 = 10$. Since each line represents one clink, **total clinks = 10.**	
Check	Use a similar problem to check your graphic solution: Form a group of five persons and have each person shake hands with every other person. Each of the five persons will shake hands four times. This gives a total of 20. But 20 is twice the number of unique handshakes, since each handshake is between two persons. Thus, the number of unique handshakes is 10. Since each handshake represents a clink, the result checks.	

Output Section

The output section of the IPO worksheet states the result you are trying to find. The actual result is given in the solution section. The output section just describes the result. In the example problem, the output section states that the number of clinks is what is to be found. Later, in the solution section, that number is found to be 10.

The problem statement is often a jumble of information. The first step in solving any problem is to extract the information about the problem output from the given problem statement and write it down. In problems such as the example, the output may be a quantity that can be expressed as a single value. In other problems, several values may be required. In general, the output includes many types of information presented in various forms. Here are a few examples of problem outputs in the information technology field:

- Printed reports

- Computer programs to produce reports

- Graphs or drawings

- Procedures or algorithms

- Computer with malfunctioning component to be identified and replaced.

In this text you will solve problems with a variety of outputs. You have already solved the string handcuffs problem by writing a procedure. However, for the most part, problems in the early chapters involving mathematical tools have output of a single value (or values). Later chapters, especially those involving algorithms, have more general outputs.

Physical Units

When output involves physical quantities such as distance or dollars, the result must be expressed in one of several appropriate units. Distance, for example, can be expressed in feet, yards, miles, meters, kilometers, and so on. Amounts of money may be expressed in U.S. dollars, Canadian dollars, European euros, British pounds, and so forth. Within one currency, for example, the U.S. dollar, money can be expressed as dollars or thousands of dollars or millions of dollars.

The units of a given quantity are just as important as the value of that quantity. Students often omit this essential information by giving only the value and not the units. The output to a problem is not complete without the proper units.

 Be sure to check your output to ensure that it includes the proper units.

Order of IPO in the Worksheet

IPO is in the order that the computer processes information. Input is provided to the computer before it is processed. Only after the processing is the output displayed. However, in general problem solving, this is not the most appropriate order. The processing needed to convert the input into the output can be defined only after the input and the output have been defined. In the IPO worksheet, output is defined first, then input, followed by process: OIP. This is a convenient order to approach the solution to a problem. The problem statement is usually not in this OIP order. So, care is needed to break apart the statement into the appropriate sections of the worksheet.

Input

Problem Statement	A group of five friends are having dinner, and someone proposes a toast. Following the toast, glasses are clinked together, two at a time. What is the total number of clinks?
Output	Total number of clinks among the group
Input	Group of five persons clink glasses two at a time

Process	Notation	In the diagram below: • Each person is represented by a dot. • Each clink is represented by a line that connects two dots.
	Additional Information	None needed
	Diagram	This problem is a natural for a graphic solution. Draw five dots equally spaced around a circle. Connect the dots to form a pentagon. Also, connect the dots to form a five-sided star.
	Approach	Interpret the diagram as follows: Each dot represents one person. Each connecting line represents one clink. Determine the number of clinks by counting the lines: Count the lines that form the pentagon. Count the lines that form the star. Add the two counts to get the total clinks.

Solution	There are 5 lines forming the pentagon. There are 5 lines forming the star. The total number of lines is 5 + 5 = 10. Since each line represents one clink, **total clinks = 10.**
Check	Use a similar problem to check your graphic solution: Form a group of five persons and have each person shake hands with every other person. Each of the five persons will shake hands four times. This gives a total of 20. But 20 is twice the number of unique handshakes, since each handshake is between two persons. Thus, the number of unique handshakes is 10. Since each handshake represents a clink, the result checks.

Input Section

The input section of the IPO worksheet identifies the information you are given to derive the desired output. This information must be extracted from the problem statement and written down. Problems are not always written in a concise way. In some cases unnecessary information may be included in the problem statement. You must sort through the problem statement and determine the essential information needed as input. This is not always clear in the beginning. You may need to revise the input section as the solution proceeds and things become clearer.

In other cases there may not be enough information. When this happens, additional information must be obtained before the solution can proceed. This additional information is covered in the process section that follows. In solving information technology problems, the analyst usually has to go back to the client many times, each time collecting additional information before the problem is solved.

On first reading, some problems may seem to be missing essential input. Here the context of the problem may offer a hint on how to proceed with the solution. If the problem is an exercise in a textbook, it is likely that a solution can be found using only the given input. Puzzle problems often appear to be missing necessary input, and the challenge is to obtain the needed information in an indirect manner. In other problems, especially real-world problems, it may be necessary to obtain the missing input before the solution can proceed.

Physical Units

Again, as in the output section, physical units of input quantities must be defined and included with their values. Chapter 4 has a section that covers the manipulation and conversion of physical units.

Process: Notation

Problem Statement	A group of five friends are having dinner, and someone proposes a toast. Following the toast, glasses are clinked together, two at a time. What is the total number of clinks?	
Output	Total number of clinks among the group	
Input	Group of five persons clink glasses two at a time	
Process	Notation	In the diagram below: • Each person is represented by a dot. • Each clink is represented by a line that connects two dots.
	Additional Information	None needed
	Diagram	This problem is a natural for a graphic solution. Draw five dots equally spaced around a circle. Connect the dots to form a pentagon. Also, connect the dots to form a five-sided star.
	Approach	Interpret the diagram as follows: Each dot represents one person. Each connecting line represents one clink. Determine the number of clinks by counting the lines: Count the lines that form the pentagon. Count the lines that form the star. Add the two counts to get the total clinks.
Solution	There are 5 lines forming the pentagon. There are 5 lines forming the star. The total number of lines is $5 + 5 = 10$. Since each line represents one clink, **total clinks = 10.**	
Check	Use a similar problem to check your graphic solution: Form a group of five persons and have each person shake hands with every other person. Each of the five persons will shake hands four times. This gives a total of 20. But 20 is twice the number of unique handshakes, since each handshake is between two persons. Thus, the number of unique handshakes is 10. Since each handshake represents a clink, the result checks.	

Process Section

The process section of the IPO worksheet defines how the input is to be converted into the output. It is the most challenging part of problem solving. To make this section clearer, we divide it into several subsections:

- Notation

- Additional information

- Diagram

- Approach

Notation

Problems almost always benefit from using a suitable notation to refer to data. This "shorthand" greatly simplifies the description of the solution plan and makes it less cumbersome to write and easier to read.

When a notation is introduced, be sure to state it in this section. For example, if you are trying to find the time to travel a given distance, it is traditional to let the letter t represent time in any equation. However, before you use the letter t in an equation, you must first state

Let t = time in hours

Although one-letter notation is conventional in mathematics, a meaningful word or short phrase is traditional notation in the computer field. So you might state

Let total earnings = sum of monthly earnings

Computer software may require limitations on the type of notation allowed. Most computer languages do not allow named quantities, called variables, to include spaces. There are several ways to avoid spaces. For example:

total_earnings = sum_of_monthly_earnings

TotalEarnings = SumOfMonthlyEarnings

It is always worth taking care when selecting a notation. It may take several drafts to come up with one that best fits the situation.

> **!** It is essential to use consistent notation throughout the various parts of the IPO worksheet.

The notation introduced in this section should match that used in the additional information section, the diagram section, the approach section, the solution section, and the check section.

In the example problem in the IPO worksheet on the opposite page the notation is graphic. Dots and lines are used to represent the input and output of the problem. However, in most problems the notation will be letters and words.

Process: Additional Information

Problem Statement	A group of five friends are having dinner, and someone proposes a toast. Following the toast, glasses are clinked together, two at a time. What is the total number of clinks?
Output	Total number of clinks among the group
Input	Group of five persons clink glasses two at a time

Process	Notation	In the diagram below: • Each person is represented by a dot. • Each clink is represented by a line that connects two dots.
	Additional Information	None needed
	Diagram	This problem is a natural for a graphic solution. Draw five dots equally spaced around a circle. Connect the dots to form a pentagon. Also, connect the dots to form a five-sided star.
	Approach	Interpret the diagram as follows: Each dot represents one person. Each connecting line represents one clink. Determine the number of clinks by counting the lines: Count the lines that form the pentagon. Count the lines that form the star. Add the two counts to get the total clinks.

Solution	There are 5 lines forming the pentagon. There are 5 lines forming the star. The total number of lines is $5 + 5 = 10$. Since each line represents one clink, **total clinks = 10.**
Check	Use a similar problem to check your graphic solution: Form a group of five persons and have each person shake hands with every other person. Each of the five persons will shake hands four times. This gives a total of 20. But 20 is twice the number of unique handshakes, since each handshake is between two persons. Thus, the number of unique handshakes is 10. Since each handshake represents a clink, the result checks.

Process Section (Continued)

Additional Information

Often some important piece of information is lacking in the problem statement. Without this information the solution is not possible. Perhaps this information is intentionally omitted to make the problem more challenging. It may be an equation or a constant value. For example, it may be the equation for the volume of a sphere, or the constant value of π (pi). It may be something you already know or something you have to look up. It may be something you need to calculate as an intermediate step in arriving at the solution. Use this section of the IPO worksheet to identify such additional information and write it down.

 Identifying the need for additional information is a major step in solving the problem.

Using Formulas

Although not needed for the example problem, you will often want to provide formulas as additional information. For example, the area of a rectangle is found by multiplying the height by the width. This relationship can be expressed using letters to represent the various quantities:

$$a = hw$$

This expression is the *formula* for computing the area of a rectangle. In the formula the letter a represents area, the letter h represents height, and the letter w represents width. The combination hw indicates that height is multiplied by width. Using the notation for multiplication employed in computers, the area formula could also be written:

Either as $\quad\quad a = h * w \quad\quad\quad$ (where * means multiply)

Or as $\quad\quad\quad a = \text{height} * \text{width}$

Should you want to use this formula in an IPO worksheet for some other problem, you must first fill in the notation and the additional information as follows:

Notation: $\quad\quad\quad\quad\quad a = \text{area}, h = \text{height}, w = \text{width}$

Additional Information: $\quad\quad a = hw$ (or perhaps $a = h * w$)

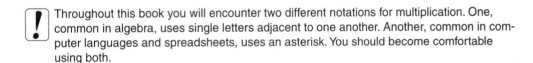 Throughout this book you will encounter two different notations for multiplication. One, common in algebra, uses single letters adjacent to one another. Another, common in computer languages and spreadsheets, uses an asterisk. You should become comfortable using both.

Image You may have to consult a reference book to find the appropriate formula for the problem you are trying to solve. (Many geometric formulas are given in Appendix C of this book.)

Process: Diagram

Problem Statement	A group of five friends are having dinner, and someone proposes a toast. Following the toast, glasses are clinked together, two at a time. What is the total number of clinks?	
Output	Total number of clinks among the group	
Input	Group of five persons clink glasses two at a time	
Process	Notation	In the diagram below: • Each person is represented by a dot. • Each clink is represented by a line that connects two dots.
	Additional Information	None needed
	Diagram	This problem is a natural for a graphic solution. Draw five dots equally spaced around a circle. Connect the dots to form a pentagon. Also, connect the dots to form a five-sided star.
	Approach	Interpret the diagram as follows: Each dot represents one person. Each connecting line represents one clink. Determine the number of clinks by counting the lines: Count the lines that form the pentagon. Count the lines that form the star. Add the two counts to get the total clinks.
Solution	There are 5 lines forming the pentagon. There are 5 lines forming the star. The total number of lines is $5 + 5 = 10$. Since each line represents one clink, **total clinks = 10.**	
Check	Use a similar problem to check your graphic solution: Form a group of five persons and have each person shake hands with every other person. Each of the five persons will shake hands four times. This gives a total of 20. But 20 is twice the number of unique handshakes, since each handshake is between two persons. Thus, the number of unique handshakes is 10. Since each handshake represents a clink, the result checks.	

Process Section (Continued)

Diagrams

The saying goes, "A picture is worth a thousand words." This is especially true in problem solving. Diagrams, charts, figures, graphs, tables, and other such techniques for displaying information in an essentially visual, nonverbal format are excellent problem-solving tools. They present a simple way to clarify a complex problem by displaying a large number of details in a manner that can be understood at a glance. Not all problems benefit from diagramming tools, but most do. It is always worth the time to examine a problem for ways to organize and best present its data.

Application to Information Technology

The field of information technology is rich in diagramming tools. We have already seen IPO charts. Later in this book, truth tables (a logical relationship tool) and flow charts (a programming design tool) will be introduced. In other information technology courses you will encounter tools such as data flow diagrams (a systems analysis tool) and entity relationship diagrams (a database design tool).

While some diagramming tools are used only in a special niche, other tools can be applied more generally. For problems involving equations, graphing is useful. When large amounts of data need to be presented, charts and tables work well. For many other problems, a simple line drawing is appropriate.

 Labels are an essential part of any diagram, chart, or graph.

Process: Approach

Problem Statement	A group of five friends are having dinner, and someone proposes a toast. Following the toast, glasses are clinked together, two at a time. What is the total number of clinks?	
Output	Total number of clinks among the group	
Input	Group of five persons clink glasses two at a time	
Process	Notation	In the diagram below: • Each person is represented by a dot. • Each clink is represented by a line that connects two dots.
	Additional Information	None needed
	Diagram	This problem is a natural for a graphic solution. Draw five dots equally spaced around a circle. Connect the dots to form a pentagon. Also, connect the dots to form a five-sided star.
	Approach	Interpret the diagram as follows: Each dot represents one person. Each connecting line represents one clink. Determine the number of clinks by counting the lines: Count the lines that form the pentagon. Count the lines that form the star. Add the two counts to get the total clinks.
Solution	There are 5 lines forming the pentagon. There are 5 lines forming the star. The total number of lines is $5 + 5 = 10$. Since each line represents one clink, **total clinks = 10.**	
Check	Use a similar problem to check your graphic solution: Form a group of five persons and have each person shake hands with every other person. Each of the five persons will shake hands four times. This gives a total of 20. But 20 is twice the number of unique handshakes, since each handshake is between two persons. Thus, the number of unique handshakes is 10. Since each handshake represents a clink, the result checks.	

Process Section (Continued)

Approach

The approach is the heart of the process section. Here the step-by-step procedure needed to convert the input into the output is given. Only later, in the solution section, will you carry out the steps in the approach section to actually arrive at a solution.

Depending on the problem, the approach may be simple or complex. In the example problem the approach is simple. The problem was solved using the diagram, and the approach just restates how this was done. But in most cases the approach is the most difficult part of the process. It is essential to be able to express the approach as a set of simple steps, in which each step is completely understood.

Most problems cannot be solved in a single step. Many relatively easy problems require at least two steps, with the first step providing information needed to perform the second step. In general, problems often require many steps to solve, and the approach section is where you lay out the plan for what steps are required and in what order they must be done.

The subdivision of big problems into several smaller problems is an idea that works for solving any type of problem. To do this you should first outline the major tasks. Then divide each major task into several smaller tasks. Continue dividing until all tasks are understood.

 Divide and conquer is the most powerful tool in problem solving. Use it with swagger!

Application to Information Technology

In computer programs, subroutines are used to divide a complex task into several smaller, and hopefully simpler, tasks. Subroutines that remain complex have to be further broken down into their own subroutines. Dividing the problem into smaller tasks continues until all the tasks are clearly defined.

Using Prior Experience

At each stage of the subdivision process look for problems that are familiar. Prior experience comes in handy at this stage. You will often find that part of a problem is similar to another problem you have solved before. The prior problem may not be identical, just similar to the current problem. The essential thing is to recognize the similarity and use it to your advantage.

 For each subprocess, recognizing the similarity between the current problem and a problem you have solved previously is the key to finding the appropriate approach.

Solution

Problem Statement	A group of five friends are having dinner, and someone proposes a toast. Following the toast, glasses are clinked together, two at a time. What is the total number of clinks?	
Output	Total number of clinks among the group	
Input	Group of five persons clink glasses two at a time	
Process	Notation	In the diagram below: • Each person is represented by a dot. • Each clink is represented by a line that connects two dots.
	Additional Information	None needed
	Diagram	This problem is a natural for a graphic solution. Draw five dots equally spaced around a circle. Connect the dots to form a pentagon. Also, connect the dots to form a five-sided star.
	Approach	Interpret the diagram as follows: Each dot represents one person. Each connecting line represents one clink. Determine the number of clinks by counting the lines: Count the lines that form the pentagon. Count the lines that form the star. Add the two counts to get the total clinks.
Solution	There are 5 lines forming the pentagon. There are 5 lines forming the star. The total number of lines is 5 + 5 = 10. Since each line represents one clink, **total clinks = 10.**	
Check	Use a similar problem to check your graphic solution: Form a group of five persons and have each person shake hands with every other person. Each of the five persons will shake hands four times. This gives a total of 20. But 20 is twice the number of unique handshakes, since each handshake is between two persons. Thus, the number of unique handshakes is 10. Since each handshake represents a clink, the result checks.	

Solution Section

In the solution section of the IPO worksheet, the steps outlined in the approach are carried out to yield the desired output in its final form. This output may be a value, a statement, a diagram, a computer program, a printed report, or whatever. For solutions that are mathematical, do the math. For solutions that use a graphical approach, describe how to interpret the graph or diagram. For solutions that are reasoned, carefully follow the reasoning steps to come to the appropriate conclusion.

Many students confuse the approach section with the solution section. In the approach section you describe the steps involved to arrive at a solution. Then in the solution section you actually carry out these steps to arrive at the solution. If formulas are part of the approach, wait to carry out the calculations in the solution section.

Check

Problem Statement	A group of five friends are having dinner, and someone proposes a toast. Following the toast, glasses are clinked together, two at a time. What is the total number of clinks?	
Output	Total number of clinks among the group	
Input	Group of five persons clink glasses two at a time	
Process	Notation	In the diagram below: • Each person is represented by a dot. • Each clink is represented by a line that connects two dots.
	Additional Information	None needed
	Diagram	This problem is a natural for a graphic solution. Draw five dots equally spaced around a circle. Connect the dots to form a pentagon. Also, connect the dots to form a five-sided star.
	Approach	Interpret the diagram as follows: Each dot represents one person. Each connecting line represents one clink. Determine the number of clinks by counting the lines: Count the lines that form the pentagon. Count the lines that form the star. Add the two counts to get the total clinks.
Solution	There are 5 lines forming the pentagon. There are 5 lines forming the star. The total number of lines is $5 + 5 = 10$. Since each line represents one clink, **total clinks = 10.**	
Check	Use a similar problem to check your graphic solution: Form a group of five persons and have each person shake hands with every other person. Each of the five persons will shake hands four times. This gives a total of 20. But 20 is twice the number of unique handshakes, since each handshake is between two persons. Thus, the number of unique handshakes is 10. Since each handshake represents a clink, the result checks.	

Check Section

The check section of the IPO worksheet verifies that the solution is correct.

Checking your solution is essential. Mistakes are prevalent for beginners and professionals alike. The main difference between the professional and the beginner is that the professional can usually detect and correct mistakes quickly and without intimidation. In the information technology field, especially in writing computer programs, mistakes are so frequent that they are not even called mistakes. Instead they are called bugs. (See the sidebar on page 34.)

When problems involve calculations, it is important to check both the approach and the calculation. Here are several methods ranging from *least* effective to *most* effective, for checking the solution to a problem:

1. Skip the check altogether.

    This is the least effective check of all. Alas, its use is all too common.

2. Look over the approach and the calculation.

    This is a little better, but it won't find most mistakes.

3. Verify the approach.

    Go over the approach carefully to confirm that it makes sense.
    Confirm that you are using only appropriate formulas.

4. Verify the reasonableness of the calculation.

    Is the answer in the ballpark you expected?
    Is the decimal point in the correct position?
    Are the units correct? (More about this in Chapter 4.)
    Redo the calculation with approximate values to simplify the calculation.
    For example: $0.679 * 3.01 = 2.04379$ is approximately $\frac{2}{3} * 3 = 2$.

5. Redo the calculation.

    Do the various steps of the calculation in a different order.
    This works best if you put the original solution aside and redo the calculation on a separate piece of paper. At the end, compare the final result of the check with the result of the original solution.

6. Redo the problem using a different approach.

    This is the best check of all. Just checking the calculation will not detect an error in reasoning, an improper approach, or the use of an inappropriate formula. It is always best to check with an alternative approach, if you can.

Notes:

- Be sure to keep the original solution and the check solution separate. You don't want mistakes in the original solution to influence the check.

- It is not enough to just describe *how* you will check your solution. You must carefully carry out the check in such a way as to maximize your confidence that your solution is correct. Too often students with invalid solutions go through the motions of a check only to falsely prove that their invalid solution is correct.

Systems Development Life Cycle

Now that you have been introduced to the IPO method, you may like to compare it to a classic methodology for designing software systems, the Systems Development Life Cycle or SDLC. The life cycle is used extensively to guide developers through the maze of an information technology project.

Five SDLC Phases	IPO Method Equivalents
Planning: Identify the problem.	Problem Statement
Analysis: Study the problem.	Input, Output, Notation, Additional Information, Diagram
Design: Define a detailed approach.	Approach
Implementation: Build, test, and deliver the new system.	Solution and Check
Operation and Maintenance: Use the system and keep it current.	Use this solution to solve a different problem.

Notice the similarity of the phases of the SDLC and the steps of the IPO method.

Computer Folklore

A bit of information technology folklore involves the origin of the word *bug*. It concerns Grace Hopper, one of the pioneers of modern computing and a key person involved in developing the COBOL computer language in the early 1950s.

The story goes that Grace Hopper was working the night shift in a computer facility and found a note in the log informing her that the computer was malfunctioning. On opening the computer cabinet, she discovered a large moth attached to one of the circuit boards. She removed the moth and the computer started working again. Her log entry for the morning shift stated:

I have **debugged** the computer and it is working again.

Whether or not this story is exactly true, the words **bug** and **debug** have become a standard part of the vocabulary used in the computer industry. *Bug* refers to the problem and *debug* refers to the process of fixing the problem. These terms are most often associated with instructions in computer programs. But they are also used to refer to a wider set of problems involving both computer hardware and software.

Tips on Using the IPO Method

Stepwise Refinement

It is not necessary to complete each section of the IPO method before moving on to the next. It is unlikely that the correct approach will emerge all at once. A cyclic approach is best:

- Draft an approach.

- See what solution this draft produces.

- Check the draft solution.

- Go back and refine the approach.

- Repeat until the problem is solved.

If the draft solution is not fully correct, which is the likely case, then the process section needs to be refined. Perhaps a new diagram is necessary, or it may become clear that some critical information is missing and must be obtained before you can proceed. Make corrections and refinements to the approach, or try a completely different approach. Then complete and check the solution again.

Hopefully, with each cycle of the process, you will get wiser and more informed about the best way to solve the problem. Sometimes this method of problem solving is called step-wise refinement.

Stepwise refinement is especially effective for problems that do not have a single solution. Life is full of problems that have many solutions, each with good points and bad points. In the field of information technology, the systems analyst has the job of sorting out many of these messy problems and finding solutions that give the most benefit for the least cost.

 Stepwise refinement is the best way to solve many complex problems.

Big and Little Problems

As a general rule, big problems require big solution plans, while small problems require small solution plans. For simple problems, the ones you know how to solve, you can skip much (if not all) of the solution planning. Obviously, you do not need to use the IPO method with diagrams and charts to determine that 1 plus 1 equals 2. Problems that involve trial and error may not benefit from the IPO method initially. However, if trial and error fails to produce a result, then the IPO method can be used to plan an alternative approach. Algebraic word problems usually benefit from the IPO method. If the problem is simple, however, then the plan for its solution will also be simple.

As problems become more complex, so must the details of the IPO method. The essential thing is to break the problem down into steps, each of which you understand. Any steps that are unclear need to be broken down into smaller steps that are clear. So, even though you did not know how to solve the problem when you started, by applying the IPO method, you can finally arrive at an approach in which each step is understood.

The solution of some problems is so complex, however, that even after the process is broken into many smaller subprocesses, some of those subprocesses remain complex. These problems may require several levels, each with their own IPO method outlining that part of the solution.

Tools for Problem Solving

The IPO method is a framework for applying a variety of tools to solve problems. We have already mentioned a few of these tools such as graphs, charts, diagrams, figures, and mathematics. In later chapters a variety of these tools will be introduced. Mathematical concepts from arithmetic, algebra, and geometry that are useful in the field of information technology will be covered. Topics such as unit analysis and number systems will be introduced. Finally, an introduction to algorithms and computer programming will be presented.

Once these tools become part of your problem-solving toolbox, the IPO method will become all the more useful for solving the problems found in information technology.

Inside Out

Where to start? It is nice when you are given a problem that can be solved 1-2-3. But often the solution does not present itself in such a simple, linear manner. Maybe the problem is more easily understood if you work backwards, 3-2-1. In other cases, if the middle part is the easiest, then work the middle part first, either 2-1-3 or 2-3-1.

Don't get locked into a conceptual box. Many problems are really quite simple once you have unraveled them. For example, some problems are craftily worded to mislead you and make them seem much harder to solve than they really need to be. Problems like this are called puzzles. Life and work are full of puzzles. So, learning to solve puzzles of all types is a life-expanding skill.

Extending Your Reach

The IPO method presents a way to organize your thoughts as you plan the solution to any problem. This method will help you solve problems that previously baffled you. Of course, the solution to many problems will remain out of reach until you develop the background and tools needed to solve them. The IPO method is intended to extend your reach by helping you see more clearly through the mental fog that makes problem solving so difficult.

The IPO method is flexible and adaptable to almost any new problem. Even as the technical world evolves, and the new replaces the old, this method will remain current, ready to assist you and empower you with greater understanding.

Making Mistakes

Making mistakes is an important part of problem solving. All successful problem solvers make mistakes. They learn from their mistakes and proceed. Don't be afraid to do the same.

Example Problems

Example Problem 1

Problem	A computer display has a rectangular viewing area 12 inches wide and 8 inches high. Express this area in square inches. A square inch is a unit of area equivalent to a square that is 1 inch on each side.	
Output	Viewing area expressed in square inches	
Input	Rectangle 12 inches wide and 8 inches high	
Process	Notation	Let h = height Let w = width Let a = area
	Additional Information	You may recall the formula for the area of a rectangle from a previous math course. If you don't remember, you could look it up in a reference book or figure it out for yourself. On graph paper draw a rectangle 2 units high and 3 units wide. Include horizontal and vertical grid lines one unit apart. Notice that there are 2 rows, each with 3 unit squares (1 unit on a side). Each of these unit squares has an area of 1 square unit. If a unit on the diagram represents an inch, then the area is in square inches. If it represent a meter, then the area is in square meters. Thus, the total number of unit squares is 2 * 3 and the area is 6 square units. Try drawing rectangles of other sizes and counting the number of unit squares. If you are adventuresome, try one with fractional dimensions. For example, try $2\frac{1}{2}$ by $3\frac{1}{2}$. Area of a rectangle is height times width: $a = h * w$
	Diagram	Rectangular viewing area: 8 in. 12 in.
	Approach	Use the formula for the area of a rectangle: $a = h * w$. Substitute 8 inches for h and 12 inches for w. Then, $a = 8 * 12$.
Solution	Carry out the calculation in the area formula: $a = 8\text{ (in.)} * 12\text{ (in.)} = 96\text{ (in.}^2)$ **Thus the viewing area is 96 square inches.**	

Check	Construct a grid in the diagram above by drawing:
	     7 horizontal lines and 11 vertical lines inside the box.
	 Count the number of small squares, each representing 1 square inch.
	     There should be 96 small squares.
	 Thus the viewing area = 96 square inches. The solution checks.	

Example Problem 2

Problem	Computer screens have nondisplay borders surrounding their viewing area. If a screen has a 2-inch wide border surrounding the rectangular viewing area of 8 inches by 12 inches, what is the area of the border in square inches?	
Output	Area of border expressed in square inches	
Input	Viewing area is 8 inches high and 12 inches wide. Border is 2 inches thick on all sides.	
Process	Notation	Let h = height Let w = width Let a = area
	Additional Information	$a = h * w$ Since border is added to both ends of each dimension, the outside area is 12 inches by 16 inches. Area inside the border is 96 square inches (from example 1-4.1).
	Diagram	Border is 2 inches thick. 12 in. 16 in.
	Approach	This is an example of using the solution from a previous problem to solve a similar problem. Here are two approaches: Approach 1: • Compute the area of the outside rectangle. • Subtract from this the area of the inside rectangle. • The remainder is the border area. Approach 2: • Divide the border into four rectangular strips: top, bottom, left side, and right side. • Compute the area of each strip, making sure not to double count the ends. • Add the four-strip areas to get the border area. Use the first approach for the solution and the second approach for the check.

Solution	Following approach 1:
	$a = h * w$
	Outside area = 16 in. * 12 in. = 192 in.^{2}
	Inside area (viewing area) = 96 in.^{2} (from example 1-4.1)
	Border area = outside area − inside area
	Border area = 192 − 96 in.^{2}
	Border area = 96 square inches
	Notice that the viewing area is only half of the total screen area.
Check	Check using approach 2:
	Area of rectangle = height * width
	Divide the border into four rectangular strips.
	Outside dimensions: 12 in. *16 in.
	Inside dimensions: 8 in. * 12 in.
	Strips are all 2 inches wide.
	Avoiding double counting the ends, the area of each strip is:
	Top: $2 * (16 − 2) = 28$ in.^{2}
	Bottom: $2 * (16 − 2) = 28$ in.^{2}
	Left side: $2 * (12 − 2) = 20$ in.^{2}
	Right side: $2 * (12 − 2) = 20$ in.^{2}
	The area of the border is the sum of the areas of the four strips:
	Border area = 28 + 28 + 20 + 20 in.^{2}
	Border area = 96 square inches
	This result matches the previous solution.

Example Problem 3

Problem	Bill leaves Seattle at 2 P.M., traveling east on Interstate 90 at 60 miles per hour. One hour later, Jim follows the same route traveling 70 miles per hour. Assuming they maintain a constant speed, how long must Jim travel before he passes Bill?
Output	How long will it take Jim to pass Bill?
Input	Bill leaves Seattle at 2 P.M. traveling east on Interstate 90 at 60 miles per hour. Jim leaves one hour later traveling at 70 miles per hour. Assume they both maintain a constant speed.

Process	Notation	None needed
	Additional Information	This is a problem that is usually solved with algebra. However, it can also be easily solved without algebra by using careful reasoning.
	Diagram	Although no diagram is needed, it may be helpful to draw a chart showing the total distance covered by Bill and Jim at the end of each hour of travel. Hour: | 0 | 1 | 2 | 3 | 4 | 5 | 6 | 7 | Time: |2:00|3:00|4:00|5:00|6:00|7:00|8:00|9:00| Bill: | 0 | 60 | 120 | 180 | 240 | 300 | 360 | 420 | Jim: | 0 | 0 | 70 | 140 | 210 | 280 | 350 | 420 |
	Approach	Using the fact that Bill covers 60 miles each hour traveled and Jim covers 70 miles each hour traveled, compute the cumulative distance covered by each at the end of each hour. Find the time when each has traveled the same distance. Then determine Jim's travel time.

Solution				
	Hour	**Time**	**Bill's Distance (miles)**	**Jim's Distance (miles)**
	0	2:00 P.M.	0	0
	1	3:00 P.M.	60	0
	2	4:00 P.M.	120	70
	3	5:00 P.M.	180	140
	4	6:00 P.M.	240	210
	5	7:00 P.M.	300	280
	6	8:00 P.M.	360	350
	7	9:00 P.M.	420	420

The distance traveled is the same for both after Bill has traveled for 7 hours and Jim has traveled for 6 hours. Thus, it will take Jim 6 hours of travel to pass Bill.

Check	An alternative approach can be used to check the result. During the hour before Jim starts, Bill travels 60 miles. Thus, when Jim starts, he has a 60-mile gap to close before passing Bill. The gap is closed at a rate of 10 miles per hour, the difference in the two speeds. Since the 60-mile gap will be closed at the rate of 10 miles each hour, it will take 6 hours to close the entire gap. Thus, it will take Jim 6 hours to pass Bill. The solution checks.

Practice Problems

Use the IPO method to plan, execute, and check the solution of the following problems. These problems should be solved using *only simple arithmetic and careful reasoning*.

Introductory Problems

1-4.1 How many 150-foot laps must you swim in a 75-foot pool in order to swim a mile?

1-4.2 A box contains 30 cookies. If each cookie weighs $\frac{1}{2}$ ounce, what is the net weight of the box when it is full? Net weight is the weight of the contents and excludes the weight of the container.

1-4.3 A computer-generated report includes inventory information for 200 items. The information for each item is given on a separate line. Up to 50 lines will fit on a page. The pages are numbered. The first page is number 1. What is the final page number?

1-4.4 Compute the floor area (in square feet) of a computer laboratory that is a rectangular room 40 feet in length and 30 feet in width using the formula: area = length * width.

1-4.5 When a thermometer measures a temperature of 90 degrees on the Fahrenheit scale, what is the equivalent temperature on the Celsius scale? The formula that converts degrees Fahrenheit (F) to degrees Celsius (C) is

$$C = \left(\frac{5}{9}\right) * (F - 32)$$

1-4.6 A box contains 30 cookies. If the empty box weighs 2 ounces and each cookie weighs $\frac{1}{2}$ ounce, what is the total weight of the box full of cookies?

1-4.7 A box of cookies contained 30 cookies when it was bought. The empty box weighs 2 ounces and each cookie weighs $\frac{1}{2}$ ounce. If one third of the cookies have been eaten, what is the weight of the box and the remaining cookies?

1-4.8 A computer-generated report includes inventory information for 220 items. The information for each item is given on a separate line. Up to 45 lines will fit on a page. The pages are numbered. The first page is number 1. What is the final page number?

Intermediate Problems

1-4.9 The volume of a sphere is given by the formula: $v = \frac{3}{4}(\pi r^3)$. In the formula, v represents volume, r represents the radius of the sphere, and π is a constant with an approximate value of 3.14159. Compute the volume of a sphere with a radius of 2 feet. Round your answer to two decimal places.

1-4.10 Students in the CIS department can receive internship credit for work experience in a position related to information technology. At the beginning of the term, each student with an approved internship position enrolls for a specified number of credits. At the end of the term each student submits a time sheet showing the number of hours worked. Credit is awarded at the rate of one credit for each 30 hours worked. Credits are rounded to the nearest half credit. (For example, values in the range 1.75 and up to but not including 2.25 will round to 2.0. Values in the range 2.25 and up to but not including 2.75 will round to 2.5.) The credits awarded can never exceed the credits enrolled, regardless of the number of hours worked. Compute the number of credits awarded to each student from the following enrollment data:

Name	Credits Enrolled	Hours Worked
Adams, Jane	1	30
Beck, Carl	2	15
Jones, Michael	2	90
Smith, Mary	2	50
Walker, William	5	75

Note: For more details on rounding see Section A-5, Decimal Numbers, in Appendix A. For tips on rounding to the nearest half see the sidebar in the same section, Rounding at Points Other Than Powers of Ten.

1-4.11 How many 2-foot-square floor tiles are needed to cover the floor of a rectangular room 32 feet in length and 30 feet in width? Use the formula: area = length * width.

1-4.12 A report generated by a computer includes information for 1000 customers. There are two types of customers, regular and preferred. The information for each regular customer takes one line on the report. The information for each preferred customer takes two lines on the report. Each page can contain up to 50 lines of customer information. If the number of regular customers is three times the number of preferred customers, how many pages will the report require?

1-4.13 Place a penny in the center of the corner cell of a tic-tac-toe grid as shown in the diagram. Slide the penny so its center passes through the center of every cell in the grid. Each move may be of any length, but must be along a horizontal, vertical, or diagonal straight line. The diagram shows how to do it in five moves. Can you do it in four moves?

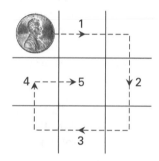

Hint: Do you know the solution to a similar problem?

1-4.14 Determine the area of a square with a diagonal dimension of 2 inches.
   Hint: Cut a square out of paper. Cut the square along both diagonals to form four right triangles. Rearrange the triangles to form two small squares. Recall that the area of a square is determined by the formula $a = s * s$, where s is the length of the side of the square.

1-4.15 Together, a bottle and a cork cost a dollar and five cents. The bottle costs a dollar more than the cork. Use trial and error to determine the cost of the cork.

1-4.16 How many feet of wallpaper must be cut from a roll one yard wide to cover a wall 18 feet wide and 9 feet high?

1-4.17 Write the step-by-step instructions for tying a shoelace into a bow. Have someone else test your instructions to make sure they are clear and correct.

More Difficult Problems

1-4.18 How many rolls of wallpaper should you purchase to cover the walls and ceiling of a room 18 feet long, 9 feet wide, and 9 feet high? Each roll is $\frac{1}{2}$ yard wide and 12 yards long. Ignore the cutouts for doors and windows.

1-4.19 If two men can paint two rooms in two hours, how long will it take for one man to paint one room? Assume all rooms are the same size.

1-4.20 Eight drawers salvaged from a kitchen remodel are to be used to make a workbench. The drawers were originally organized as two stacks of equal heights, but information on how they were stacked has been lost. The heights, in inches, of the drawers are 9, 9, 8, 7, 6, 4, 2, and 1.

   Divide the drawers into two stacks of equal height. Arrange the drawers in each stack in ascending order with the largest at the bottom and the smallest at the top. Following these rules, how many different ways can you stack the drawers? What are the ways?

Note: The number of drawers in each stack need not be the same, but the total height of both stacks must be the same.

1-4.21 At noon on Saturday a hiker starts up the trail to the cabin, arriving at 3:00 P.M. At noon the following day he hikes from the cabin to the trailhead down the same trail. On both trips the hiker never leaves the trail.

Is there some point along the trail that the hiker passes on the way down at exactly the same time of day as he passed on the way up?

There are three possible outcomes to this question:

1. The answer must always be true.

2. The answer may sometimes be true, depending on the speed of the hiker.

3. The answer can never be true.

Show which of these outcomes fits.

1-4.22 A group of friends are having dinner, and someone proposes a toast. Following the toast, glasses are clinked together, two at a time. Let the letter n represent the number of persons in the group. Find a formula that computes the number of clinks (c), given a value for the number of persons (n).

Hint: You already have seen a simpler version of this problem, an example in Section 1-3, that uses a pentagon to represent a group of five persons. Now use other polygons to represent different sized groups. For example, use a triangle for three persons, a square for four persons, a hexagon for six persons, and so on. Create a table that gives the number of clinks for each value of n from 1 to 7. Notice how the solutions for the different polygons are similar. Then generalize the solutions into a formula.

1-4.23 Bill leaves Seattle at 2 P.M. traveling east on Interstate 90 at 45 miles per hour. Two hours later, Jim follows the same route, traveling 75 miles per hour. Assuming they maintain a constant speed, how long will Jim travel before passing Bill?

Chapter Summary

Problem solving benefits from a systematic methodology that clearly presents all the essential information needed to solve the problem. The methodology presented here is called IPO method. The IPO worksheet is used to present all the essential information about the problem:

Problem Statement:	State the problem in clearly written sentences. Before you start the solution, you must understand the problem.
Output:	State what you are trying to find.
Input:	State the information you are given.
Process:	The heart of the problem-solving method, it involves the following sections:
Notation:	Define the "shorthand" notation that will be used to represent various quantities in your solution.
Additional Information:	Identify what information is lacking in the input statement that is essential to obtaining the solution, for example, equations, constant values, conversion factors, etc.
Diagram:	Use a graphic presentation to restate the problem, for example, diagrams, figures, charts, graphs, etc., that clarify the problem.
Approach:	List the specific steps needed to convert the input into the output. If a step is not clear, break it down into several substeps that are clear. Continue doing this until all steps are completely understood.
Solution:	Carry out the steps in the approach section to yield the solution. All values must be accompanied by their appropriate units. (Only at this point will you do any calculations.)
Check:	Before you can have confidence in your results, the solution must be checked. When possible, use an alternative solution.

This methodology can be used both formally and informally.

As you are learning the IPO method, write all information clearly in each section of the worksheet. Later, when you understand the method, you can do many of the sections mentally, leaving only the essential parts of the solution to be written.

Practice is an essential part of learning to solve problems. The best way to learn to solve problems is to solve a lot of problems. The solution of a simple problem becomes a building block that can be used later to solve more difficult problems.

Review Questions

1. Why is it important to be familiar with a formal problem-solving method?
2. Discuss the similarities between the Polya, Rubenstein, and IPO problem-solving methods.
3. Describe each of the six main sections of the IPO worksheet.
4. Describe each of the components of the process section in the IPO worksheet.
5. When a formula is part of a problem's solution, where in the IPO worksheet is it introduced?
6. Discuss the difference between the approach and the solution sections of the IPO worksheet.

7. Why should you use both a formal and informal IPO method? When is each appropriate?
8. Why is checking the solution of a problem so important?
9. Describe several ways to check a solution.
10. Give several examples of how notation can be used to solve problems.
11. List a few reasons you have failed to complete the solution to problems in the past.
12. List a few things to do if you get stuck when solving a problem.

Problem Solutions

1-2.2 Data: A box contains 30 cookies.
       Condition: Each cookie weighs $\frac{1}{2}$ ounce.
       Unknown: The weight of the cookies is unknown.
1-2.4 Data: A box contains 30 cookies.
                   The empty box weighs 2 ounces.
                   Each cookie weighs $\frac{1}{2}$ ounce.
       Condition: One third of the cookies have been eaten.
       Unknown: The weight of the box and the remaining cookies is unknown.
1-2.6 Data: Report contains information on 1000 customers.
                   There are two types of customers: regular and preferred.
                   Information for each regular customer takes one line.
                   Information for each preferred customer takes two lines.
                   Each page of the report can contain 50 lines.
       Condition: Regular customers are three times the number of preferred customers.
       Unknown: Number of pages required to print the report is unknown.
1-2.8 White (Do you know why?)
1-2.10 Condition 2 (Do you know why?)
1-2.12 It can be done in 17 minutes with no tricks. (Try all possible combinations.)

1-4.2 15 oz   1-4.4 1200 ft^2
1-4.6 17 oz   1-4.8 The final page number is 5.
1-4.10 1.0, 0.5, 2.0, 1.5, 2.5 credits   1-4.12 25 pages
1-4.14 2 in.^{2} 1-4.16 54 ft.
1-4.18 12 rolls   1-4.20 three ways: (1, 4, 9, 9 and 2, 6, 7, 8)
1-4.22 $c = n(n - 1) / 2$ (1, 6, 7, 9 and 2, 4, 8, 9)
                                                   (1, 2, 4, 7, 9 and 6, 8, 9)

Exponents

As far as the laws of mathematics refer to reality, they are not certain; as far as they are certain they do not refer to reality.
—Albert Einstein

2-1 ■ Introduction

How to Use This Chapter

Before you begin this chapter, you should look over the material in Appendix A on arithmetic and ensure that you have mastered the concepts presented there. A quick way to determine this is by taking the self-test at the beginning of Appendix A.

The first computational tool that will be covered in this book is that of exponential numbers. This foundation will prepare you for the next chapter on number systems.

Background

Before you can understand number systems, you must first understand exponential numbers. Multiplying a base number by itself one or more times forms what is called an exponential number. The simplest exponential number is the square. This is where a number is multiplied by itself one time. Since this calculation involves two like factors, the square of a number is sometimes called the second power.

The chapter begins with squares and the opposite operation, the square root. Following this, the general rules of exponents are covered. With this foundation, understanding number systems will be a snap.

Chapter Objectives

In this chapter you will learn:

- How to use squares and square roots

- How to use the rules for exponents in general

- How to multiply like bases with exponents

- How to divide like bases with exponents

- How to raise an exponential number to a power

- That negative exponents are the same as reciprocals
- That fractional exponents are the same as roots (square roots, for example)
- How to write numbers in scientific notation

Introductory Problem

Products of the same base number can be represented with exponent notation as follows:

$$10 = 10^1$$
$$10 * 10 = 10^2$$
$$10 * 10 * 10 = 10^3$$
$$10 * 10 * 10 * 10 = 10^4$$

$$2 = 2^1$$
$$2 * 2 = 2^2$$
$$2 * 2 * 2 = 2^3$$
$$2 * 2 * 2 * 2 = 2^4$$

Using this background information, answer the following questions:

1. The product $10^2 * 10^3$ can be expressed as the base 10 with what exponent?

2. The division $10^4 / 10^2$ can be expressed as the base 10 with what exponent?

3. The product $2^2 * 2$ can be expressed as the base 2 with what exponent?

4. The division $2^5 / 2^2$ can be expressed as the base 2 with what exponent?

5. Can you discover a rule for multiplying common base numbers that have exponents? Test your rule using the base number 3.

6. Can you discover a rule for dividing common base numbers that have exponents? Test your rule using the base number 3.

Before exponents are covered in general, the simplest and most frequently encountered exponent, the exponent 2, will be introduced. This is more commonly called the square of a number. The reverse operation, the square root, will also be covered.

Squares

Definition: The square of a number is that number multiplied by itself.

$$b \text{ squared} = b * b$$

Here are some examples using numbers:

5 squared	$= 5 * 5$	or	25
10 squared	$= 10 * 10$	or	100
2 squared	$= 2 * 2$	or	4
-2 squared	$= (-2) * (-2)$	also	4

 Squares of real numbers[1], whether positive or negative, are always positive. This isbecause multiplying two negative numbers gives a positive result.

Exponent Notation
Another way to write the square of a number is by using the superscript 2. In this case the superscript is called an *exponent.* Thus

$$b^2 = b * b$$

Because there are *two* factors involved, this is also called the *second* power of b.

Example Using Geometry
The area of the geometric figure called a square, with four equal sides, is found by multiplying the length of the side by itself:

$$\text{area} = \text{side} * \text{side}$$

Let the letter a represent the area of the square and letter s represent the length of the side. Then the formula for the area of a square figure is

$$a = s^2$$

Larger or Smaller
Depending on the value being squared, the result may be

- Larger than the original number
- Smaller than the original number
- Equal to the original number

---

[1] If you are not clear about the definition of "real" numbers, consult Section A-3 in Appendix A.

Squaring a positive number that is

- **Greater than 1** will give a result that is *larger* than the original number.

$$(1.1)^2 = 1.21$$
$$(2.2)^2 = 4.84$$

- **Between zero and 1** will give a result *less than* the original number.

$$(0.9)^2 = 0.81$$
$$(0.5)^2 = 0.25$$

- **Exactly equal to 1** will give a result that is *unchanged.*

$$1^2 = 1$$

- **Exactly equal to zero** will give a result that is *unchanged.*

$$0^2 = 0$$

Perfect Squares

Perfect squares are whole numbers formed as the square of some other whole number. Following are examples of the first few perfect squares:

Perfect Square	Formed by	Expressed as Square
1	1 * 1	1^2
4	2 * 2	2^2
9	3 * 3	3^2
16	4 * 4	4^2
25	5 * 5	5^2
36	6 * 6	6^2
49	7 * 7	7^2
64	8 * 8	8^2
81	9 * 9	9^2
100	10 * 10	10^2

Perfect squares are important because they are the only numbers that have square roots that are also whole numbers. This will be made clear in the next section.

Square Roots

While the concept of squaring a number is straightforward, the reverse operation of taking the square root is more difficult to grasp. This is because the definition must be stated backwards.

Definition: The square root of a number is that value which when multiplied by itself results in the original number.

Stated in another way:

If some number, represented by the letter *a*, is separated into two equal factors, *b * b*, then the factor *b* is said to be the square root of the number *a*.

In mathematical terms, if

$$a = b * b$$

then *b* is the square root of *a*. Only in a few special cases will the factor *b* (the square root) be a whole number. In most cases the square root will be a never-ending decimal number. This is explained in the next section.

Square Roots of Perfect Squares

Most numbers have square roots that are *not* whole numbers. However, a few special numbers, called perfect squares, do have square roots that are whole numbers. The square roots of perfect squares, because of their simplicity, present a good place to start in the discussion of square roots.

Recall that a perfect square is a whole number that is formed by squaring another whole number. It then follows that taking the square root of a perfect square gives the whole number that was used to form it in the first place.

Perfect Square	*Formed by*	*Square Root*
1	1 * 1	1
4	2 * 2	2
9	3 * 3	3
16	4 * 4	4
25	5 * 5	5
36	6 * 6	6
49	7 * 7	7
64	8 * 8	8
81	9 * 9	9
100	10 * 10	10

Notice that multiplying the square root of a number by itself (squaring) gives the original number.

Square Roots Come in Pairs

Numbers have two square roots, with the same absolute value and opposite signs. Here are a few examples of square roots of perfect squares:

Perfect Square	*Positive Square Root*	*Negative Square Root*	*Notation for Both Square Roots*
1	$+1$	-1	± 1
4	$+2$	-2	± 2
9	$+3$	-3	± 3
16	$+4$	-4	± 4
25	$+5$	-5	± 5

Note: When a *positive* square root is multiplied by itself, the result is the original number. When a *negative* square root is multiplied by itself, the result is *also* the original number.

(This is because the multiplication of two negative numbers gives a positive number.)

The Radical Sign

The symbol $\sqrt{}$ is one way to indicate the *positive* square root. When

$$b^2 = a$$

then

$$\sqrt{a} = b$$

The positive and negative square roots can be expressed using the radical as follows:

$$+\sqrt{a} = +b$$
$$-\sqrt{a} = -b$$

Remember, the radical symbol denotes only the positive square root.

Irrational Numbers

You have seen a few special square roots that are whole numbers. However, the vast majority of square roots are not whole numbers, but rather a special type of decimal number called an *irrational* number.

Before irrational numbers are defined, let's review rational numbers. Decimal numbers formed from fractions involve the ratio of two whole numbers. These numbers have decimal parts that either end ($\frac{1}{8} = 0.125$) or repeat indefinitely ($\frac{2}{3} = 0.66666 \ldots$). Because these numbers are formed from a ratio, they are called *rational* numbers. Integers are also rational because they may be formed by the ratio of a whole number and the number 1.

It turns out that square roots that are not whole numbers are never-ending decimals that do not repeat. Therefore, they cannot be formed from fractions and are *not* rational. They belong to a new class of numbers called *irrational* numbers. Because irrational numbers have decimal parts that never end and do not repeat, they can only be calculated approximately.

Calculating Approximate Square Roots

Before computers and calculators, high school students were taught to calculate square roots by hand. The process is much like doing long division, but a little harder. Luckily, you now have computers and calculators that do this work for you.

Although a few numbers (the perfect squares such as 4, 9, and 16) have whole number square roots, most numbers have square roots that are endless decimals. Thus, except for the perfect squares, you can never calculate a square root exactly. You can compute the approximate value of a square root to as many decimal places as you are willing to take the time to find.

As you have learned, the radical symbol $\sqrt{}$ can be used to represent the positive square root of a number. The radical symbol implies the *exact* value of the square root. However, when you press the key on your calculator to find the square root of a number (other than a perfect square), the result displayed is approximate. Most scientific calculators give the result to 10 significant figures.

Here are a few examples of positive square roots computed on a calculator:

Number	Square Root Using Radical	Approximate Square Root
2	$\sqrt{2}$	1.414213562
3	$\sqrt{3}$	1.732050808
5	$\sqrt{5}$	2.236067977
7	$\sqrt{7}$	2.645751311

> **!** Thus, for square roots of numbers that are not perfect squares, the radical can be used to represent the exact value. However, when the radical is evaluated as a decimal number, the result is approximate.

Inverse Operations

Taking the square root of a number is the reverse or opposite operation of taking the square. Mathematicians call these opposite operations *inverse* operations. An inverse operation undoes what another operation has just accomplished, leaving the original value unchanged.

Although you may not have realized it, you are already familiar with other inverse operations. Addition and subtraction are inverse operations:

Start with some value a, add another value b, then subtract the same value b. The result of the addition and subtraction of the same number "cancel" each other, leaving the original value a as the final result.

$$a + b - b = a$$

Multiplication and division are inverse operations:

Start with some value a, multiply it by another value b, then divide the result by the same value b. When both operations are combined, they "cancel" each other, leaving the original value a unchanged.

$$\frac{(a * b)}{b} = a$$

(Of course this will not work if the value of b is zero, because division by zero is undefined.)

Now look at the square and square root operations in the same way. Squares and square roots are inverse operations:

Start with some value, b. Square it. Then take the square root of the result. The operations cancel each other, leaving the value b unchanged.

$$\sqrt{b^2} = b$$

Try this inverse operation on a few numbers using your calculator:

Number	Square Root	Square of Square Root	Exactly
2	1.41421 . . .	1.999989 . . .	2
3	1.73205 . . .	2.999997 . . .	3
5	2.23606 . . .	4.999964 . . .	5

The preceding calculations have been rounded to five decimal places.

 Because calculating a square root is approximate, so is the square of the square root.

The inverse operations do not depend on which operation is done first:

$$\sqrt{b} * \sqrt{b} = b \qquad \text{and} \qquad \sqrt{b^2} = b$$

Example from Geometry
Recall that the area of a square figure is the length of the side squared:

$$a = s^2$$

Therefore, the square root of the area is equal to the length of the side:

$$\sqrt{a} = s$$

Thus, using the formula:

- If the area of a square is 4 square feet, the length of the side is 2 feet.

- If the area of a square is 2 square feet, the length of a side is $\sqrt{2}$ feet.

Square Root of Negative Numbers

The square root of negative numbers is a topic that will not be covered in this book. A brief definition is given in the sidebar.

Aside: Imaginary Numbers

The square root of negative numbers is a little trickier. Of the numbers we have discussed so far, the so-called *real numbers,* none will serve as the square root of a negative number. Therefore, an entire branch of mathematics involving *complex numbers* that combine real numbers and *imaginary numbers* have been created to deal with this.

The imaginary number called i is defined such that:

$$i^2 = -1$$

Therefore

$$i = \sqrt{-1}$$

Imaginary numbers are useful, but will not be covered further in this text.

Note: Don't confuse a negative square root with taking the square root of a negative number. When we take the square root of a positive number, the result is a positive square root and a negative square root. For example, the square root of $+16$ is either $+4$ or -4. In this case -4 is called the *negative square root* of 16. This is very different from taking the square root of -16, which is something that is not covered in this book.

Smaller or Larger

The square root of a number greater than 1 is *smaller* than the original number. The square root of a number between zero and 1 is *larger* than the original number.

- Numbers larger than 1 have square roots *smaller* than the original number:

$$\sqrt{2} = 1.414\ldots \qquad \sqrt{4} = 2 \qquad \sqrt{9} = 3$$

- Numbers between zero and 1 have square roots *larger* than the original number:

$$\sqrt{0.3} = 0.547\ldots \qquad \sqrt{0.5} = 0.707\ldots \qquad \sqrt{0.9} = 0.948\ldots$$

- Zero and 1 have square roots equal to themselves:

$$\sqrt{1} = 1 \qquad \sqrt{0} = 0$$

Using a Calculator to Demonstrate Inverse

You will need a calculator with the square root key $\sqrt{}$ to do this exercise.

1. Enter any positive number larger than 1.

2. Compute the square of the number. Do this by either multiplying the number by itself, or if your calculator has a x^2 key, you can do this with one keystroke. Notice that the result is larger than the original number.

3. Compute the square root of the result. Notice that after the square root key is pressed, the display has been reduced by exactly the amount needed to become the

original number. The inverse operations of square and square root have canceled each other out; for example:

2 $\sqrt{}$ × 2 $\sqrt{}$ =   displays 2

Or try:

2 x^2 $\sqrt{}$ displays 2

Try the same thing in reverse. Take the square root first, then square the result. Notice that the square root is smaller than the original number, but the square makes it larger again by exactly enough to return it to the original number.

Try the same procedure using an original number between zero and 1, for example, 0.5. Notice that the smaller and larger parts are reversed, but the end result is the same.

Application to Information Technology

Computer languages do not recognize the radical sign $\sqrt{}$ and use alternative ways to indicate the square root in calculations.

Functions are one way to calculate a square root on a computer. Functions in computer languages and spreadsheets accept input, process that input in a specified way, and return the result as output. Many different functions are available to the computer user.

The square root function is written in different ways, depending on the language. Two examples are SQR(x) or SQRT(x). The value x inside the parentheses is the input and can be assigned any positive value up to some maximum value that the computer can handle. The function then calculates the positive square root of x and returns the result.

The square root function can appear inside a more complicated calculation such as

$$(4 + SQR(4))/2$$

which is evaluated as

$$(4 + 2)/2 = 3$$

Another example is

$$2 * SQR(2)$$

which is evaluated as

$$2 * 1.4142136 = 2.8284271$$

Manipulating Radicals

Separating Factors Inside Radicals

Multiplied factors inside a radical $\sqrt{}$ can be written as the product of separated radicals:

$$\sqrt{2 * 3} = \sqrt{2} * \sqrt{3}$$
$$\sqrt{a * b} = \sqrt{a} * \sqrt{b}$$

This works only for multiplied factors, not for additive terms.

Removing Radicals

When separate radicals contain like terms, the radicals can be removed to leave only the term inside. This is because the square and the square root are inverse operations.

$$\sqrt{a^2} \;=\; \sqrt{a} * \sqrt{a} \;=\; a$$

Once you understand this, you can use a shortcut: Two like factors inside a radical can be replaced with one on the outside.

Simplifying Radicals

Combining the two concepts of separating radicals and removing radicals provides a way to simplify radicals when they contain like factors. Like factors can be removed from inside the radical, and in some cases the radical can be eliminated altogether. Always start by factoring the quantity inside the radical as much as possible.

▶ *Example:*

$$\sqrt{12} \;=\; \sqrt{2 * 2 * 3}$$

Use the shortcut to replace the pair of 2s inside the radical with a single 2 on the outside:

$$\sqrt{2 * 2 * 3} \;=\; 2\sqrt{3}$$

Thus, $\sqrt{12}$ can be simplified to

$$\sqrt{12} \;=\; 2\sqrt{3}$$

▶ *Other Examples:*

Expression	*Separated Radicals*	*Simplified Expression*
$\sqrt{2 * 2}$	$= \sqrt{2 * 2}$	$= 2$
$\sqrt{3^2}$	$= \sqrt{3 * 3}$	$= 3$
$\sqrt{32}$	$= \sqrt{4 * 4 * 2}$	$= 4\sqrt{2}$
$\sqrt{75}$	$= \sqrt{5 * 5 * 3}$	$= 5\sqrt{3}$
$\sqrt{a^2 b}$	$= \sqrt{a * a * b}$	$= a\sqrt{b}$

This simplification applies only to factors that are multiplied, not to terms that are added or subtracted. The following cannot be simplified:

$$\sqrt{a^2 + b^2}$$
$$\sqrt{a} + \sqrt{b}$$

Removing Radicals from the Denominator of Fractions

It is customary to remove radicals from the denominator of a fraction when presenting an expression in final form. It is easy to do this. Simply multiply the fraction by the ratio of the radical to itself. Since that ratio is equal to 1, the value of the fraction is unchanged. In the process, the radical is moved from the denominator to the numerator of the fraction. Here are some examples:

To remove the radical from the denominator of $1/\sqrt{2}$, multiply by the ratio of $\sqrt{2}$ to itself. That ratio, $(\sqrt{2} / \sqrt{2})$, is equal to 1. Thus, it will not change the fraction.

$$\frac{1}{\sqrt{2}} \;=\; \left(\frac{1}{\sqrt{2}}\right) * \left(\frac{\sqrt{2}}{\sqrt{2}}\right) \;=\; \frac{\sqrt{2}}{\left(\sqrt{2} * \sqrt{2}\right)} \;=\; \frac{\sqrt{2}}{2}$$

To remove the radical from the denominator of $3/\sqrt{3}$, multiply by the ratio of $\sqrt{3}$ to itself. That ratio, $(\sqrt{3}/\sqrt{3})$, is equal to 1. Thus, it will not change the fraction.

$$\frac{3}{\sqrt{3}} = \left(\frac{3}{\sqrt{3}}\right) * \left(\frac{\sqrt{3}}{\sqrt{3}}\right) = \frac{\left(3\sqrt{3}\right)}{\left(\sqrt{3}*\sqrt{3}\right)} = \frac{3\sqrt{3}}{3} = \sqrt{3}$$

Practice Problems

Simplify the following expressions by factoring as much as possible. Remove like pairs of factors from inside the radical. Where possible, eliminate the radical altogether. When it is not possible to eliminate the radical, leave each remaining factor inside one or more individual radicals.

▶ **Examples:**

$\sqrt{2*2} = 2$	The radical was eliminated.
$\sqrt{12} = \sqrt{2*2*3} = 2\sqrt{3}$	The radical was simplified.
$\sqrt{24} = \sqrt{2*2*2*3} = 2\sqrt{2}*\sqrt{3}$	The radical was simplified.
$1/\sqrt{8} = 1/\sqrt{2*2*2} = 1/2\sqrt{2} = \sqrt{2}/2$	The radical was simplified.

Introductory Problems

2-2.1 $\sqrt{3*3}$

2-2.2 $\sqrt{2*3*2}$

2-2.3 $\sqrt{3*3*3}$

2-2.4 $\sqrt{3*3*3*2*2}$

2-2.5 $\sqrt{2*2*2*2*3*3*5}$

Intermediate Problems

2-2.6 $\sqrt{4^2}$

2-2.7 $\sqrt{2^2*3^2}$

2-2.8 $\sqrt{2^2*3}$

2-2.9 $\sqrt{a^2}$

2-2.10 $\sqrt{a^2*a}$

More Difficult Problems

2-2.11 $\sqrt{12}$

2-2.12 $\sqrt{24}$

2-2.13 $\sqrt{(60)}$

2-2.14 $9/\sqrt{27}$

2-2.15 $8/\sqrt{128}$

2-3 ■ Powers and Exponents

Powers

The product of equal base numbers (b) is called a power.

$b * b$ is the second power of the number b (or b squared)
$b * b * b$ is the third power of the number b (or b cubed)
$b * b * b * b$ is the fourth power of the number b

 The power is equal to the number of like factors being multiplied.

Exponent Notation

Exponents provide a shorthand notation for writing powers.

$$b^3 = b * b * b$$

The third power of b is written using the superscript 3. The superscript is the exponent. For example:

$$3^2 = 3 * 3$$

or

$$10^6 = 10 * 10 * 10 * 10 * 10 * 10$$

Note: Exponents apply only to the base quantity immediately on the *left*. For example, ab^2 indicates that a is multiplied by b squared. However, parentheses may be used to indicate that exponents apply to an entire expression. For example, $(ab)^2$ indicates that the product of a and b is squared.

Here are examples of base 2 raised to different powers:

$2^4 =$ $2 * 2 * 2 * 2$ $= 16$
$2^3 =$ $2 * 2 * 2$ $= 8$
$2^2 =$ $2 * 2$ $= 4$
$2^1 =$ 2 $= 2$
$2^0 =$ 1 $= 1$ (The zero exponent will be explainted later.)

Computer Notation

Computers use an alternative notation for exponents with the $^\wedge$ symbol instead of a superscript. In the following examples the value following the circumflex symbol $^\wedge$ is the exponent.

$b \wedge 3$ represents $b^3 = b * b * b$

$3 \wedge 2$ represents $3^2 = 3 * 3 = 9$

Except when discussing computer calculations, this book will use the superscript notation.

The Exponent 1

The first power of any number is the number itself. Thus the exponent 1 is usually not written.

$$5^1 = 5$$
$$4^1 = 4$$
$$3^1 = 3$$
$$2^1 = 2$$
$$1^1 = 1$$
$$0^1 = 0$$

Expressions raised to the power of 1 are equal to themselves:

$$(2 + 5)^1 = (2 + 5)$$
$$(a + b)^1 = (a + b)$$
$$\left[\frac{(a * b)}{c} \right]^1 = \left[\frac{(a * b)}{c} \right]$$

The Exponent Zero

Any number (except zero) raised to the zero power is equal to 1.

$$b^0 = 1 \quad \text{for all } b \neq 0$$

Students often insist that the zero power should equal zero. But this is *not* the case. We can see this for powers of 2 by noticing that each higher power of 2 is exactly *twice* the preceding power. The reason it is *twice* the preceding power is that the higher power has been multiplied by the base, 2. Conversely, each lower power of two is exactly *half* of the preceding power. Again, the reason it is *half* is that the lower power has been divided by the base, 2.

$$2^3 = 2 * 2 * 2 = 8$$
$$2^2 = 2 * 2 = 4$$
$$2^1 = 2$$
$$2^0 = 1$$

Notice that 2^0 follows the pattern, being half of 2^1.

Similarly, for powers of 3, 3^0 is one third of 3^1 or 1. (The one third comes about because the base is 3.) In fact, any base number (with the exception of zero) raised to the power of zero is equal to 1. Here are a few examples:

$$5^0 = 1 \qquad 4^0 = 1 \qquad 3^0 = 1 \qquad 2^0 = 1 \qquad 1^0 = 1$$

Expressions raised to the power of zero are also equal to 1:

$$(2 + 5)^0 = 1$$
$$(a + b)^0 = 1 \qquad \text{as long as } (a + b) \neq 0$$
$$\left[\frac{(a * b)}{c} \right]^0 = 1 \qquad \text{as long as} \left[\frac{(a * b)}{c} \right] \neq 0 \text{ and } c \neq 0$$

Using the rules of exponents, it is easy to prove that any number except zero, raised to the zero power, gives 1. We will introduce these rules next and then return to state the proof.

Rules of Exponents

Exponent Rule 1: Multiplying Like Bases (b) with Exponents
To compute the product of b^n and b^m, add exponents:

$$b^n * b^m = b^{n+m}$$

For example:

$$2^3 * 2^2 = 2^{3+2}$$
$$= 2^5$$

Check using another method:
Write 2^3 as $(2 * 2 * 2)$
Write 2^2 as $(2 * 2)$
Multiply $(2 * 2 * 2) * (2 * 2)$
$$= (2 * 2 * 2 * 2 * 2)$$
Which is also $= 2^5$

Another example:

$$3 * 3^2 = 3^1 * 3^2$$
$$= 3^{1+2}$$
$$= 3^3$$

Check using another method:
Write 3^1 as (3)
Write 3^2 as $(3 * 3)$
Multiply $(3)*(3 * 3)$
$$= (3 * 3 * 3)$$
Which is also $= 3^3$

Remember: exponent rule 1 *cannot* be applied when there is no common base. Neither of the following can be simplified using exponent rule 1:

$$2^2 * 3^2$$
$$x^3 * y^2$$

Practice Problems
Where possible, apply exponent rule 1 to simplify the following expressions:

2-3.1 $2^2 * 2$

2-3.2 $3^2 * 3^3$

2-3.3 $4^3 * 4^6$

2-3.4 $y^7 * y^8$

2-3.5 $2^4 * 2^8 * 3^5$

2-3.6 $5^4 * 5^7$

2-3.7 $3x^5 * (4y^3)$

2-3.8 $5y^2 * y(-2y^4)$

2-3.9 $a^3 * b^2$

2-3.10 $(2^3) * (2^3)$

Exponent Rule 2: Dividing Like Bases (b) with Exponents
To compute b^n divided by b^m, subtract exponents:

$$\frac{b^n}{b^m} = b^{n-m} \qquad \text{for all b} \neq 0$$

For example:

$$\frac{3^3}{3^2} = 3^{3-2}$$

$$= 3^1$$

Check using another method:
Write 3^3 as $(3 * 3 * 3)$
Write 3^2 as $(3 * 3)$
Divide: $\dfrac{(3 * 3 * 3)}{(3 * 3)}$

$$= (3) * \frac{(3 * 3)}{(3 * 3)}$$

$$= 3 * 1$$

Which is also $= 3^1$

Another example:

$$\frac{2^5}{2^2} = 2^{5-2}$$

$$= 2^3$$

Check using another method:
Write 2^5 as $(2 * 2 * 2 * 2 * 2)$
Write 2^2 as $(2 * 2)$
Divide: $\dfrac{(2 * 2 * 2 * 2 * 2)}{(2 * 2)}$

$$= (2 * 2 * 2) * \frac{(2 * 2)}{(2 * 2)}$$

$$= (2 * 2 * 2) * 1$$

Which is also $= 2^3$

Remember: exponent rule 2 *cannot* be applied when there is no common base. The following cannot be simplified using exponent rule 2:

$$\frac{2^2}{3^2}$$

$$\left(\text{Thus, } \frac{2^3}{3^2} \text{ is not } 2^{3-2} \text{ nor is it } 3^{3-2} \right).$$

Practice Problems

Where possible, apply exponent rule 2 to simplify the following expressions:

2-3.11 $\dfrac{2^2}{2^1}$

2-3.12 $\dfrac{3^3}{3^2}$

2-3.13 $\dfrac{4^6}{4^5}$

2-3.14 $\dfrac{y^8}{y^6}$

2-3.15 $\dfrac{2^8}{2^4}$

2-3.16 $\dfrac{5^7}{5^3}$

2-3.17 $\dfrac{3x^5}{(4y^3)}$

2-3.18 $\dfrac{5y^6}{(y * y^4)}$

2-2.19 $\dfrac{a^3}{b^2}$

2-3.20 $\dfrac{(2^3)}{\left(\dfrac{2^2}{2}\right)}$

Exponent Rule 3: Powers of Bases Themselves Raised to a Power

To compute $(b^n)^m$, multiply exponents.

$$(b^n)^m = b^{n*m}$$

For example:

$(2^3)^2 = 2^{3*2}$
$\qquad = 2^6$

Check using another method:
$\quad$ Write 2^3 as $(2 * 2 * 2)$
$\quad$ Then $(2^3)^2 = (2 * 2 * 2) * (2 * 2 * 2)$
$\quad$ Combine as $(2 * 2 * 2 * 2 * 2 * 2)$
$\quad$ Which is also 2^6

Another example:

$(3^2)^3 = 3^{2*3}$
$\qquad = 3^6$

Check using another method:
$\quad$ Write 3^2 as $(3 * 3)$
$\quad$ Then $(3^2)^3 = (3 * 3) * (3 * 3) * (3 * 3)$
$\quad$ Combine as $(3 * 3 * 3 * 3 * 3 * 3)$
$\quad$ Which is also 3^6

Practice Problems

Apply exponent rule 3 to simplify the following expressions:

2-3.21 $(3^2)^3$

2-3.22 $(y^3)^5$

2-3.23 $(x^2)^5$

2-3.24 $(a^5)^6$

2-3.25 $(2^2)^3$

2-3.26 $(10^3)^2$

2-3.27 $(2^3)^4$

2-3.28 $(x^b)^4$

2-3.29 $[(x^3)^3]^4$

2-3.30 $[(x^2)^4]^2$

Exponent Rule 4: Product Raised to a Power

A product of several factors $(a * b)$ raised to the power n is the same as the power applied to each factor individually.

$$(a * b)^n = (a^n) * (b^n)$$

The reverse is also true:

$$(a^n) * (b^n) = (a * b)^n$$

▶ Example:

$$(2 * 3)^2 = (2^2) * (3^2)$$

Check using another method:
Write $(2 * 3)^2$ as $(2 * 3) * (2 * 3)$
Rearrange as $(2 * 2) * (3 * 3)$
Which is also $(2^2) * (3^2)$

Work the same problem in reverse:

$$(2^2) * (3^2) = (2 * 3)^2$$

Check using another method:
Write 2^2 as $(2 * 2)$
Write 3^2 as $(3 * 3)$
Multiply $(2^2) * (3^2) = (2 * 2) * (3 * 3)$
Rearrange as $(2 * 3) * (2 * 3)$
Which is also $(2 * 3)^2$

Another example:

$$(x * y)^3 = x^3 * y^3$$

Check using another method:
Write $(x * y)^3$ as $(x * y) * (x * y) * (x * y)$
Recombine as $(x * x * x) * (y * y * y)$
Which is also $x^3 * y^3$

Note: This can also be written without the asterisk as $(xy)^3 = x^3y^3$

Practice Problems
Use exponent rule 4 to give an expression with a single exponent:
2-3.31 $2^3 * 4^3$

2-3.32 $4^2 * 3^2$

2-3.33 $(3x)^3 * 2^3$

2-3.34 $a^2 * c^2$

2-3.35 $(2x)^3 * (3y)^3$

Use the reverse of exponent rule 4 to expand the following expressions:
2-3.36 $(3 * 5)^4$

2-3.37 $(a * b)^n$

2-3.38 $(2 * 2 * 2)^2$

2-3.39 $(x * y * z)^2$

2-3.40 $(x^2 * y^3)^2$

Exponent Rule 5: Quotient or Fraction Raised to a Power
A fraction or quotient $(\frac{a}{b})$ raised to the power n is the same as the power applied to both numerator and denominator individually.

$$\left(\frac{a}{b}\right)^n = \frac{(a^n)}{(b^n)} \qquad \text{for all } b \neq 0$$

The reverse is also true:

$$\frac{(a^n)}{(b^n)} = \left(\frac{a}{b}\right)^n \qquad \text{for all } b \neq 0$$

► Example:

$$\left(\frac{2}{3}\right)^2 = \frac{(2^2)}{(3^2)}$$

Check using another method:

Write $\left(\frac{2}{3}\right)^2$ as $\left(\frac{2}{3}\right) * \left(\frac{2}{3}\right)$

Rearrange as $\dfrac{\left(\dfrac{2}{2}\right)}{\left(\dfrac{3}{3}\right)}$

Which is also $\dfrac{(2^2)}{(3^2)}$

Work the same problem in reverse:

$$\frac{(2^2)}{(3^2)} = \left(\frac{2}{3}\right)^2$$

Check using another method:

Write 2^2 as $(2 * 2)$
Write 3^2 as $(3 * 3)$

Divide as $\dfrac{(2^2)}{(3^2)} = \dfrac{(2 * 2)}{(3 * 3)}$

Recombine as $\dfrac{\left(\dfrac{2}{3}\right)}{\left(\dfrac{2}{3}\right)}$

Which is also $\left(\dfrac{2}{3}\right)^2$

Another example:

$$\left(\frac{x}{y}\right)^3 = \frac{x^3}{y^3}$$

Check using another method:

Write $\left(\dfrac{x}{y}\right)^3$ as $\left(\dfrac{x}{y}\right) * \left(\dfrac{x}{y}\right) * \left(\dfrac{x}{y}\right)$

Recombine as $\left(\dfrac{x * x * x}{y * y * y}\right)$

Which is also $\left(\dfrac{x^3}{y^3}\right)$

Practice Problems

Use exponent rule 5 to give an expression with a single exponent:

2-3.41 $\dfrac{2^2}{3^2}$

2-3.42 $\dfrac{3^2}{4^2}$

2-3.43 $\dfrac{a^3}{b^3}$

2-3.44 $\dfrac{x^2}{y^2}$

2-3.45 $\left(\dfrac{x^2}{y^2}\right)^2$

Use the reverse of exponent rule 5 to expand the following expressions:

2-3.46 $\left(\dfrac{2}{3}\right)^3$

2-3.47 $\left(\dfrac{5}{7}\right)^2$

2-3.48 $\left(\dfrac{3}{4}\right)^n$

2-3.49 $\left(\dfrac{2x}{5}\right)^2$

2-3.50 $\left(\dfrac{3}{2y}\right)^3$

Zero Power Proof

Use the rules of exponents to show that $b^0 = 1$ for all $b \neq 0$.

First:	$b^1 / b^1 = 1$	A number divided by itself is unity.
Also:	$b^1 / b^1 = b^{(1-1)} = b^0$	Exponent rule 2: Subtract exponents.
Therefore:	$b^0 = 1$	Two expressions that both equal to b^1 / b^1 are themselves equal.

The same thing can be shown using a specific number:

First:	$3^2 / 3^2 = 9 / 9 = 1$
Also:	$3^2 / 3^2 = 3^{(2-2)} = 3^0$
Therefore:	$3^0 = 1$

First Power Proof

Use the rules of exponents to show that $b^1 = b$:

First:	$b^3 / b^2 = (b * b * b) / (b * b) = b$	Dividing out the matched b's
Then:	$b^3 / b^2 = b^{(3-2)} = b^1$	Applying the rules of exponents
Therefore:	$b^1 = b$	Two expressions both equal to b^3 / b^2 are themselves equal.

The same thing can be shown using a specific number:

First:	$3^3 / 3^2 = 9 / 3 = 3$
Also:	$3^3 / 3^2 = 3^{(3-2)} = 3^1$
Therefore:	$3^1 = 3$

Reciprocals and Negative Exponents

Reciprocals

The reciprocal of a number is 1 divided by that number. All numbers have reciprocals except zero, since division by zero is undefined. The reciprocal of some value b is $\frac{1}{b}$ for all $b \neq 0$.

Here are some consequences of this definition:

1. A quantity multiplied by its reciprocal is unity:

$$b * \left(\dfrac{1}{b}\right) = 1 \quad \text{for all } b \neq 0$$

▶ Examples:

$$2 * \left(\frac{1}{2}\right) = 1$$

$$3 * \left(\frac{1}{3}\right) = 1$$

2. Multiplying by the reciprocal of a number is the same as dividing by that number.

▶ Example: Multiplying 6 by the reciprocal of 2 is

$$6 * \left(\frac{1}{2}\right) = 3$$

This is the same as dividing 6 by 2:

$$\frac{6}{2} = 3$$

3. The reciprocal of a fraction is the same as exchanging its numerator and denominator. This can be shown for the fraction (a / b) as

$$= \frac{1}{\left(\dfrac{a}{b}\right)}$$

$$= \frac{(b * 1)}{\left(b * \dfrac{a}{b}\right)} \quad \text{(multiplying numerator and denominator by } b\text{)}$$

$$= \frac{b}{a}$$

▶ Examples: The reciprocal of 1/3 is 3/1 or just 3. The reciprocal of 3/4 is 4/3.

4. Taking the reciprocal of the reciprocal of a number results in the original number.

▶ Example: The reciprocal of the reciprocal of 2 is 2. Since the reciprocal of 2 is 1/2, then the reciprocal of 1/2 is 2/1 or 2.

Practice Problems
Apply the rules for reciprocals to the following problems:

2-3.51 Take the reciprocal of 5.

2-3.52 Take the reciprocal of $3x$.

2-3.53 Take the reciprocal of $\frac{x}{y}$.

2-3.54 Take the reciprocal of $\frac{1}{5}$.

2-3.55 Take the reciprocal of $\frac{1}{6}$.

2-3.56 Multiply 4 by the reciprocal of 2.

2-3.57 Take the reciprocal of the reciprocal of 3.

2-3.58 Express the reciprocal of $\frac{8}{7}$ as a fraction.

2-3.59 Express the reciprocal of $\frac{5}{3}$ as a fraction.

2-3.60 Multiply the number 12,345 by its reciprocal.

Negative Exponents

An alternative way to express a reciprocal is by using a negative exponent. Why this is true is shown here:

The reciprocal of some value b (for any value except zero) is $\frac{1}{b}$. Then

$$\frac{1}{b} = \frac{b^0}{b^1} \qquad \text{Because } b^0 \text{ is equivalent to 1 for all } b \text{ except zero}$$

$$= b^{0-1} \qquad \text{Using the second rule of exponents}$$

$$= b^{-1} \qquad \text{After subtracting exponents}$$

▶ Example:

Reciprocal of 1000 is $\frac{1}{1000}$, which also can be written as 1000^{-1}.

It follows that taking the reciprocal of an expression having a positive exponent will yield a result with a negative exponent. Here are two examples:

$$\text{Reciprocal of } 10^3 \qquad = \frac{1}{10^3} \qquad = 10^{-3}$$

$$\text{Reciprocal of } 2^2 \qquad = \frac{1}{2^2} \qquad = 2^{-2}$$

$$\text{Reciprocal of } (x+y)^2 \qquad = \frac{1}{(x+y)^2} \qquad = (x+y)^{-2}$$

Also, taking the reciprocal of a number with an expression having a negative exponent will give a result with a positive exponent. Here are two examples:

$$\text{Reciprocal of } 10^{-3} \qquad = \frac{1}{(10^{-3})} \qquad = 10^3$$

$$\text{Reciprocal of } 2^{-2} \qquad = \frac{1}{(2^{-2})} \qquad = 2^2$$

$$\text{Reciprocal of } (x+y)^{-2} \qquad = \frac{1}{(x+y)^{-2}} \qquad = (x+y)^2$$

All the rules of exponents also apply to negative exponents. Here are three examples:

$$10^3 * 10^{-2} = 10^{3+(-2)} = 10^1 \qquad = 10$$

$$\frac{10^3}{10^{-2}} = 10^{3-(-2)} \qquad = 10^5$$

$$(10^{-3})^2 \qquad = 10^{-6}$$

Reciprocals by Flipping the Sign of an Exponent

Taking the reciprocal of a base number (b) with an exponent (n) is the same as multiplying the exponent by -1. More simply put, taking the reciprocal is accomplished by flipping the sign of the exponent. Thus, the reciprocal of b^n is b^{-n}, and the reciprocal of b^{-n} is b^n. Here are two examples:

The reciprocal of 3^2 is 3^{-2} Exponent 2 flips to -2 (The + sign in implied.)
The reciprocal of 5^{-3} is 5^3 Exponent -3 flips to 3 (Again, the + sign is implied.)

Practice Problems

Determine the reciprocal of the following expressions by flipping the sign of the exponent. Remember that expressions without an exponent have an implied exponent of 1.

2-3.61 5

2-3.62 3^2

2-3.63 $(x + y)^2$

2-3.64 5^{-1}

2-3.65 $(x + y)^{-2}$

Double Flipping

Double flipping incorporates the two methods for taking reciprocals introduced previously. It provides a way to change the *sign* of an exponent of a factor within an expression without changing the *value* of the expression. This is accomplished following these steps:

1. Flip the *sign* of the exponent of the factor.

2. Flip the *position* of the factor above or below the fraction bar.

When both flips are done at the same time, the *value* of the expression is *unchanged*. Here are some examples:

5^{-2}	is the same as	$\frac{1}{5^2}$
x^{-5}	is the same as	$\frac{1}{x^5}$
$\frac{1}{x}$	is the same as	x^{-1}
$\frac{1}{x^{-3}}$	is the same as	x^3

Note: As you can see, exponents and fraction bars are sometimes *implied* rather than shown *explicitly*. Thus, 5 implies 5^{+1}. Also, 5 implies $\frac{5}{1}$.

The reason double flipping works is that each flip represents one way to take the reciprocal of the factor. When you do both flips together, you are in effect taking the reciprocal of the reciprocal. Both reciprocals cancel each other, leaving the value of the expression unchanged.

Note: In addition to working on single terms, double flipping will also work on any factor that makes up a larger expression.

$$xy^{-2} \quad \text{is the same as} \quad \frac{x}{y^2}$$

Practice Problems

Use the double flip method to write equivalent expressions having only positive exponents.

2-3.66 3^{-4}

2-3.67 $\dfrac{1}{3^{-2}}$

2-3.68 $\dfrac{3y}{x^{-2}}$

2-3.69 $2a^2b^{-1}$

2-3.70 $(2x)^{-2}$

Signs of Exponents vs. Signs of Base Numbers

Negative exponents are used to indicate reciprocals and do not indicate negative values. Reciprocals of positive integers are not less than zero, but are between zero and plus one.

5^{-2} evaluates as $\dfrac{1}{25}$ This is a positive value between zero and one.

2^{-3} evaluates as $\dfrac{1}{8}$ This is a positive value between zero and one.

A minus sign in front of an expression causes the value of the expression to be multiplied by -1. Thus, if the expression is positive, a leading minus sign will make the result less than zero.

$$- (5^2) \text{ evaluates as } -25 \quad \text{This is a negative value.}$$

However, when the minus sign is inside parentheses and is acted upon by the exponent, the sign of the resulting expression can be either positive or negative.

$$(-5)^2 \text{ evaluates as } 25 \quad \text{This is a positive value.}$$

This is because the $(-5) * (-5) = +25$, since the product of two minuses makes a plus. Changing the exponent in the preceding expression from 2 to 3 gives a negative result.

$$(-5)^3 \text{ evaluates as } -625 \quad \text{This is a negative value.}$$

This is because $(-5) * (-5) * (-5) = -625$, since the product of three minuses makes a minus.

These last two examples provide a rule for simplifying expressions with minus signs inside parentheses that are acted on by integer exponents.

- If the integer exponent is *even,* the minus sign can be dropped.

- If the integer exponent is *odd,* the minus sign can be moved outside the parentheses.

Here are some examples:

$(-1)^0$	$= 1$	Even exponent, drop minus (consider zero even)
$(-2)^1$	$= -2$	Odd exponent, drop minus
$(-3)^2$	$= 3^2$	Even exponent, drop minus
$(-3)^{-3}$	$= -3^{-3}$	Odd exponent, retain minus
$(-2xy)^{-3}$	$= -(2xy)^{-3}$	Odd exponent, retain minus

Note: The odd–even rule is the same regardless of the sign of the exponent.

Practice Problems
Remove any negative signs from inside parentheses according to the preceding rule.

2-3.71 $(-5)^1$

2-3.72 $(-2x)^3$

2-3.73 $(-100)^2$

2-3.74 $\dfrac{1}{(-3^2)}$

2-3.75 $(-x)^{-3} * (-y)^{-3}$

2-3.76 $(x)^3 * (-y)^3$

Roots and Fractional Exponents

Roots
A root is the inverse operation of raising a number to a power. An inverse operation can be thought of as an operation that undoes the effect of a previous operation. Thus, the square root is the inverse operation of squaring a number (raising it to the second power). The cube root is the inverse operation of cubing a number (raising it to the third power). Higher roots are the inverses of their corresponding powers.

This can be easily demonstrated for the square root by using a calculator.

- Enter any positive number.

- Square it by pressing the x^2 key.

- Take the square root of the result by pressing the $\sqrt{}$ key.
- The final result will be the number you entered first.

In other words, the square root is the inverse operation of squaring a number. Thus, the square root of the square of a number is the original number.

$$\sqrt{b^2} = b$$

where b is positive.

> **Note:** Since the radical operator returns only the positive square root, the expression above is for positive values of b. But remember, square roots come in positive and negative pairs. Therefore, since b^2 is positive for both positive and negative values of b; the negative square root allows for negative values of b as well (as long as the radical is not used).

Fractional Exponents

Fractional exponents can be used to represent roots. An alternative notation for the square root radical is the fractional exponent $\frac{1}{2}$. For example:

$$\sqrt{2} \text{ can be written as } 2^{\frac{1}{2}}.$$

One way to convince yourself of this is to take the square root several times. Start with the fourth power of 2. Each time you take the square root, the exponent will be halved. Eventually you will arrive at a fractional exponent.

16	$= 2^4$	Start with 16, the fourth power of 2.
4	$= 2^2$	Taking the square root 16 gives 4, the second power of 2. (Notice that taking the square root halves the exponent.)
2	$= 2^1$	Taking the square root of 4 gives 2, the first power of 2. (Again, taking the square root halves the exponent.)
$\sqrt{2}$	$= 2^{\frac{1}{2}}$	Taking the square root of 2 gives $\sqrt{2}$. (Halving the exponent 1 gives the exponent $\frac{1}{2}$.)

> **Note:** Students who want a more formal proof may wish to examine the sidebar that follows.

Aside: Proof that $\sqrt{2}$ can be written as $2^{\frac{1}{2}}$

To show that $\sqrt{2} = 2^{\frac{1}{2}}$, replace the exponent $\frac{1}{2}$ with the letter x. Then, use the rules of exponents and a little algebra to show that x must equal $\frac{1}{2}$.

$\sqrt{2} = 2^x$	Start with this equation. Then find the value x.
$\sqrt{2} * \sqrt{2} = 2^x * 2^x$	Multiply both sides by themselves. This does not change things.
$2 = 2^x * 2^x$	This is because $\sqrt{2} * \sqrt{2} = 2$ (from Section 2-2).
$2 = 2^{(x+x)}$	Using exponent rule 1 (from Section 2-3).
$2^1 = 2^{(x+x)}$	Since 2 is the same as 2^1
$2^1 = 2^{2x}$	Since $x + x = 2x$.
$1 = 2x$	Since both bases are equal, both the exponents must be equal.
$x = \frac{1}{2}$	Since $\frac{1}{2}$ is the only value of x that will make $1 = 2x$ true

Therefore, $\sqrt{2} = 2^{\frac{1}{2}}$.

In a similar fashion, the cube root can be written as the exponent $\frac{1}{3}$, and the fourth root can be written as the exponent $\frac{1}{4}$.

The inverse nature of powers and roots can be seen clearly using fractional exponents together with exponent rule 3 (exponents raised to a power):

$$(b^{\frac{1}{2}})^2 = b^{\frac{1}{2}*2} = b^1 = b$$

Note: All the rules for exponents also apply to fractional exponents.

Thus

$$10^{\frac{1}{2}} * 10^{\frac{1}{2}} = 10^{(\frac{1}{2}+\frac{1}{2})} = 10^1 = 10$$

$$\frac{10}{10^{\frac{1}{2}}} = \frac{10^1}{10^{\frac{1}{2}}} = 10^{(1-\frac{1}{2})} = 10^{\frac{1}{2}}$$

$$(10^2)^{\frac{1}{2}} = 10^{(2*\frac{1}{2})} = 10^1 = 10$$

Application to Information Technology

When doing calculations on a computer, exponents are written using the $^\wedge$ symbol rather than with a superscript. Thus, 5 squared can be written as $5 \wedge 2$ and evaluates to 25.

Similarly, square roots can be calculated using a fractional exponent 1/2. The square root of 4 can be written as $4 \wedge (1/2)$ or as $4 \wedge 0.5$. Both of these expressions evaluate to 2.

Note: The use of parentheses around the exponent (1/2) is essential. Without the parentheses the order of precedence would evaluate the $4 \wedge 1$ first and then divide 2. In this case the result would be 2 rather than the square root of 2.

The exponent 1/2 presents an alternative to using a square root function to calculate the square. (See Application to Information Technology in Section 2-2.)

Practice Problems

Where possible, apply the rules of exponents to simplify the following expressions to a base number with a single exponent. Note that there are three parts to each problem.

2-3.77 $10^{\frac{1}{2}} * 10^{\frac{1}{2}}$ $10^{\frac{1}{3}} * 10^{\frac{1}{3}} * 10^{\frac{1}{3}}$ $10^{\frac{1}{4}} * 10^{\frac{1}{4}} * 10^{\frac{1}{4}} * 10^{\frac{1}{4}}$

2-3.78 $\dfrac{10^{\frac{1}{2}}}{10^{\frac{1}{2}}}$ $\dfrac{10^{\frac{1}{3}}}{10^{\frac{1}{3}}}$ $\dfrac{10^{\frac{1}{4}}}{10^{\frac{1}{4}}}$

2-3.79 $2^2 * 2^{\frac{1}{2}}$ $2^{\frac{1}{2}} * 2^{\frac{1}{3}}$ $2^{\frac{1}{2}} * \sqrt{2}$

2-3.80 $10^{\frac{1}{2}} * 10^{\frac{1}{3}}$ $\dfrac{10^{\frac{1}{2}}}{10^{\frac{1}{3}}}$ $10^{\frac{1}{2}} * 10$

2-3.81 $(2^{\frac{1}{2}})^2$ $(\sqrt{2})^2$ $2^{\frac{1}{2}} * 3^{\frac{1}{2}}$

2-3.82 $2 \wedge (\frac{1}{2})$ $2 \wedge 0.5$ $2 \wedge 1.5$ (computer notation)

2-3.83 $(7^{\frac{1}{2}})^2$ $(6^{\frac{1}{2}})^2$ $(5^{\frac{1}{3}})^2$

2-3.84 $2^{\frac{1}{2}} * 2^{\frac{1}{2}}$ $2^{\frac{1}{2}} * (\frac{1}{2})^{\frac{1}{2}}$ $\dfrac{1}{2^{\frac{1}{2}}}$

Exponent Rules Summary

Rule 1 applies to like bases with exponents that are **multiplied:** $\quad b^n * b^m = b^{n+m}$.

Rule 2 applies to like bases with exponents that are **divided:** $\quad \dfrac{b^n}{b^m} = b^{n-m}$.

Rule 3 applies to a base with an exponent **raised to a power:** $\quad (b^n)^m = b^{n*m}$.

Rule 4 applies to a **product of bases** raised to a power: $\quad (ab)^n = a^n * b^n$.

Rule 5 applies to the **quotient of bases** raised to a power: $\quad \left(\dfrac{a}{b}\right)^n = \dfrac{a^n}{b^n}$.

Note: Rules 1 and 2 apply to expressions with like *bases*.
Rules 4 and 5 apply to expressions with like *exponents*.

Combining the Rules of Exponents

The rules for exponents work in combination with each other. Complex expressions that include like bases with exponents can be simplified using the various rules.

The following examples were specially selected to demonstrate how certain terms that are initially not in exponent notation can be replaced with equivalent terms that use exponent notation and have the correct base as well. Only after this has been done can the rules of exponents be fully applied.

Note: To work these examples, 1 is replaced with 2^0 and $\sqrt{2}$ is replaced with $2^{\frac{1}{2}}$.

a) $\dfrac{2}{2}$ $\qquad = \dfrac{2^1}{2^1}$ $\qquad = 2^{(1-1)}$ $\qquad = 2^0$

b) $\dfrac{2}{1}$ $\qquad = \dfrac{2^1}{2^0}$ $\qquad = 2^{(1-0)}$ $\qquad = 2^1$

c) $\dfrac{1}{2}$ $\qquad = \dfrac{2^0}{2^1}$ $\qquad = 2^{(0-1)}$ $\qquad = 2^{-1}$

d) $\sqrt{2} * \sqrt{2}$ $\qquad = 2^{\frac{1}{2}} * 2^{\frac{1}{2}}$ $\qquad = 2^{(\frac{1}{2}+\frac{1}{2})}$ $\qquad = 2^1$

e) $(2^{\frac{1}{2}}) * (2^{\frac{1}{2}})$ $\qquad = 2^{\frac{1}{2}} * 2^{\frac{1}{2}}$ $\qquad = 2^{(\frac{1}{2}+\frac{1}{2})}$ $\qquad = 2^1$

f) $(2^2)^3$ $\qquad = 2^{(2*3)}$ $\qquad = 2^6$

g) $2^3/2^6$ $\qquad = 2^{(3-6)}$ $\qquad = 2^{-3}$

h) $\sqrt{2^2}$ $\qquad = (2^{\frac{1}{2}})^2$ $\qquad = (2^{\frac{1}{2}*2})$ $\qquad = 2^1$

i) $\left(\dfrac{2^5}{2}\right)^2$ $\qquad = \left(\dfrac{2^5}{2^1}\right)^2$ $\qquad = (2^{(5-1)})^2$ $\qquad = 2^{4*2}$ $\qquad = 2^8$

j) Reciprocal of 2^3 $\qquad = \dfrac{1}{2^3}$ $\qquad = \dfrac{2^0}{2^3}$ $\qquad = 2^{(0-3)}$ $\qquad = 2^{-3}$

k) Reciprocal of 2^{-2} $\qquad = \dfrac{1}{2^{-2}}$ $\qquad = \dfrac{2^0}{2^{-2}}$ $\qquad = 2^{0-(-2)}$ $\qquad = 2^2$

l) $\dfrac{1}{2^{-4}}$ $\qquad = \dfrac{2^0}{2^{-4}}$ $\qquad = 2^{0-(-4)}$ $\qquad = 2^4$

m) $\dfrac{1}{\left(\dfrac{1}{2}\right)}$ $= \dfrac{2^0}{\left(\dfrac{2^0}{2^1}\right)}$ $= \dfrac{2^0}{(2^{0-1})}$ $= 2^{0-(-1)}$ $= 2^1$

n) $\dfrac{1}{\left(\dfrac{1}{2}\right)^2}$ $= \dfrac{2^0}{\left(\dfrac{2^0}{2^1}\right)^2}$ $= \dfrac{2^0}{(2^{0-1})^2}$ $= \dfrac{2^0}{2^{(-1*2)}}$ $= 2^{0-(-2)}$ $= 2^2$

Note: Following the rules for order of precedence, terms that are inside parentheses are evaluated first.

The following additional example uses the rules of exponents to simplify a messy expression:

$$\frac{(2 * 10^{-3} * 2^2 * 2^{\frac{1}{2}} * 10^6)}{(2^3 * 10^2 * 2^{-\frac{1}{2}})}$$

1. Rearrange to group like bases together:

$$\left(\frac{10^{-3} * 10^6}{10^2}\right) * \frac{(2^1 * 2^2 * 2^{\frac{1}{2}})}{(2^3 * 2^{-\frac{1}{2}})}$$

Note: The rearrangement of parentheses is is permitted by the associative law of arithmetic, which allows multiplication (or division) of several factors to be carried out in any order you wish. (See Appendix A, Section A-3.)

2. Combine like bases that are multiplied by adding exponents:

$$\left(\frac{10^{(-3+6)}}{10^2}\right) * \frac{2^{(1+2+\frac{1}{2})}}{2^{(3-\frac{1}{2})}} = \left(\frac{10^3}{10^2}\right) * \left(\frac{2^{3\frac{1}{2}}}{2^{2\frac{1}{2}}}\right)$$

3. Combine like bases that are divided by subtracting exponents:

$$10^{(3-2)} * 2^{(3\frac{1}{2}-2\frac{1}{2})} = 10^1 * 2^1$$

Practice Problems
Use the rules of exponents to combine like bases and simplify the following expressions.

2-3.85 $\dfrac{(2^3 * 5^3 * 2^{-2} * 5)}{(5^6 * 5^{-2} * 2)}$

2-3.86 $(10^3 * 3^2 * \sqrt{3}) * \left(\dfrac{10^2}{3^{\frac{1}{2}}}\right)$

2-3.87 $\dfrac{(10^2 * 10^4 * 10^3 * 10)}{(10^2 * 10^5 * 10^4)}$

2-3.88 $\dfrac{(2^2 * 3^2 * 2^3 * 3^5 * 2^4 * 3^{-4})}{(2^5 * 3)}$

2-3.89 $\dfrac{(2^{\frac{1}{2}} * 2^{\frac{1}{2}} * 3^{\frac{1}{2}} * 3^2)}{(3^{-\frac{1}{2}} * 2)}$

Scientific Notation

Calculations using values that range from the very small to the very large sometimes present a problem. For example, suppose you need to multiply

$$123,000,000 * 0.000000123$$

The fields of engineering and science use a scientific notation to simplify these difficulties. In scientific notation a number is expressed as the product of

- a decimal value between 1 and 10 (but not equal to 10), and

- an appropriate power of 10.

Converting Decimal Notation into Scientific Notation

When a number is expressed in decimal notation, it can be thought of as being multiplied by 10^0. This changes nothing because $10^0 = 1$. Starting with this thought, move the decimal point of the original number to create a new value that has exactly one nonzero digit to the left of the decimal point. Discard any leading zeros.

To keep the new value equivalent to the original, adjust the exponent zero by an appropriate amount:

- Add 1 to the exponent zero for each decimal place moved to the *left*.

- Subtract 1 from the exponent zero for each decimal place moved to the *right*.

▶ *Examples:*

Express 2500 in scientific notation:
Write 2500 as $2500 * 10^0$.
Move the decimal point three places to the left to get 2.500.
Add 3 to the exponent to give $2.5 * 10^3$ in scientific notation.

Note: Whether or not you discard the trailing zeros depends on the number of significant digits in the original number. In this example it is assumed to be 2, so drop the trailing zeros. (Significant digits are discussed in a later section.)

Express -0.000027 in scientific notation:
Write -0.000027 as $-0.000027 * 10^0$.
Move the decimal point five places to the right to get -2.700.
Subtract 5 from the exponent to give $-2.7 * 10^{-5}$ in scientific notation.

Note: Since the original number had two significant digits, you drop the trailing zeros.

In scientific notation the sign of the exponent and the sign of the value are very different things:

- A leading minus sign indicates that the original number is less than zero.

- No leading sign indicates that the original number is greater than or equal to zero.

- Negative exponents show that the *absolute value* of the original number is less than 1.

- Positive exponents show that the *absolute value* of the original number is greater than 1.

Converting Scientific Notation into Decimal Notation

Move the decimal point the appropriate number of places needed to adjust the exponent back to zero:

- Move the decimal point to the *right* to remove *positive* exponents.

- Move the decimal point to the *left* to remove *negative* exponents.

Remember that right-positive and left-negative are the same on the number line. Finally, drop the 10^0.

▶ *Examples:*

Express $2.500 * 10^3$ in decimal notation.
Move the decimal point three places to the right to get $2500 * 10^0$.
Drop the 10^0 to get 2500 in decimal notation.

Express $-2.7 * 10^{-5}$ in decimal notation.
Move the decimal point five places to the left to get $-0.000027 * 10^0$.
Drop the 10^0 to get -0.0000027 in decimal notation.

Application to Information Technology

Scientific Notation on Computers

Computers use an E to represent 10 to some power.

$-1.2345 * 10^8$	is expressed on a computer as	$-1.2345 \text{ E} + 08$
$6.7890 * 10^{-5}$	is expressed on a computer as	$6.7890 \text{ E} - 05$

Note: The appropriate sign always precedes the exponent, even if it is positive.

Scientific Notation on Calculators

Calculators usually place the exponent off to the right. However, there is no E on most computers.

$-1.2345 * 10^8$	appears on a calculator as	$-1.2345 \quad 08$
$6.7890 * 10^{-5}$	appears on a calculator as	$6.7890 \ -05$

Note: Negative *values* display a sign while positive values do not.
Negative *exponents* display a sign while positive exponents do not.

Significant Digits

All digits in a number expressed in scientific notation are significant. Therefore, you will sometimes find zeros padded on the right side of the value to imply a certain level of significance. (See Appendix A for more information about significant digits.)

▶ *Examples:*

$2.0 * 10^3$	Implies the original number 2000 is known only to two significant digits.
$2.000 * 10^3$	Implies the original number 2000 is known to all four significant digits.
$2.5 * 10^{-3}$	Implies the original number 0.0025 is known to two significant digits.
$2.500 * 10^{-3}$	Implies the original number 0.0025 is known to four significant digits.

Practice Problems

Convert the following numbers from decimal notation to scientific notation. Round the result to *five significant digits.*

2-4.1   123456.7

2-4.2   12345.67

2-4.3   1234.567

2-4.4   123.4567

2-4.5   12.34567

2-4.6   1.234567

2-4.7   0.1234567

2-4.8   0.01234567

2-4.9   0.001234567

2-4.10 100000

2-4.11 1000000000

2-4.12 1000000001

2-4.13 0.000001234567

2-4.14 0.00000123

Convert the following numbers from scientific notation to decimal notation. Maintain the same number of significant digits.

2-4.15 $3.21 * 10^4$

2-4.16 $3.2 * 10^{-3}$

2-4.17 $3.210 * 10^{-3}$

2-4.18 $1.57 * 10^5$

2-4.19 $9.99 * 10^1$

2-4.20 $8.123 * 10^{-2}$

Evaluating Expressions Using Scientific Notation

When expressions involve multiplying and dividing values of widely differing magnitudes, scientific notion makes the evaluation easier. Follow these steps:

1. Express each number in scientific notation as a value multiplied by 10 with an exponent.

2. Multiply or divide the values.

3. Round the resulting value to the desired number of significant digits.

4. Add or subtract the exponents (using the rules of exponents).

5. Convert the result from scientific notation into conventional notation (if desired).

▶ *Example:*

Evaluate 0.000000123 * 342,000,000 / 0.02 to three significant digits.

$$\frac{(1.23 * 10^{-7}) * (3.42 * 10^8)}{(2.0 * 10^{-2})}$$

$$\left(\frac{1.23 * 3.42}{2.0}\right) * 10^{-7 + 8 - (-2)}$$

$$2.1033 * 10^3$$

$$2.10 * 10^3 \quad \text{(scientific notation rounded to three significant digits)}$$

$$2,100 \quad \text{(expressed in conventional notation)}$$

Practice Problems

Evaluate the following expressions by writing each number in scientific notation. Combine the decimal part and their exponents to express the result in scientific notation, rounded to four significant digits.

2-4.21 $\dfrac{10,002,000 * 0.000234}{1.5678}$ $\quad 1.0002 \times 10^7 \quad 2.34 \times 10^{-4}$

$\quad 1.5678 \times 10^0$

2-4.22 $\dfrac{2.5678 * 0.00001987}{234,000}$

2-4.23 $0.00024689 * 0.00000001 * 58,000,000,000$

2-4.24 $\dfrac{365.1 * (\frac{1}{2}) * 0.00001 * 0.0000023}{0.00000000045}$

Exponential Curiosities

Exponents present a way to express extremely large numbers using a simple notation. For example:

The age of the universe in seconds is roughly 10^{17} or 1 followed by 17 zeros.

The weight of the earth in grams is about 10^{27} or 1 followed by 27 zeros.

An estimate of the number of elementary particles (neutrons, protons, and electrons) in the universe is about 10^{80} or 1 followed by 80 zeros.

A number even larger than all of these examples added together is the googol, 10 raised to the power of 100 or

$$10^{100}.$$

The googol can be written in conventional notation as 1 followed by 100 zeros. Although the 101 digits of the googol can be written on a single line of this text, it is much larger than the age of the universe in seconds or the weight of the earth in grams. In fact, it is over one hundred million, million, million times larger than the estimated number of elementary particles in the universe given in the preceding example.

By using exponents that themselves have exponents, numbers immensely larger than a googol can be written. One such number is called the googolplex, 10 raised to the power of a googol. When written in exponent notation, a googolplex looks deceptively simple:

$$10^{\left(10^{100}\right)}$$

If the googolplex were written in conventional notation, it would be 1 followed by a googol of zeros. Not 100 zeros, but zeros numbering 10 raised to the hundreth power. While a googol can be written in conventional form on one line of this textbook, the googolplex is a very different matter. In fact, it is impossible to write all the zeros of the googolplex on a printed page or even in an entire book. There is not enough matter in the universe to make the pages needed to hold all the zeros in a googolplex. Even if there were, the time needed to get the job done would far exceed the age of the universe.

Order of Precedence with Exponents

While the googol and the googolplex are curiosities that you are not likely to encounter in the world of information technology, it is likely you will have to do calculations involving exponents. These curiosities present a way to demonstrate the importance of carrying out calculations in the proper order and the consequences if you do not. It is easier to see this using the computer notation for exponents that employs the ^ symbol.

A googol can be written as 10 ^ 100. Another way to write a googol is 10 ^ (10 ^ 2). The parentheses are important since they dictate the order of precedence in carrying out the calculation. Recall that expressions inside parentheses are evaluated before expressions that are not inside parentheses.

Suppose the parentheses are moved to indicate that 10 is to be raised to the tenth power before the result is raised to the second power. This can be written as (10 ^ 10) ^ 2. Using exponent rule 3 and multiplying exponents, this new number can be rewritten as 10 ^ (10 * 2) or 10 ^ 20.

10 ^ 20 is not a googol, but a number much smaller. Not a thousand times smaller or a million times smaller or even a quadrillion times smaller but smaller than a googol by a factor of 10 followed by 80 zeros. Clearly, 10 ^ 20 is not as large as 10 ^ 100. The difference is huge.

 The order of carrying out exponential calculations can change the result.

Practice Problems

2-4.25 This famous problem involves placing grains of wheat on the squares of a chess-board. A single grain of wheat is placed on the first square. Two grains are placed on the next square and four grains on the next. The board is filled with wheat, each square receiving twice that of the preceding square, until all 64 squares are filled. Of course, the number of grains of wheat become far too great to fit on a square, or on the entire board, or even the room that holds the board.

  a) Express the number of grains on the first square as 2 raised to some power.

  b) Express the number of grains that belong on the final square as 2 raised to a power.

  c) Express the total number of grains that belong on the board in terms of 2 raised to some power.

2-4.26 The National Collegiate Athletic Association (NCAA) men's Division I basketball tournament, played in the United States each March, involves the best college teams in the nation, playing an elimination-style tournament. Games are played in rounds; winning teams in each round continue, while losers are eliminated. There are the following rounds: first round, second round, regional semifinal, regional final, national semifinal, and national final. Many teams pair up to play games in the first round. Only the winners go on to play games in the second round, and only two teams survive to play in the final round. The winner of the final round becomes the national champion. From this information, determine the following:

  a) How many teams begin the tournament?

  b) How many games are played in the tournament?

2-4.27 The following problem is adapted from one presented as a weekly puzzler on the National Public Radio Program *Cartalk*.

      Your math class is studying probability by flipping a coin and seeing how many times in a row you can call the flip correctly. Your instructor breaks the class of 32 students into teams of two persons each. One member of each team flips a coin while the other calls heads or tails. If a flip is called correctly, the caller wins; otherwise, the flipper wins. You can form new teams whenever you like, and each person keeps track of his or her individual wins.

Your instructor offers a wager. Any student who can win five flips in a row will receive an A on the next quiz. But if they fail even once, they receive an F. There are no takers. So another wager is offered. An A will go to the student who can devise a method that guarantees someone in the class will win five flips in a row. Can you come up with such a method? (*Hint:* A look at problem 2-4.26 will be helpful.)

2-4.28 Contestants on the television quiz show *Who Wants to Be a Millionaire* can win $1 million if they answer 15 questions correctly. Each question is multiple choice with four possible answers. What are the odds of winning the million by guessing the answers to all 15 questions? (*Hint:* The odds of guessing the correct answer to one question are 1 in 4 or $\frac{1}{4}$. The odds of guessing the correct answer to two questions are 1 in 16 or $\frac{1}{4} * \frac{1}{4}$.)

The producers of the show want to make it more likely to succeed. They make the first five questions much easier than the later questions. Also, each contestant has three lifelines that can often help the contestant to answer correctly. Suppose the contestant gets the first five questions correct and of the remaining ten questions, three are answered correctly using lifelines. Of the remaining questions, the contestant knows the answer to three and for each of the others can identify two of the possible answers as being incorrect. What are the odds of winning the million dollars by guessing these two-part questions.

2-4.29 Consider the sequence of terms $t_1, t_2, t_3, t_4 \ldots, t_{(n-1)}, t_n$, where $t_1 = 1, t_2 = 2, t_3 = 4$, and $t_4 = 8$. Thus, the sequence is 1, 2, 4, 8 . . .

    a) What are the next three terms in the sequence?

    b) What formula finds any term t_n for all $n > 1$, given the preceding term, t_{n-1}?

    c) What formula finds any term t_n for all $n > = 1$, given 2 raised to some power?

Extra credit: Answer a), b), and c) for the sequence: 2, 4, 16. . . *Hint:* To answer part c) you need to think of the googolplex.

Chapter Summary

Rules of Exponents

- When multiplying exponential numbers with like bases, add exponents.
- When dividing exponential numbers with like bases, subtract exponents.
- When raising an exponential number to a power, multiply exponents.
- A product of several factors raised to a power can be rewritten with the power applied to each factor individually.
- A fraction or quotient raised to a power can be rewritten with the power applied to both numerator and denominator individually.

Reciprocals

- The reciprocal of a number is that number divided into 1.
- Taking the reciprocal of an exponential number causes the exponent to be multiplied by -1.

Roots

- Taking a root of a number is the same as raising it to a fractional power.
- For example, the square root of a number is the same as the $\frac{1}{2}$ power of that number.

Review Questions

1. Define the square and square root of a number.
2. How are the square and square root operations related?
3. What is the result of taking the square root of the square of a number?
4. What is a perfect square?
5. How many square roots does a number have?
6. What is a radical?
7. To which root does the radical apply?
8. How do you remove like factors from inside a radical?
9. Does taking the square root of a positive number between zero and 1 give a result larger or smaller than the original number? How about square roots of numbers greater than 1?
10. Give the five rules of exponents. Tell when they do and don't apply.
11. What is the result of raising a number to the zero power? Are there any exceptions?
12. What is the difference between a power and an exponent?
13. Define a reciprocal.
14. What is the relationship between reciprocals and negative exponents?
15. What is the relationship between reciprocals and division?
16. When a positive number is raised to a negative power, is the result positive or negative?
17. When is scientific notation useful?
18. Describe the two parts of a number expressed in scientific notation.
19. What does it mean when a number in scientific notation has a negative exponent?
20. What does it mean when a number in scientific notation has a negative value?

Problem Solutions

Section 2-2

2-2.2	$2\sqrt{3}$	2-2.4	$6\sqrt{3}$
2-2.6	4	2-2.8	$2\sqrt{3}$
2-2.10	$a\sqrt{a}$	2-2.12	$2\sqrt{2} * \sqrt{3}$
2-2.14	$\sqrt{3}$		

Section 2-3

2-3.2	3^5	2-3.4	y^{15}
2-3.6	5^{11}	2-3.8	$-10y^7$
2-3.10	2^6	2-3.12	3
2-3.14	y^2	2-3.16	5^4
2-3.18	$5y$	2-3.20	2^2
2-3.22	y^{15}	2-3.24	a^{30}
2-3.26	10^6	2-3.28	x^{4b}
2-3.30	x^{16}	2-3.32	12^2
2-3.34	$(a * c)^2$	2-3.36	$3^4 * 5^4$
2-3.38	$2^2 * 2^2 * 2^2$	2-3.40	$(x^2)^2 * (y^3)^2$
2-3.42	$(\frac{3}{4})^2$	2-3.44	$(x / y)^2$
2-3.46	$2^3/3^3$	2-3.48	$3^n / 4^n$
2-3.50	$3^3 / 8y^3$	2-3.52	$1 / (3x)$
2-3.54	5	2-3.56	2
2-3.58	$\frac{7}{8}$	2-3.60	1
2-3.62	3^{-2}	2-3.64	5
2-3.66	$1 / 3^4$	2-3.68	$3yx^2$
2-3.70	$1 / (2x)^2$	2-3.72	$-(2x)^3$
2-3.74	$1 / 3^2$	2-3.76	$-(x^3y^3)$
2-3.78	$10^0, 10^0, 10^0$	2-3.80	$10^{\frac{5}{6}}, 10^{\frac{1}{6}}, 10^{\frac{3}{2}}$
2-3.82	$2^{\frac{1}{2}}, 2^{\frac{1}{2}}, 2^{\frac{3}{2}}$	2-3.84	$2^{\frac{3}{2}}$ (see note), $2^0, 2^{-\frac{1}{2}}$
2-3.86	$10^5 * 3^2$	2-3.88	$2^4 * 3^2$

Note: The rules of exponents cannot be applied to the first part of problem 2-3.84 as originally written. However, by rewriting $2^{\frac{1}{2}} + 2^{\frac{1}{2}}$ as $2(2^{\frac{1}{2}})$, the rules of exponents can be applied.

Section 2-4

2-4.2	$1.2346 * 10^4$	2-4.4	$1.2346 * 10^2$
2-4.6	$1.2346 * 10^0$	2-4.8	$1.2346 * 10^{-2}$
2-4.10	$1.0000 * 10^5$	2-4.12	$1.0000 * 10^9$
2-4.14	$1.2300 * 10^{-6}$	2-4.16	0.0032
2-4.18	157000	2-4.20	0.08123
2-4.22	$2.180 * 10^{-10}$	2-4.24	$9.330 * 10^0$
2-4.26	64 and 63	2-4.28	1 in 1,073,741,824 and 1 in 16

Number Systems

The mind that is not baffled is not employed.
—Wendell Berry

3-1 ■ Introduction

How to Use This Chapter

As an information technology professional, you will sometimes need to deal with the internal number systems employed by computers. The electronic circuitry that runs inside each computer is based on a transistor. A transistor is a two-state device that works like a switch in that it is either on or off. Thus, it is natural to find that at the electronic level computers do not use our familiar decimal numbers, but rather a number system based on the number 2. This system is called the binary number system. Another number system, called the hexadecimal system, is a companion to the binary number system.

Computers employ both binary and hexadecimal numbers. There will come a time when you will need knowledge of these new number systems to solve problems. This chapter will give you the tools to be able to make sense of both binary and hexadecimal numbers.

Background

Persons working with computers need to be at least aware of the binary and hexadecimal numbers lurking inside. Computer programmers and technical support persons must understand these number systems in order to make conversions between them and our familiar decimal system.

All of these number systems (decimal, binary, and hexadecimal) are positional number systems, which means they employ a set of symbols to represent zero and the first few counting numbers. Our decimal system has 10 such symbols (0, 1, 2, 3, 4, 5, 6, 7, 8, 9) to represent the numbers up to nine. Numbers larger than nine are represented by several symbols in combination.

When several symbols are combined to represent a number, each symbol contributes to the overall value according to its position. The value of a symbol is multiplied by the magnitude of the position it occupies. Positional magnitudes in the decimal system are all powers of the base number, 10. For numbers 1 and larger, the rightmost place has a positional magnitude of 10 to the zero power (the units place). The next place to the left is 10 times larger, with a magnitude of 10 to the first power (the tens place). The value at each position counts 10 times more than the position on its right.

Other number systems work in this same way except that they use a different base number and a different set of symbols.

 Note: The discussion throughout this chapter of presenting numbers in different number systems is limited to integer numbers. Therefore, when you read the word *number*, think *integer*.

Chapter Objectives

In this chapter you will learn:

- The rules of a positional number system
- How the binary number system works
- How to convert from binary to decimal and vice versa
- How the hexadecimal number system works
- How to convert from hexadecimal to decimal and vice versa
- How to convert from hexadecimal to binary and vice versa
- How numbers are represented in computer memory
- How memory is addressed in computers

Introductory Problem

In the days before radio was developed, a signaling system between the shore and ships entering a harbor at night might have involved three lanterns placed next to one another. Each lantern may have had either a red filter or a green filter placed over it. The various combinations of colors represented different messages. Red-red-red represented one message, red-red-green another, green-red-green still another. Both the harbormaster and the ship's captain had a signaling manual that listed all of the possible messages.

1. How many different messages could be represented by this system?

2. Suppose 4 two-color lanterns were used. How many messages would be possible?

3. Suppose three colors were used with three lanterns. How many messages would be possible?

Several Examples

What is 345? Most people would say that 345 is the number three hundred and forty-five. For everyday use, this definition would be true enough. But as we are about to introduce number systems, we need to be more precise with our definitions.

More precisely, 345 is a coded representation of the number we call three hundred and forty-five. In addition, the coded representation uses a positional notation based on the powers of the number 10. The rightmost position is called the units place. The next place to the left is called the tens place. Moving left one more place reaches the hundreds place. Thus, the three-digit number 345 is

$$(3 * 100) + (4 * 10) + (5 * 1) = 345$$

This system of coding numbers is more commonly called the *decimal number system*. For most people it is the only number system they will ever use. But many other number systems are possible. For example:

CCCVL	is 345 using the Roman numerals, a nonpositional number system.
159	is 345 using hexadecimal notation, a base 16 number system.
531	is 345 using octal notation, a base 8 number system.
101011001	is 345 using binary notation, a base 2 number system.

The remainder of this chapter will examine different number systems. First, the decimal number system, the one we use every day, will be examined. Then, two other number systems that are used by computers, the binary number system and the hexadecimal number system, will be examined. As an information system professional you are likely to come across binary and hexadecimal numbers and will need to be able to convert them to decimal numbers so you can understand what you are seeing.

The Decimal Number System

Number systems involve raising a base number to different powers. The number system with which we are most familiar, the decimal system, has a base of 10. The word *decimal* means "ten."

The decimal number system will be used to explain how a number system works. In the sections that follow, you will see that other number systems work exactly like the decimal number system except that they each have a different bases.

Decimal Basics

Base:	10
Number of symbols:	Ten
Symbols:	0, 1, 2, 3, 4, 5, 6, 7, 8, 9 (zero and the first nine counting numbers)
Positional weighting factors:	Based on powers of 10
Symbol name:	Digit

Notice that there is no symbol for the number 10. Numbers larger than 9 are formed by combining one or more symbols in what is called a positional notation.

Positional Notation

In positional notation, the value of any symbol depends on

- the position of the symbol and

- the weighting factor assigned to that position.

Positional weighting factors are all powers of the base, which is 10. The rightmost position, the units position, is weighted by 10 to the zero power. Recall that any number except zero raised to the zero power is 1. Moving left, the next position is weighted by the first power of 10, or 10. Next is the second power of 10, or 100, and so on. Each subsequent position is 10 times the preceding one.

Position weights for the first six decimal positions are shown in the following table:

Position Name	Hundred Thousands	Ten Thousands	Thousands	Hundreds	Tens	Units
Power of 10	10^5	10^4	10^3	10^2	10^1	10^0
Weighting factor	100,000	10,000	1,000	100	10	1

Evaluating Decimal Numbers

Any number can be evaluated as the sum of the contributions of the symbols at each position. The contribution of any position is the product of two things:

- The value of the symbol

- The weighting factor for that position

The leftmost symbol in any number is called the most significant position. Positions to the left of this are all multiplied by zero and therefore do not contribute to the evaluation of the number.

The following table shows how positional notation can be used to evaluate the digits 345 (in the shaded row) as a base 10 number:

Power of 10	10^5	10^4	10^3	10^2	10^1	10^0
Weighting factor	100,000	10,000	1,000	100	10	1
Example digits	0	0	0	3	4	5
Digit * weighting factor				3 * 100	4 * 10	1 * 5
Contribution at each place (product)	0	0	0	300	40	5

Summing the products gives

$$(3 * 100) + (4 * 10) + (1 * 5) = 300 + 40 + 5 = 345$$

Least Significant and Most Significant Digit

Within any number the rightmost digit is the least significant and the leftmost digit is the most significant. These terms are abbreviated LSD and MSD.

▶ Example for the number 10234:

- The LSD is the symbol 4 in the units place.

- The MSD is the symbol 1 in the ten thousands place.

Binary numbers, as their name implies, have a base of 2. Binary numbers are formed from combinations of just two symbols. It is important to know about binary numbers because of the transistor, an electronic device that has two states.

Digital computers are built from millions of transistors, an electronic device that acts like a switch. Like the light switch in your home, transistors have only two states: off and on. While your home light switch must be flipped manually, transistors change states electronically. This permits them to cycle from off to on millions of times faster than a light switch can.

Because transistors have two states, it is convenient to use a number system with a base of 2 to represent numbers inside the computer. If computers were required to do internal computations in the decimal number system, they would be much slower. Therefore, internal calculations and storage is in binary, and only the final output, like that displayed on the screen or printed out, is given in familiar decimal numbers.

Binary Basics

Base:	2
Number of symbols:	Two
Symbols:	0, 1
Positional weighting factors:	Based on powers of 2
Symbol name:	Bit

Position weights for the first six binary positions are shown in the following table:

Position name	Thirty-two	Sixteen	Eight	Four	Two	Unit
Power of 2	2^5	2^4	2^3	2^2	2^1	2^0
Weighting factor	32	16	8	4	2	1

 Notice the similarity between the binary number system and the decimal number system. Like the decimal positions, each binary position has a weighting factor but in this case the factor is a power of 2 rather than a power of 10.

The First Few Binary Numbers
The following table shows the first few integers written in binary notation. Notice that the decimal equivalent of any binary number can be found by summing the products of each bit (0 or 1) with its associated weighting factor. A subscript following a value indicates its base.

Binary	*Decimal Equivalent*
000_2	$(0 * 4) + (0 * 2) + (0 * 1) = 0_{10}$
001_2	$(0 * 4) + (0 * 2) + (1 * 1) = 1_{10}$
010_2	$(0 * 4) + (1 * 2) + (0 * 1) = 2_{10}$
011_2	$(0 * 4) + (1 * 2) + (1 * 1) = 3_{10}$
100_2	$(1 * 4) + (0 * 2) + (0 * 1) = 4_{10}$
101_2	$(1 * 4) + (0 * 2) + (1 * 1) = 5_{10}$
110_2	$(1 * 4) + (1 * 2) + (0 * 1) = 6_{10}$
111_2	$(1 * 4) + (1 * 2) + (1 * 1) = 7_{10}$

Identifying Numbers in Different Systems
When values expressed in different number systems are used in close proximity, it may not be obvious which system belongs to which value. Context may provide a clue. The symbols

employed may also provide a clue. But in certain situations it is easy to be misled. For example, what does the following expression mean?

$$101 < 100$$

To avoid confusion, a subscript following a number is used to indicate its base. Thus

$$101_2 < 100_{10} \text{ (which means } 5_{10} < 100_{10})$$

When a number does not have a subscript, its base may be made clear in some other way. For example, large decimal values are often written with commas separating groups of three digits:

1,000,000 is one million in decimal notation.

Large binary values are often written with spaces between groups of four bits:

1111 0100 0010 0100 0000 is one million in binary notation.

When no explicit clue is given, it is best to assume that a value uses the decimal system. For example:

$$100_2 = 2^2$$

The value 2^2 would be even more confusing if it were written with both a superscript and a subscript. In this situation, it is natural to assume that 2^2 is a decimal value.

Finally, the symbols 0 and 1 *alone* represent the values zero and 1 in most positional systems.

$$0_2 = 0_{10} = 0_{16}$$
$$1_2 = 1_{10} = 1_{16}$$

Important Powers of 2

It is useful to memorize all of the powers of 2 from zero to nine:

$$
\begin{array}{ll}
2^0 = 1 & 2^5 = 32 \\
2^1 = 2 & 2^6 = 64 \\
2^2 = 4 & 2^7 = 128 \\
2^3 = 8 & 2^8 = 256 \\
2^4 = 16 & 2^9 = 512
\end{array}
$$

In addition, you should know these three very important powers of 2:

2^{10}	1024	1 K	Approximately one thousand
2^{20}	1048576	1 Meg	Approximately one million
2^{30}	1073741824	1 Gig	Approximately one billion

Converting Binary to Decimal

Binary numbers follow all the rules of the more familiar decimal number system, except that the weighting factors for each position is a power of 2, not 10.

To convert any binary number to decimal, use the following steps:

1. Multiply the decimal value (0 or 1) of each bit by the power of two associated with that bit (the weighting factor) to give the decimal contribution of each bit.

2. Sum the decimal contributions from all bits to get the decimal equivalent of the entire number.

The following table shows the conversion of the binary number 110101 (in the shaded row) to a base 10 number:

Power of 2	2^5	2^4	2^3	2^2	2^1	2^0
Weighting factor	32	16	8	4	2	1
Example bits	1	1	0	1	0	1
Bit * weighting factor	1 * 32	1 * 16	0 * 8	1 * 4	0 * 2	1*1
Decimal contribution of each bit (product)	32	16	0	4	0	1

The decimal equivalent of 110101_2 is found by summing the products in the final row of the table:

$$32 + 16 + 0 + 4 + 1 = 53$$

Thus

$$11\ 0101_2 = 53_{10}$$

More examples:
Convert 111_2 to decimal.

Multiply: $1 * 4 = 4$, $1 * 2 = 2$, $1 * 1 = 1$
Sum: $4 + 2 + 1 = 7$
The result is 7_{10}.

Convert 10001_2 to decimal.

Multiply: $1 * 16 = 16$, $0 * 8 = 0$, $0 * 4 = 0$, $0 * 2 = 0$, $1 * 1 = 1$
Sum: $16 + 0 + 0 + 0 + 1 = 17$
The result is 17_{10}.

Convert 10011_2 to decimal.

Multiply: $1 * 16 = 16$, $0 * 8 = 0$, $0 * 4 = 0$, $1 * 2 = 2$, $1 * 1 = 1$
Sum: $16 + 0 + 0 + 2 + 1 = 19$
The result is 19_{10}.

Convert 11101_2 to decimal.

Multiply: $1 * 16 = 16$, $1 * 8 = 8$, $1 * 4 = 4$, $0 * 2 = 0$, $1 * 1 = 1$
Sum: $16 + 8 + 4 + 0 + 1 = 29$
The result is 29_{10}.

Practice Problems
Convert the following binary numbers to decimal numbers.
3-3.1 1001

3-3.2 1110

3-3.3 1010

3-3.4 1111

3-3.5 10 1010

3-3.6 11 0110

3-3.7 1 0010 0100

3-3.8 1 1111 1111

Converting Binary Numbers Composed of All 1s

A special case of finding the decimal equivalent of binary numbers occurs when the binary number is composed of all 1s. Using the standard method of conversion is time consuming because the power of 2 for each bit must be added to get the total decimal value. As the number of bits grows, so does the complexity of the conversion.

$$
\begin{aligned}
111_2 &= 1 + 2 + 4 &&= 7_{10} \\
1111_2 &= 1 + 2 + 4 + 8 &&= 15_{10} \\
1\,1111_2 &= 1 + 2 + 4 + 8 + 16 &&= 31_{10} \\
11\,1111_2 &= 1 + 2 + 4 + 8 + 16 + 32 &&= 63_{10}
\end{aligned}
$$

There is, however, a much easier way to convert these numbers. Notice that each decimal number above is exactly one less than a power of 2.

$$
\begin{aligned}
7 &= 8 - 1 &&= 2^3 - 1 \\
15 &= 16 - 1 &&= 2^4 - 1 \\
31 &= 32 - 1 &&= 2^5 - 1 \\
63 &= 64 - 1 &&= 2^6 - 1
\end{aligned}
$$

To see why this is so, you first have to know that any number that is an exact power of 2 is represented in binary notation as a 1 followed by a string of zeros. The number of zeros is the same as the power of 2.

$$
\begin{aligned}
2^1 &= 10_2 \\
2^2 &= 100_2 \\
2^3 &= 1000_2
\end{aligned}
$$

Subtracting 1 from the binary numbers above results in a string of all 1s. Each zero becomes a 1. The number of 1s is the same as the power of 2.

$$
\begin{aligned}
2^1 - 1 &= 1_2 \\
2^2 - 1 &= 11_2 \\
2^3 - 1 &= 111_2
\end{aligned}
$$

> **!** To find the decimal equivalent of any binary number composed of all 1s, just raise 2 to a power equal to the number of bits and subtract 1.

▶ Examples:

$$
\begin{aligned}
11\,1111\,1111_2 \text{ (ten 1s)} &= 2^{10} - 1 &&= 1024 - 1 &&= 1{,}023_{10} \\
111\,1111\,1111\,1111_2 \text{ (fifteen 1s)} &= 2^{15} - 1 &&= 32768 - 1 &&= 32{,}767_{10} \\
1111\,1111\,1111\,1111\,1111_2 \text{ (twenty 1s)} &= 2^{20} - 1 &&= 1048576 - 1 &&= 1{,}048{,}575_{10}
\end{aligned}
$$

Note: All binary numbers that end in 1 are odd. Those that end in zero are even.

Converting Decimal to Binary

The easiest method to convert decimal (base 10) numbers to binary numbers proceeds bit by bit from the least significant bit (LSB) to the most significant bit (MSB), as a series of divisions by 2:

1. Divide the base 10 number by 2. The result will be a whole number and a remainder (either 0 or 1). The remainder (abbreviated R) becomes the binary digit for this position.

2. The whole number replaces the decimal number in step 1, and the process is repeated until the whole number is zero. The final remainder is the MSB.

When dividing any integer by 2, the remainder will be either zero or 1:

- Zero when the integer is even
- 1 when the integer is odd

The following table shows the conversion of the base 10 number 13_{10} to a binary number.

Division by 2	Whole Number	Remainder	Place	Bit
$\dfrac{13}{2} =$	6	R 1	2^0	1 (LSB)
$\dfrac{6}{2} =$	3	R 0	2^1	0
$\dfrac{3}{2} =$	1	R 1	2^2	1
$\dfrac{1}{2} =$	0	R 1	2^3	1 (MSB)

When the division yields a whole number of zero, the remainder is the most significant bit (MSB) and the conversion is complete. Now reverse the bits to give the result: 13_{10} is equivalent to 1101_2.

Note: Remember to reverse the bits.

▶ *Examples.* Convert 21_{10} to binary.

21 / 2 = 10 R1	LSB is	1
10 / 2 = 5 R 0	Next bit is	0
5 / 2 = 2 R 1	Next bit is	1
2 / 2 = 1 R 0	Next bit is	0
1 / 2 = 0 R 1	MSB is	1

Reversing bits gives $21_{10} = 10101_2$.

Convert 27_{10} to binary.

$27/2 = 13$ R1	LSB is	1
$13/2 = 6$ R 1	Next bit is	1
$6/2 = 3$ R 0	Next bit is	0
$3/2 = 1$ R 1	Next bit is	1
$1/2 = 0$ R 1	MSB is	1

Reversing bits gives $27_{10} = 11011_2$.

Convert 29_{10} to binary.

$29/2 = 14$ R 1	LSB is	1
$14/2 = 7$ R 0	Next bit is	0
$7/2 = 3$ R 1	Next bit is	1
$3/2 = 1$ R 1	Next bit is	1
$1/2 = 0$ R 1	MSB is	1

Reversing bits gives $2910 = 11101_2$.

Practice Problems

Convert the following decimal numbers to binary numbers.

3-3.9 7

3-3.10 15

3-3.11 31

3-3.12 63

3-3.13 100

3-3.14 200

3-3.15 300

3-3.16 400

Another important number system when working with computers is the hexadecimal system. *Hex* means "six" and *decimal* means "ten." Thus, the base of the hexadecimal system is 16. Hexadecimal numbers are often called hex numbers for short.

As we have seen in the previous section, the on–off nature of transistors requires that numbers stored in the memory of a computer must be expressed in the binary number system. While binary numbers are the most efficient system for the computer, they are very difficult for people to read. To express the base 10 number 1000 in binary requires ten bits. To express the base 10 number 1 million requires 20 bits. Any human, even an expert mathematician, is likely to miss a bit when reading a 20-bit number.

To make it easier for us humans, computers are designed to make binary numbers readable. Most computer output (to the screen or printer) is converted to the familiar decimal system. However, the conversion from binary to decimal requires a fair amount of processing, while conversion from binary to hex is very fast. Therefore, some of the internal workings of computers, such as memory allocation, are displayed in hex rather than decimal.

Hexadecimal numbers offer the following advantages:

- Large hexadecimal numbers require fewer places (digits) than decimal numbers.

- The conversion from binary to hexadecimal is much faster than the conversion from binary to decimal.

- Reading hexadecimal numbers is much easier than reading binary numbers.

Hexadecimal Basics

Base:	16
Number of symbols:	Sixteen
Symbols:	0, 1, 2, 3, 4, 5, 6, 7, 8, 9, A, B, C, D, E, F
Positional weighting factors:	Based on powers of 16
Symbol name:	Hex digit

Position weights for the first six hex positions are shown in the following table:

Power of 16	16^5	16^4	16^3	16^2	16^1	16^0
Weighting factor	1048576	65536	4096	256	16	1

The Hexadecimal Digits

Hex Digit	Decimal Value	Hex Digit	Decimal Value
0	0	8	8
1	1	9	9
2	2	A	10
3	3	B	11
4	4	C	12
5	5	D	13
6	6	E	14
7	7	F	15

Ways to Label Hex Numbers

Whenever hexadecimal numbers are used in a context in which they might be confused with numbers using another base, they must be labeled in some way that makes it clear they are hex numbers. The fact that they may contain letters as symbols is not enough since many hex numbers use only the usual decimal digits. For example, the number 100 might be 100 decimal or 100 binary or 100 hex. Thus, you will sometimes see hex numbers with the letter H at the end. This method of labeling is often used in situations in which subscripts are not convenient, such as on computers. For example, hex 100 would be written 100H.

 Remember, the H is not a hex digit, since the only letters that can serve as hex digits are A–F.

A better way to indicate a hex number is by using a subscript to indicate the base, for example, 100_{16}.

To show that the hex number 100 is equivalent to the decimal number 256, you might write

$$100_{16} = 256_{10}$$

Important Hexadecimal Numbers

10_{16}	$= 16_{10}$	$= 2^1$	(10 in any system = base)
400_{16}	$= 1024_{10}$	$= 2^{10}$	(1 K)
$FFFF_{16}$	$= 65535_{10}$	$= 2^{16} - 1$	(1 less than 64 K)
10000_{16}	$= 65536_{10}$	$= 2^{16}$	(64 K)
$FFFFF_{16}$	$= 1048575_{10}$	$= 2^{20} - 1$	(1 less than 1 Meg)
100000_{16}	$= 1048576_{10}$	$= 2^{20}$	(1 Meg)

Converting Hexadecimal to Decimal

Hexadecimal numbers follow the same rules as binary and decimal numbers except that the weighting factors for each position is a power of 16.

To convert any hexadecimal number to decimal, follow these steps:

1. Multiply the decimal value of each hex digit (0–15) by the power of 16 for that position (weighting factor) to get the decimal contribution at each position.

2. Sum the decimal contributions for all positions to get the decimal equivalent for the entire number.

The following table shows the conversion of the hexadecimal number 1A5F (in the shaded row) to a base 10 number:

	16^5	16^4	16^3	16^2	16^1	16^0
Power of 16	16^5	16^4	16^3	16^2	16^1	16^0
Weighting factor	1048576	65536	4096	256	16	1
Example hex digits	0	0	1	A	5	F
Decimal value of each hex digit	0	0	1	10	5	15
Decimal value of hex digit times weighting factor	0 * 1048576	0 * 65536	1 * 4096	10 * 256	5 * 16	15 * 1
Product	0	0	4096	2560	80	15

The decimal equivalent is found by summing the products:

$$4096 + 2560 + 80 + 15 = 6751_{10}$$

Thus

$$1A5F_{16} = 6751_{10}$$

▶ Examples:
Convert 93_{16} to decimal.

Multiply: $(9 * 16) = 144,$ $(3 * 1) = 3$
Sum: $144 + 3 = 147$
The result is 147_{10}.

Convert $2B_{16}$ to decimal.

Multiply: $(2 * 16) = 32,$ $(11 * 1) = 11$
       (Notice that B is replaced with 11)
Sum: $33 + 11 = 43$
The result is 43_{10}.

Convert ABC_{16} to decimal.

Multiply: $(10 * 256) = 2560,$ $(11 * 16) = 176,$ $(12 * 1) = 12$
       (Notice that A is replaced with 10, B is replaced with 11, C is replaced with 12.)
Sum: $2560 + 176 + 12 = 2748$
The result is 2748_{10}.

Convert $2B3F_{16}$ to decimal.

Multiply: $(2 * 4096) = 8192,$ $(11 * 256) = 2816,$ $(3 * 16) = 48,$
       $(15 * 1) = 15$
       (Notice that B is replaced with 11; F is replaced with 15.)
Sum: $8192 + 2816 + 48 + 15 = 11071$
The result is 11071_{10}.

Practice Problems
Convert the following hexadecimal numbers to decimal numbers.
3-4.1 10

3-4.2 123

3-4.3 CA2

3-4.4 ABCD

3-4.5 FFFF

Decimal Equivalent of Hexadecimals Numbers with all Fs
Hexadecimal numbers composed of all Fs are always 1 less than the next higher power of 16.

F	(one F)	$= 16^1 - 1$
FF	(two Fs)	$= 16^2 - 1$
FFF	(three Fs)	$= 16^3 - 1$
FFFF	(four Fs)	$= 16^4 - 1$
FFFFF	(five Fs)	$= 16^5 - 1$

To find the decimal equivalent of any hexadecimal number composed of all Fs, simply subtract 1 from the next higher power of 16. The exponent will be the same as the number of hex digits in the original number.

Converting Decimal to Hexadecimal

The easiest method to convert decimal numbers to hexadecimal numbers proceeds hex digit by hex digit, from the least significant digit (LSD) to the most significant digit (MSD) as a series of divisions by 16:

1. Divide the base 10 number by 16.

2. The result will be a whole number and a decimal part. The decimal part needs to be converted into a remainder. (See the sidebar that follows for instructions on how to compute remainders on your calculator.) Once you know the remainder, convert it into a hex digit for this position. The remainder of the first division becomes the least significant hex digit (LSD).

3. The whole number replaces the decimal number in step 1 and the process is repeated digit by digit until the whole number is zero. The final remainder is the most significant hex digit (MSD).

Using a Calculator to Find a Remainder When Dividing by 16

Division of any number by 16 gives a result that is made up of a whole number and a decimal part.

$$\text{any number} / 16 = \text{result}$$
$$\text{result} = \text{whole number} + \text{decimal part}$$

To find the remainder:

Subtract the whole number from the result to get the decimal part alone.
Then multiply the decimal part by 16 to get the remainder.

▶ Example: $150 \div 16 = 9.375$ Result is whole number + decimal part.
             $9.375 - 9 = 0.375$ Subtract whole number to get decimal part alone.
             $0.375 * 16 = 6$ Multiply decimal part by 16 to get remainder.

Thus, 150 divided by 16 gives a whole number of 9 and a remainder of 6. This result can be abbreviated as:

$$9\ R6$$

Remainders for divisions by other numbers can be found by this same method, just replace 16 with the other number.

The following table shows the conversion of the base 10 number 743_{10} to a hexadecimal number using the steps listed above:

Division by 16	Whole Number	Remainder	Place	Hex Digit
$\dfrac{743}{16} = 46.4375$	46	R 7	16^0	7 (LSD)
$\dfrac{46}{16} = 2.875$	2	R 14	16^1	E ($14_{10} = E_{16}$)
$\dfrac{2}{16} = 0.125$	0	R 2	16^2	2 (MSD)

When the division yields a whole number of zero, this is the most significant digit (MSD), and the conversion is complete. Now reverse the hex digits to get the result $2E7_{16}$.

$$743_{10} = 2E7_{16}$$

Note: This method is identical to the conversion from decimal to binary except that the division is now by 16. The remainders that result from each division may range from 0 to 15. Values above 9 must be converted to the appropriate hex digit A–F.

▶ Examples
Convert 216_{10} to hexadecimal.

216 / 16	= 13 R 8	LSD is	8	
13 / 16	= 0 R 13	MSD is	D	(13_{10} becomes D_{16})

Reversing the digits gives $D8_{16}$.

Convert 549_{10} to hexadecimal.

549 / 16	= 34 R 5	LSD is	5
34 / 16	= 2 R 2	Next digit is	2
2 / 16	= 0 R 2	MSD	2

Reversing the digits gives 225_{16}.

Convert 1020_{10} to hexadecimal.

1020 / 16	= 63 R 12	LSD is	C	(12_{10} becomes C_{16})
63 / 16	= 3 R 15	Next digit is	F	(15_{10} becomes F_{16})
3 / 16	= 0 R 3	MSD	3	

Reversing the digits gives $3FC_{16}$.

Convert 3030_{10} to hexadecimal.

3030 / 16	= 189 R 6	LSD is	6	
189 / 16	= 11 R 13	Next digit is	D	(13_{10} becomes D_{16})
11 / 16	= 0 R 11	MSD	B	(11_{10} becomes B_{16})

Reversing the digits gives $BD6_{16}$.

Practice Problems
Convert the following decimal numbers to hexadecimal numbers using the digit-by-digit method described on page 96. Check your solution with a scientific calculator.

3-4.6   10
3-4.7   20
3-4.8   30
3-4.9   40
3-4.10 100
3-4.11 200
3-4.12 300
3-4.13 400
3-4.14 1024
3-4.15 2000
3-4.16 5000

Hexadecimal Words

For fun, some hexadecimal numbers can be read as English language words. To work, the hex number must be restricted to only certain symbols that are valid letters in the English language or look like them. The following are valid symbols:

A, B, C, D, E, F   which represent the corresponding letters
0 (zero)   which looks like the letter *O*
1 (one)   which looks like the letter *I*

An example is the hex number DECADE, which has a decimal value of 14,600,926.

    Use a scientific calculator to convert the following decimal numbers into hexadecimal numbers and read the words hidden inside. Note: Many calculators display the hex "letters" B and D in lowercase. So, for example, hex number DECADE would be displayed as dECAdE.

3-4.17   3,243

3-4.18   44,061

3-4.19   48,879

3-4.20   53,710

3-4.21   64,206

3-4.22   14,613,198

3-4.23   233,577,965

Complete the following sentences by converting the decimal numbers into hexadecimal "words."

3-4.24   14,596,894 looked great in her 12,496,365 11,588,046 as she entered the 51,966 to get an 7,405 14,600,958 12,648,430.

3-4.25   DNA is an 44,061 that is a form of 185,647,326.

3-4.26   My 700,638 is a sturdy 708,798 248,639,950 with a 1,027,565 16,435,934.

3-4.27   61,904 and 61,937 were 4,077 48,879 for dinner.

3-4.28   The 233,573,869 message read "The 3,098 and the 4,017 are on your trail."

3-4.29   2,827 2845 diamonds because he had the 2,766.

3-4.30   So far, in the previous problems the largest decimal number that converts into a hexadecimal "word" is 233,577,965. Can you discover a larger one?

3-4.31   Suppose you let the hex digit 5 (five) represent the letter S. Will this help you find an even larger decimal number that converts into a hexadecimal "word"?

Converting Hexadecimal to Binary

Because 16 is a power of 2 (i.e., 2^4), converting hexadecimal numbers to binary is easy. Each digit in a hex number converts to a 4-bit binary group. You simply need to know the binary equivalents for the 16 hex symbols.

Hex Digit	Binary Value	Hex Digit	Binary Value
0	0000	8	1000
1	0001	9	1001
2	0010	A	1010
3	0011	B	1011
4	0100	C	1100
5	0101	D	1101
6	0110	E	1110
7	0111	F	1111

To convert any hexadecimal number into binary, simply replace each hex digit with the corresponding 4-bit group. Then string the groups together to form the binary number.

Notice that the resulting binary number has four times the number of bits as there were digits in the equivalent hex number. Because large binary numbers are difficult to read, it is conventional to place spaces between the 4-bit groups. This spacing is analogous to using a comma to separate 3-digit groups in large decimal numbers.

Finally, drop any leading zeros from the result. The following examples will make all of this clear.

▶ **Example.** Convert hex $7AB1_{16}$ to binary. Replace each hex digit with the corresponding 4-bit group.

$$\begin{array}{cccc} 7 & A & B & 1 \\ 0111 & 1010 & 1011 & 0001 \end{array}$$

Dropping the leading zero gives $111\ 1010\ 1011\ 0001_2$.

▶ **More Examples.** Convert 12_{16} to binary.

$1_{16} = 0001_2$
$2_{16} = 0010_2$
Combining binary digits gives $0001\ 0010_2$.
Dropping leading zeros gives $1\ 0010_2$.

Convert AA_{16} to binary.

$A_{16} = 1010_2$
$A_{16} = 1010_2$
Combining binary digits gives $1010\ 1010_2$.

Convert $1B2_{16}$ to binary.

$1_{16} = 0001_2$
$B_{16} = 1011_2$
$2_{16} = 0010_2$
Combining binary digits gives $0001\ 1011\ 0010_2$.
Dropping leading zeros gives $1\ 1011\ 0010_2$.

Practice Problems
Convert the following hexadecimal numbers to binary numbers.
3-4.32 10

3-4.33 FF

3-4.34 40

3-4.35 AB

3-4.36 FFFF

3-4.37 1000

3-4.38 4000

3-4.39 ABCD

3-4.40 1234

3-4.41 10EE

3-4.42 A001

3-4.43 2C01

Converting Binary to Hexadecimal

Converting from binary to hexadecimal is just the reverse of process described in the previous section

- Starting with the LSB, separate the bits into 4-bit groups. You may need to pad the MSB with leading zeros to make the final group have four-bits.

- Then use the following table to convert each 4-bit group into its corresponding hex digit.

Binary Value	Hex Digit		Binary Value	Hex Digit
0000	0		1000	8
0001	1		1001	9
0010	2		1010	A
0011	3		1011	B
0100	4		1100	C
0101	5		1101	D
0110	6		1110	E
0111	7		1111	F

▶ *Example.* Convert 100111001012 to hexadecimal.

Group by fours gives: 0100   1110   0101
Convert each group: 4   E   5
Thus 100111001012 is equivalent to $4E5_{16}$.

▶ *More Examples.* Convert 1011011012 to hexadecimal.

Grouping by fours gives: 0001   0110   1101
Convert each group: 1   6   D
Thus the hexadecimal equivalent is $16D_{16}$.

Convert 1111011112 to hexadecimal.

Grouping by fours gives: 0001   1110   1111
Convert each group: 1   E   F
Thus the hexadecimal equivalent is $1EF_{16}$.

Convert 10110010112 to hexadecimal.

Grouping by fours gives: 0010   1100   1011
Convert each group: 2   C   B
Thus the hexadecimal equivalent is $2CB_{16}$. (*Note:* This binary number may be too large to check on a hand calculator. Try checking it on your computer's calculator.)

Practice Problems

Convert the following binary numbers to hexadecimal numbers.

3-4.44 1001

3-4.45 1010 1110

3-4.46 10 1011 1011

3-4.47 110 1101 1011

3-4.48 1111 1111 1111

3-4.49 10 0100 1010 1000

3-4.50 1000 1000 1000

3-4.51 1 0001 0001 0001

3-4.52 1 0110 1011 0101

3-4.53 10 1010 1010

Representing Numbers in Computer Memory

Numbers are represented inside a computer in a binary notation using collections of transistors. Each transistor can be electronically set to one of two states: on or off. Thus, a transistor represents the binary digit in the "on" state and a transistor in the "off" state represents the binary digit zero. Larger numbers can be represented by several binary digits, formed by several transistors grouped together.

Bits and Bytes

Different transistor groupings have the following names:

- Each binary digit is called a **bit.**

- A group of eight bits is called a **byte.**

- A group of one or more bytes is called a **word.**

Word sizes of 8, 16, 32, or 64 bits are commonly used. The more digits that need to be stored, the larger the word size needed to store them.

The Sign Bit

In order to explain how data is represented in memory, a 4-bit example will be used. Although a 4-bit unit of storage is too small to be practical, it is the perfect size to demonstrate how data is stored. Once these concepts are understood, they can be expanded to larger units of data.

The leftmost bit of a unit of storage is reserved for the sign of the number. It is called the sign bit.

- Positive numbers have the sign bit set to zero.

- Negative numbers have the sign bit set to 1.

The remaining bits are available to represent the value of the number.

Representing Positive Integers

The following positive integers can be represented in 4-bit words:

Decimal Number	Binary Representation
0	0 0 0 0
+1	0 0 0 1
+2	0 0 1 0
+3	0 0 1 1
+4	0 1 0 0
+5	0 1 0 1
+6	0 1 1 0
+7	0 1 1 1

Note: A larger unit of storage is required to store numbers larger than +7. All positive numbers have zero in the leftmost bit.

Adding Positive Integers

Numbers are added bit by bit, starting from the rightmost bit and proceeding to the left.

Add 0001 (+1) and 0001 (+1) to get 0010 (+2).

Carry bit:	1	
First number:	0 0 0 1	(+1)
Second number:	0 0 0 1	(+1)
Result:	0 0 1 0	(+2)

Notice the carry from the units place to the twos place. Full rules for the carry are given next.

Rules for Adding Bits

At each position, there are three possible bits to add:

- The bit from the first number

- The bit from the second number

- A possible carry bit (assigned from the position to the right)

As the three bits at any position are added, the following outcomes are possible:

0 + 0	Becomes 0
0 + 0 + carry	Becomes 1
0 + 1	Becomes 1
0 + 1 + carry	Becomes 0 and a carry to the next place
1 + 1	Becomes 0 and a carry to the next place
1 + 1 + carry	Becomes 1 and a carry to the next place

► *More Examples*

Add 0001 (+1) and 0011 (+3) to get 0100 (+4).

Carry bit:	1 1	
First number:	0 0 0 1	(+1)
Second number:	0 0 1 1	(+3)
Result:	0 1 0 0	(+4)

Add 0011 (+3) and 0001 (+1) to get 0100 (+4) (the reverse of the previous example).

Carry bit:	1 1	
First number:	0 0 1 1	(+3)
Second number:	0 0 0 1	(+1)
Result:	0 1 0 0	(+4)

Add 0010 (+2) and 0100 (+4) to get 0110 (+6).

Carry bit:		
First number:	0 0 1 0	(+2)
Second number:	0 1 0 0	(+4)
Result:	0 1 1 0	(+6)

Add 0011 (+3) and 0100 (+4) to get 0111 (+7).

Carry bit:		
First number:	0 0 1 1	(+3)
Second number:	0 1 0 0	(+4)
Result:	0 1 1 1	(+7)

Add 0011 (+3) and 0011 (+3) to get 0110 (+6).

Carry bit:	1 1	
First number:	0 0 1 1	(+3)
Second number:	0 0 1 1	(+3)
Result:	0 1 1 0	(+6)

Note: In the second column (from the right) 1 + 1 + carry = 1 and a carry to the next place.

Problems Representing Negative Integers

You might imagine that negative integers from -1 through -7 can be represented with a sign bit set to 1 and the three remaining bits set the same as for positive integers:

0	-1	-2	-3	-4	-5	-6	-7
1000,	1001,	1010,	1011,	1100,	1101,	1110,	1111

But this is wrong! Although this system could be made to work, it has some very undesirable characteristics. First, there are two ways to represent zero: 0000 (plus zero) and 1000 (minus zero). The second problem is even more troublesome. Addition and subtraction of positive and negative integers does not work in a consistent manner. A system of notation called 2's complement solves these problems for numbers that are integers.

The 2's Complement Notation

To avoid these problems, a different notation is employed to represent negative numbers. This notation is called 2's complement. To convert any positive integer to its respective negative value take the 2's complement of the number. Use these steps:

- Start with the binary representation of the number.
- Flip the bits (change all 1s to zeros, and all zeros to 1s)[1]
- Add 1 to the result.

To find the binary representation of -1_{10}, take the 2's complement of $+1_{10}$:

Start with +1:	0001
Flip the bits:	1110
Add 1:	1111 (This is the binary representation of -1_{10}.)

The following negative integers can be represented using 2's complement notation:

Decimal Number	Binary Representation
-1	1 1 1 1
-2	1 1 1 0
-3	1 1 0 1
-4	1 1 0 0
-5	1 0 1 1
-6	1 0 1 0
-7	1 0 0 1
-8	1 0 0 0

[1]Flipping the bits is also called taking the *1's complement*. Adding 1 to the 1's complement is given the name *2's complement.*

Note:

- The largest negative integer (-8_{10}) has an absolute value that is one more than the largest positive integer ($+7_{10}$). This is because zero is grouped with the positive integers and there is no negative zero.

- Taking the 2's complement of a binary number flips its sign. Thus, taking the 2's complement of a positive number gives a negative number with the same absolute value. Taking the 2's complement of a negative number gives a positive number with the same absolute value.

Adding 2's Complement Numbers

▶ *Examples*

Add 1111 (-1) and 1111 (-1) to get 1110 (-2).

Carry bit:	1 1 1 1	
First number:	1 1 1 1	(-1)
Second Number:	1 1 1 1	(-1)
Result:	* 1 1 1 0	(-2)

* The final carry is discarded.

Add 0001 ($+1$) and 1111 (-1) to get 0000 (0).

Carry bit:	1 1 1 1	
First number:	0 0 0 1	($+1$)
Second number:	1 1 1 1	(-1)
Result:	* 0 0 0 0	(0)

* The final carry is discarded.

Subtracting 2's Complement Numbers

The 2's complement can be used to subtract binary numbers. Suppose you have two binary numbers. To subtract the second number from the first number, convert the subtraction problem into an equivalent addition problem.

Here is an example using decimal numbers:

If the subtraction problem is:	$1 - 1 = 0$
An equivalent addition problem is:	$1 + (-1) = 0$

Another example subtracting a negative number:

If the subtraction problem is:	$1 - (-1) = 2$
An equivalent addition problem is:	$1 + (+1) = 2$

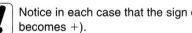 Notice in each case that the sign of the second number was flipped (+ becomes − or − becomes +).

Since taking the 2's complement of a binary number flips the sign, we can use the 2's complement to subtract binary numbers in the same way as in the previous examples. To subtract binary numbers:

- Convert the problem into an equivalent addition problem using the 2's complement to flip the sign of the number to be subtracted.

- Then perform the binary addition.

This will work even if the number to be subtracted is negative, since the 2's complement of a negative number is the positive number with the same absolute value. Once computers were designed to calculate the 2's complement, there was no need for a separate subtraction routine. Only addition was needed. This allowed computers to do arithmetic very quickly.

▶ *Examples*

From 0011 (+3) subtract 0001 (+1) to get 0010 (+2).
Change this into an equivalent addition problem: Add +3 to −1 to get +2.

Carry bit:	1 1 1 1	
First number:	0 0 1 1	(+3)
2's comp. of 2nd num:	<u>1 1 1 1</u>	(−1) 2's complement of +1 is −1.
Result:	* 0 0 1 0	(+2)

* The final carry is discarded since there is no place to store it.

From 1110 (−2) subtract 1101 (−3) to get 0001 (+1).
Change this into an equivalent addition problem: Add −2 to +3 to get +1.

Carry bit:	1 1 1	
First number:	1 1 1 0	(−2)
2's comp. of 2nd num:	<u>0 0 1 1</u>	(+3) 2's complement of −3 is +3.
Result:	* 0 0 0 1	(+1)

* The final carry is discarded.

Overflow

An error condition, called an overflow, occurs whenever the result of binary calculation becomes too large to fit in the storage unit. For the 4-bit storage unit, results greater than +7 or less than −8 will cause an overflow.

This can be seen best by taking the segment of the number line from −8 to +7 and wrapping it into a circle.

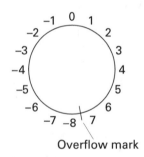

Overflow mark

Starting at any number on the circle:

- Adding a positive number (or subtracting a negative number)—moves clockwise around the circle.

- Adding a negative number (or subtracting a positive number)—moves counterclockwise around the circle.

An error occurs when calculations cross the overflow mark between +7 and −8.

▶ *Examples.* Work the following decimal addition and subtraction problems using 2's complement binary notation. Indicate where an overflow occurs.

Start at +7 and add 1 to give −8 on the circle, causing an overflow.

Carry bit:	1 1 1		
First number:	0 1 1 1	(+7)	
Second number:	0 0 0 1	(+1)	
Result:	1 0 0 0	(−8)	An overflow has occurred.

Start at -8 and subtract 1 to give $+7$ on the circle, causing an overflow.
Change this problem to an equivalent addition problem: $-8 + (-1)$

Carry bit:	1		
First number:	1 0 0 0	(-8)	
2's comp. of 2nd num:	1 1 1 1	(-1)	2's complement of $+1$ is -1.
Result:	*0 1 1 1	$(+7)$	An overflow has occurred.

* The final carry is discarded.

Note: It is not the discarded carry bit that causes the overflow. Rather, it is crossing the overflow mark as you move around the circle.

Practice Problems

3-5.1 Perform the following decimal addition problems using binary numbers in a 4-bit word. Use 2's complement notation to represent negative values. Show your work and your carry bits as in the previous examples. Indicate when an overflow occurs.
a) Add $+2$ and -2.
b) Add -2 and $+5$.
c) Add -2 and -5.
d) Add -2 and $+2$.
e) Add $+3$ and $+5$.

3-5.2 Perform the following decimal subtraction problems using binary numbers in a 4-bit word. Use 2's complement notation to convert the subtraction problem into an equivalent addition problem. Then carry out the binary addition. Show your work and your carry bits as in the previous examples. Indicate when an overflow occurs.
a) From $+2$ subtract $+2$.
b) From -2 subtract $+3$.
c) From $+5$ subtract -3.
d) From -5 subtract -3.
e) From -5 subtract -4.

3-5.3 What is the largest positive decimal integer that can be stored as binary in an 8-bit word (a) if only positive numbers are stored and (b) if both positive and negative numbers are stored using 2's complement notation?

3-5.4 What is the decimal value of the negative number with the largest absolute value that can be stored as binary in an 8-bit word using 2's complement binary notation?

3-5.5 What arrangement of bits is needed to store the following decimal values as binary in an 8-bit word using 2's complement binary notation?
a) 127   b) -127 c) 0   d) -1

3-5.6 What range of signed decimal integers can be stored as binary in a 16-bit word using 2's complement notation?

Memory Addressing in Computers

The most likely way you will use hexadecimal numbers in the field of information technology is in computer memory addressing.

Basic Definitions

One byte is eight bits. A byte is the smallest unit of memory that can be addressed.

One kilobyte (KB) is 2^{10}, which represents 1,024 bytes.

One megabyte (MB) is 2^{20}, which represents 1,048,576 bytes.

One gigabyte (GB) is 2^{30}, which represents 1,073,741,824 bytes.

Absolute Memory Addressing

The Intel Pentium® processor is currently able to address up to four gigabytes of memory. (However, the actual memory installed is usually much less than the addressable maximum.) Under the *absolute memory addressing* scheme, byte-sized addresses start at zero and count up continuously.

Memory is divided into several different areas, each with a specific function. Table 3-5.1 gives a map of the different memory areas. On the left of each area is given the total amount of memory (in decimal) used up to and including that area. On the right of each area is the beginning and ending address (hexadecimal). Because the first address is zero (and not 1), the ending hex address for each area is one less than the amount of memory used up to that point, given in decimal on the left.

Table 3–5.1 Intel Pentium® Memory Map

Amount of Memory (up to this point)	Named Memory Areas	Hex Address Range (absolute)
Up to 4 GB ($4 * 2^{30}$)	Extended Memory (includes HMA below)	FFFFFFFF *110000*
1 MB + 64 KB	High Memory Area (HMA) (first 64 K of extended memory)	*10FFFF* 100000
1 MB (2^{20})	Motherboard ROM (read only memory)	FFFFF *1048575* E0000 *917*
		DFFFF CC000
	Hard Drive ROM (read only memory)	CBFFF C8000
		C7FFF C4000
	EGA ROM (read only memory)	C3FFF C0000
	Video ROM (read only memory)	BFFFF A0000
640 KB ($640 * 2^{10}$)	Conventional Memory (640 KB)	9FFFF 00000

Intel Pentium is a registered trademark of the Intel Corporation.

Practice Problems

Use Table 3-5.1 to answer the following questions.

Introductory Problems. If the following decimal values represent bytes of memory counting from zero, what area of the memory map will contain the final byte?

3-5.7   0

3-5.8   24,000

3-5.9   64,001

3-5.10 655,360

3-5.11 1,000,000

3-5.12 1,100,000

Intermediate Problems. In what memory area will the following absolute hex addresses occur?

3-5.13 00000

3-5.14 0FFFF

3-5.15 90000

3-5.16 A1234

3-5.17 B0000

3-5.18 C9000

3-5.19 F0000

Other Problems

3-5.20 What is the highest hex address in the high memory area?

3-5.21 What is the hex address at the end of the second megabyte of memory.

3-5.22 What is the hex address at the end of the first gigabyte of memory?

Segmented Addressing

The original PC with its 8088 chip had a 16-bit processor.[1] The maximum number of bytes that can normally be addressed in 16 bits is 2^{16}. This is 64 KB or 65,536 bytes. IBM and Intel, who built the original PC, realized that this amount of memory was not enough to be practical. Thus, a scheme was devised to expand the addressable memory to 1,048,576 bytes or 1 megabyte.

This was done using a technique called *segment addressing*. A segment is a contiguous 64-KB area memory. To achieve 1 megabyte of addressable memory, they chose not 16 adjacent segments, but rather 65,536 overlapping segments, a new one starting every 16 bytes. To locate a specific address within the first megabyte of memory, two 16-bit addresses were needed. The first 16-bit address identified a specific *segment number* from among the 65,536 possibilities. This is sometimes called the *paragraph number*. The second 16-bit address identified a specific byte within that segment from among the 65,536 possibilities. This location within the segment was called the *offset* since it was a count (starting with zero) of bytes from the first byte in the segment.

Since a 16-bit binary number can be written with four hex digits, you will usually see a segmented address written as two 4-digit hex numbers separated by a colon. An example is ADCD:0001. An address like this, separated by a colon, is always a segmented hexadecimal address.

All Intel processors subsequent to the 8088 have inherited this addressing scheme. Even though the new processors are able to handle absolute addressing up to and beyond 1 megabyte in what is called *protected mode,* for reasons of backward compatibility, they also are able to switch to *real mode* to emulate the original 8088. Thus, segment offset addressing lurks inside even the most up-to-date Intel processors of today. Therefore, it is useful to learn how to convert to and from segmented addresses.

Converting from Segmented to Absolute Addresses. To process an address stored in segment notation requires it to be converted into a 20-bit absolute address (20 bits can address up to 2^{20} or 1 megabyte of memory). The conversion process is as follows:

1. Multiply the 4-digit segment (paragraph) number by 16 to get the absolute address of the beginning of the segment. (In hex this means appending a zero at the right.)

2. Add the offset to the number from step 1 to get the full absolute address. (The result is a five-digit hex number.)

▶ Examples

Convert the following segmented address to absolute addresses:

1. ABCD:0001 becomes ABCD0 + 0001 = ABCD1.

2. A000:0FFF becomes A0000 + 0FFF = A0FFF.

3. 2000:ABCD becomes 20000 + ABCD = 2ABCD.

[1]Don't confuse the original 8088 processor with its more modern relative, the Pentium, discussed in the previous section.

Converting from Absolute to Segmented Addresses

 A segmented address will map to only one absolute address. The reverse is not true.

Several different segment offset addresses can map to the same absolute address. For example, segmented addresses 0001:0010 and 0002:0000 both map to the same five-digit absolute hex address: 00020.

There is no best way to convert a five-digit absolute address into the two parts of a segmented address. The simplest method is as follows:

- If needed, add zeros on the left to make a five-digit absolute hex address.

- Take the leftmost hex digit of the absolute address (which may be zero) and add three zeros to form the segment number part.

- The remaining four-hex digits become the offset part.

For example:

> 12345 becomes 1000:2345.
> ABCDE becomes A000:BCDE.

 The segment-offset result may be checked by converting back to absolute.

Practice Problems

Introductory Problems
Convert the following segmented addresses into absolute addresses with five hex digits using the simplest method described earlier.

3-5.23 0000:0000

3-5.24 0001:0001

3-5.25 1000:0000

3-5.26 1000:0001

3-5.27 A000:0001

3-5.28 A000:A000

Convert the following absolute hex addresses into segmented addresses using the method described earlier.

3-5.29 00000

3-5.30 00001

3-5.31 000AA

3-5.32 10001

3-5.33 A0001

3-5.34 F1000

Chapter Summary

Rules of Positional Number Systems

The value of the base is the same as the number of symbols required.

The value of a symbol is multiplied by the magnitude of the position it occupies.

The rightmost position has a magnitude of the base to the zero power (always equal to 1).

The next position to the left has a magnitude of the base to the first power.

Each position to the left has a magnitude of one higher power of the base.

No single symbol can represent a number equal to the base. That always takes two digits.

The symbols 10 represent a number equal to the base in any system. For example:

$$10_2 = 2_{10} \qquad 10_3 = 3_{10} \qquad 10_8 = 8_{10} \qquad 10_{10} = 10_{10} \qquad 10_{16} = 16_{10}$$

Conversion Summary

Conversion from decimal to any base involves repeated *divisions* by the base number.

Conversion from any base to decimal involves repeated *multiplication* of the positional weighting factors by the decimal equivalent of each digit and then summing the various products.

Conversion Details

To convert from binary (or hexadecimal) to decimal:

Sum all the positional products formed by multiplying

- the symbol at a position
- the magnitude of the position the symbol occupies

To convert from decimal to binary (or hexadecimal):

Divide the decimal number by the base to get a whole number and a remainder.
The remainder becomes the least significant digit.
The whole number is fed into the next step.
The three previous steps above are repeated over and over to find the digits to the left.
The most significant digit is reached when the division gives a whole number of zero.

To convert from binary to hexadecimal:

Separate the binary digits into groups of four.
Zero-fill the leftmost group so it also contains four bits.
Left to right, each four-bit group represents one hex digit.

To convert from hexadecimal to binary:

Convert each hex digit into a four-bit group.
Combine the four-bit groups, in the same order as the hex digits.
Drop any leading zeros.

Binary and Hex Summary

The First 16 Integers in Decimal, Binary, and Hex

Decimal	Hex	Binary
0	0	0000
1	1	0001
2	2	0010
3	3	0011
4	4	0100
5	5	0101
6	6	0110
7	7	0111
8	8	1000
9	9	1001
10	A	1010
11	B	1011
12	C	1100
13	D	1101
14	E	1110
15	F	1111

Some Important Binary Numbers

Binary	Decimal		Binary	Decimal	
10_2	2^1	$= 2$	1000000_2	2^6	$= 64$
100_2	2^2	$= 4$	10000000_2	2^7	$= 128$
1000_2	2^3	$= 8$	100000000_2	2^8	$= 256$
10000_2	2^4	$= 16$	1000000000_2	2^9	$= 512$
100000_2	2^5	$= 32$	10000000000_2	2^{10}	$= 1024$

Note: All of these numbers are even since even binary numbers have a rightmost bit of zero. The power of 2 is equal to the number of zero bits in the binary number.

Binary	Decimal		Binary	Decimal	
1_2	$2^1 - 1$	$= 1$	111111_2	$2^6 - 1$	$= 63$
11_2	$2^2 - 1$	$= 3$	1111111_2	$2^7 - 1$	$= 127$
111_2	$2^3 - 1$	$= 7$	11111111_2	$2^8 - 1$	$= 255$
1111_2	$2^4 - 1$	$= 15$	111111111_2	$2^9 - 1$	$= 511$
11111_2	$2^5 - 1$	$= 31$	1111111111_2	$2^{10} - 1$	$= 1023$

Note: All of these numbers are odd since odd binary numbers have a rightmost bit of 1. The power of 2 (that 1 is subtracted from) is equal to the number of bits (all 1s) in the binary number.

Some Important Hexadecimal Numbers

Hex	Decimal	
10_{16}	16^1	$= 16$
100_{16}	16^2	$= 256$
1000_{16}	16^3	$= 4096$
10000_{16}	16^4	$= 65536$

Note: All of these decimal numbers *end in 6.* Notice that the power of 16 is equal to the number of zeros in the hex number.

Hex	Decimal		
F_{16}	$16^1 - 1$	$= 15$	
FF_{16}	$16^2 - 1$	$= 255$	
FFF_{16}	$16^3 - 1$	$= 4085$	
$FFFF_{16}$	$16^4 - 1$	$= 65535$	

Note: All of these decimal numbers *end in 5.* The power of 16 (that 1 is subtracted from) is equal to the number of Fs in the hex number.

Review Questions

1. What is the base of each of the following number systems: decimal, binary, hexadecimal?
2. Name the valid symbols of the following number systems: decimal, binary, hexadecimal.
3. What number is equivalent to the base when written in each of the above systems?
4. What are the first 10 powers of 2?
5. What are the first three powers of 16?
6. What name is given to each of the following values: 2^{10} and 2^{20}?
7. Write a procedure in your own words to convert a decimal number to binary.
8. Define the following terms: MSB, LSB, MSD, LSD.
9. Write a procedure in your own words to convert a decimal number to hexadecimal.
10. Write a procedure in your own words to convert a hexadecimal number to binary.
11. Write a procedure in your own words to convert a binary number to hexadecimal.
12. Write a procedure in your own words to find the remainder of a division by 16.
13. Fill in the missing numbers in the following table:

Decimal	Hex	Binary
469	?	?
?	1B2	?
?	?	1101110

14. Define the following terms: bit, byte, kilobyte, megabyte.
15. Write a procedure in your own words to find the 2's complement of a binary number.
16. What is the purpose of the 2's complement?
17. Describe how to add binary numbers. How is carrying handled?
18. What is the purpose of segmented addresses and when are they used?
19. In your own words, write a procedure to convert absolute addresses to segmented addresses.
20. In your own words, write a procedure to convert segmented addresses to absolute addresses.
21. List the starting and ending addresses of the Pentium processor memory areas.

Problem Solutions

Section 3-3	3-3.2	14_{10}	3-3.4	15_{10}
	3-3.6	54_{10}	3-3.8	511_{10}
	3-3.10	1111_2	3-3.12	$11\ 1111_2$
	3-3.14	$1100\ 1000_2$	3-3.16	$1\ 1001\ 0000_2$
Section 3-4	3-4.2	291_{10}	3-4.4	$43{,}981_{10}$
	3-4.6	A_{16}	3-4.8	$1E_{16}$
	3-4.10	64_{16}	3-4.12	$12C_{16}$

3-4.14 400_{16}

3-4.16 1388_{16}

3-4.18 ACID

3-4.20 DICE

3-4.22 DEFACE

3-4.24 <u>DEBBIE</u> looked great in her <u>BEADED</u> <u>BODICE</u> as she entered the <u>CAFE</u> to get an <u>ICED</u> <u>DECAF</u> <u>COFFEE</u>.

3-4.26 My <u>ABODE</u> is a sturdy <u>ADOBE</u> <u>EDIFICE</u> with a <u>FADED</u> <u>FACADE</u>.

3-4.28 The <u>DECODED</u> message read "The <u>CIA</u> and the <u>FBI</u> are on your trail."

3-4.30 233,811,181 is DEFACED

3-4.32 10000_2

3.4.34 1000000_2

3-4.36 1111111111111111_2

3-4.38 100000000000000_2

3-4.40 0001001000110100_2

3-4.42 1010000000000001_2

3-4.44 9_{16}

3-4.46 $2BB_{16}$

3-4.48 FFF_{16}

3-4.50 888_{16}

3-4.52 $16B5_{16}$

Section 3-5   3-5.2   a) 0000_2 b) 1011_2 c) 1000_2 (overflow)
d) 1110_2 e) 1111_2

3-5.4 -128_{10}

3-5.6 $+32767_{10}$ to -32768_{10}

3-5.8   conventional memory

3-5.10 video ROM

3-5.12 high memory area

3-5.14 conventional memory

3-5.16 video ROM

3-5.18 hard drive ROM

3-5.20 $10FFFF_{16}$

3-5.22 $3FFFFFFF_{16}$

3-5.24 00011_{16}

3-5.26 10001_{16}

3-5.28 $AA000_{16}$

3-5.30 0000:0001

3-5.32 1000:0001

3-5.34 F000:1000

CHAPTER

4

Unit Analysis

A handful of patience is worth a bushel of brains.
—saying from a Chinese fortune cookie

4-1 ■ Introduction

How to Use This Chapter

The unit analysis tools presented in this chapter provide a formal way to track units in a calculation. Since most calculations involve physical quantities measured in units, the tools presented here will prove especially useful. The field of information technology presents many opportunities to apply these tools. A variety of these problems are presented here as examples and for practice.

At the heart of the chapter is the concept of unit conversion. An easy method to convert one unit of measurement to another is introduced. Once understood, this same technique can be applied to a variety of problems that require unit conversions as part of a larger calculation.

In an equation, units on one side of the equal sign must balance (be equivalent to) the units on the other side of the equal sign. The technique of balancing units within an equation will help you determine which calculations are appropriate to solve a given problem. Once the solution is complete, the same method will also provide a means to help you check the solution.

This is definitely a chapter that you should master. Adding unit analysis skills to your problem-solving toolbox will empower you to solve problems that until now have been beyond your reach. You will receive good value for time invested in this chapter.

Background

When numbers represent physical quantities, units must be specified. A length of 100 has no meaning unless the units accompany the value: 100 millimeters, 100 yards, 100 miles, and 100 light-years are all lengths, but they are lengths of different magnitudes.

Many problems involve the conversion of units from one form to another. Problems in the field of information technology are no exception. The solution of these problems is often a challenge for many students.

This chapter develops a technique that makes unit conversions easy. The same technique can be applied as a general problem-solving tool for a large body of problems with unit conversions embedded within them.

Chapter Objectives

In this chapter you will learn:

- The difference between a physical quantity and a unit used to measure it
- To understand simple units and composite units, rates, and conversion factors
- How to balance units within an equation
- How to convert the measurement of a physical quantity from one unit into another
- To define the decimal and binary prefixes used to modify a given unit
- How to change prefixes on a given unit
- How to use unit analysis as an important problem-solving tool

Introductory Problem

Before we present a formal method to attack this type of problem, try your hand at solving the following problem involving units.

A computer can process 100 records of data in 200 milliseconds. If a millisecond is $\frac{1}{1000}$ of a second, how many seconds does it take to process a file of 1000 records?

Your instructor will provide the correct answer.

Physical Quantities

Physical quantities are things that can be measured, such as time, distance, mass, weight, area, volume, and money. To measure a physical quantity, you must be able to compare it to a previously defined standard. This standard is called a **unit** of measurement.

Simple Units

Each type of physical measurement has its own set of units. For periods of time, for example, units such as seconds, minutes, days, years, and centuries apply. If you require smaller measurements of time, there are milliseconds, microseconds, and nanoseconds.

Length has an equally robust set of units. Units such as inches, feet, yards, and miles are used in the English measurement system and employed throughout the United States. The metric system[1] of measurement includes lengths such as millimeters, centimeters, meters, and kilometers. The metric system is employed throughout most of the countries of the world. One of the compelling attractions of the metric system is that units of the same physical quantity are related to one another by powers of 10, making conversion from one unit to another relatively easy.

Electrical measurements involve volts, amps, ohms, and watts. Computers involve measurements of processing rates and storage capacities. In the business world, time and money seem to be at the top of the list.

Application to Information Technology

A Gary Larson cartoon of a few years past shows a white-haired professor in front of a chalkboard scrawled full of equations that culminate with the entry: time = $. Beneath the cartoon is the statement:

Einstein discovers that time equals money.

This is a fundamental axiom in the business world and in the world of information technology.

Measuring money involves different currency systems, each country having a different one. Within a given currency system there is the standard measurement unit. The U.S. dollar, the European euro, and the Japanese yen are examples. Each of these has its own set of units for measuring smaller or larger amounts of money. In the United States, for example, there are cents (hundredths of a dollar) and mills (thousandths of a dollar). Large amounts of money in the United States are often stated in million $, billion $, and trillion $. There are also conversion factors (called exchange rates) to convert the currency of one country into that of another country.

 This book will refer to any *single unit* of measurement as a *simple unit*.

Composite Units

The physical quantities we have discussed so far involve single units of measurement (a foot, an inch, a second) that we call simple units. When simple units are combined through multiplication or division, they form *composite units* of measurement.

 This book will refer to composite units as the combination of two or more simple units through multiplication or division.

[1]The metric system is the common name for what is known technically as the system international or SI.

► Examples of composite units:

feet^2	read as square feet
yards^3	read as cubic yards
pounds/inch^2	read as pounds per square inch or psi
feet/second	read as feet per second or fps

Rates

Quantities that have composite units that are ratios (one unit divided by another) are called rates.

⚠ Rates have units that relate one physical measurement with another.

The most familiar rates are those that involve time, for example:

Speed of an automobile is measured in miles per hour (mi/hr).

Speed of a runner is measured in feet per second (ft/sec).

Flow rate of a garden hose is measured in gallons per second (gal/sec).

Cost of a phone call is measured in cents per minute (cents/min).

Cost of a labor is measured in dollars per hour ($/hr).

Cost of leasing equipment can be measured in dollars per month ($/mo).

Production rates can be measured in tons of material produced per day (tons/day).

The following familiar rates do not involve time:

Fuel consumption is measured in miles per gallon (mi/gal).

Cost of bulk food is measured in dollars per pound ($/lb).

Cost of gold is measured in dollars per ounce ($/oz).

The density of text on the page is measured in characters per page (char/pg).

One of the measures of data storage on a hard disk is bytes per track (bytes/track).

Conversion Factors

When a ratio of units involves the same physical quantity, such as two lengths or two times, we will call it a conversion factor.

► Examples of conversion factors:

12 (in./ft)	60 (sec/min)
3 (ft/yd)	60 (min/hr)
5280 (ft/mi)	24 (hr/day)

These factors will be used to convert a measurement of some physical quantity from one unit of measurement to another. Conversion factors are covered in more detail in Section 4-3, Unit Conversions.

Units in Equations

⚠ Just as the numeric value of both sides of an equation must balance, so must the units.[2]

---

[2]Some equations may appear to have units that do not balance. However, on close inspection, they include derived units or factors with hidden units that allow them to balance. This is explained in the next section.

To ensure that units balance, it is helpful to include the units of each term in parentheses following the term. Here are some important rules for combining terms with units in an equation:

- Quantities that add or subtract must have the same units.
- Quantities that multiply or divide form composite units.
- Like units that divide one another, cancel out and can be removed from a composite unit.

Adding and Subtracting Quantities. Before quantities can be added or subtracted, they must be expressed in like units.

▶ **Example Problem**

Compute the total time T in hours required to do three tasks, t_1, t_2, and t_3:

$$T = t_1 + t_2 + t_3$$

If some of the times are given in minutes and some in hours, they must all be converted into the same units.

- Convert the times given in hours to minutes. Then add the times and convert the result to hours.
- Or, convert the times given in minutes to hours. Then add the times to get the result in hours.

Multiplying and Dividing Quantities. Quantities with different units can be multiplied or divided. The result will have composite units combined from the calculation. If the same unit appears on both the top and bottom of a ratio, that pair of units may be removed (canceled), thus simplifying the ratio. When quantities are multiplied or divided within an equation, the combined units on one side of an equation must equal the combined units on the other side, after like units are canceled.

▶ **Example Problem**

How many miles can you travel on 50 gallons of fuel if your vehicle consumes fuel at the rate of 25 miles per gallon?

Let x = the number of miles traveled.

The rate is given as 25 miles per gallon or $25 \left(\dfrac{mi}{gal} \right)$

$x \text{ (mi)} = 50 \text{ gal} * 25 \left(\dfrac{mi}{gal} \right)$

$x \text{ (mi)} = 50 \cancel{\text{ gal}} * 25 \left(\dfrac{mi}{\cancel{gal}} \right)$ Gal divides out (cancels)

$x \text{ (mi)} = 1250 \text{ (mi)}$ leaving miles on both sides.

Derived Units

The fields of science and engineering make use of units that are derived from other units. Newton's law of motion, which relates units of force, mass, and acceleration, is a good example of this:

$$\text{force} = \text{mass} * \text{acceleration}$$

$$f = ma$$

In the metric system, physicists measure force in a unit called a newton, mass in units of kilograms (kg), and acceleration in units of meters per second squared (m/sec^2). Thus

$$\text{force (newton)} = \text{mass (kg)} * \text{acceleration}\left(\frac{m}{\sec^2}\right)$$

At first glance, it appears that the units in this equation do not balance, since the units on the left are in newtons and the units on the right combine to kg * m/sec^2. However, the newton has been derived from kilograms, meters, and seconds so that 1 newton is defined as 1 kg m/sec^2. Substituting the definition for a newton into the equation makes the units balance:

$$\text{force}\left(kg\,\frac{m}{\sec^2}\right) = \text{mass (kg)} * \text{acceleration}\left(\frac{m}{\sec^2}\right)$$

> **!** If you work in science and engineering, you will have to take derived units into account when balancing units in an equation.

Hidden Units

The formula that allows you to convert a temperature (t_F) measured on the Fahrenheit scale to a temperature (t_C) measured on the Celsius scale is:

$$t_C = \left(\frac{5}{9}\right) * (t_F - 32)$$

At first glance it appears that the units of this equation do not balance, t_C being measured in units of °C and t_F being measured in units of °F:

$$t_C\,(°C) = \left(\frac{5}{9}\right) * (t_F - 32)\,(°F)$$

However, the factor $\frac{5}{9}$ is the ratio of *degrees* on the Celsius scale to *degrees* on the Fahrenheit scale. When these hidden units (°C/°F) are included, the equation does, in fact, balance:

$$t_C\,(°C) = \left(\frac{5}{9}\right)\left(\frac{°C}{°F}\right) * (t_F - 32)\,(°F)$$

Notice the °F units cancel out, leaving only the °C units on both sides.

> **Remember:** Both temperatures (t_C and t_F) represent a number of degrees, but the units in which these degrees are mesasured are different. Temperature t_C is measured in units of °C, while temperature t_F is measured in units of °F.

Weight vs. Mass

Outside the fields of science and engineering, there is a great deal of confusion about weight and mass. Mass is a measure of the matter that makes up a body. Weight is the force of gravity that pulls on that body. Therefore, how much things weigh depends on your location. You have no doubt heard that astronauts weigh only one sixth of their usual earth weight when walking on the moon. This is because the moon has only one sixth the gravitational pull of the earth. The mass of the astronaut, however, is the same on earth as it is on the moon.

On earth, at places where gravity is more or less the same, mass and weight have a fixed relationship that depends on the units used to measure them. At markets in the United States you purchase items by the pound, a unit of weight. At markets in most other countries, you purchase items by the kilogram, a unit of mass. If you were to weigh one kilogram of bananas on a scale calibrated in pounds, you would find they weigh about 2.2 pounds. The exact weight would vary slightly from place to place because the force of gravity is not exactly the same at all places on the surface of the earth. However, keep in mind that this relationship applies only here on earth. If markets are ever built on the moon, this relationship won't work.

 Even though mass and weight are different quantities, for practical purposes we can say that a kilogram mass is equivalent to about 2.2 pounds.

Practice Problems

4-2.1 Which of the following units can be used to measure length?
millimeters, miles/hour, kilometers, feet/meter, feet, foot-pounds

4-2.2 Which of the following units can be used to measure time?
seconds, miles/hour, hours, $/day, weeks, centuries

4-2.3 Which of the following units are used to measure rates?
miles, miles/hour, feet, feet/second, seconds, furlongs/fortnight

4-2.4 Which of the following units are used with conversion factors?
miles/hour, miles/kilometer, gallons/minute, hours/second

4-2.5 What are the units of x in each of the following equations?
a) $x = 5$ (ft) $+ 10$ (ft)
b) $x = 1$ (yd) $* 2$ (yd) $* 3$ (yd)
c) $x = 50$ (mi)/2 (hr)
d) $x = 10$ (mi) $* 1/10$ (1/hr)
e) $x = 20$ (gal/min) $* 10$ (min)

Equivalence Expressions and Conversion Factors

Equivalence Expressions

An equivalence expression is a statement that relates two measures of the same physical quantity in different units. Physical quantities such as length, time, or weight, have different units of measurement. An equivalence expression can be used to relate two times or two lengths.

The following equivalence expressions between feet and yards (both lengths) are different ways of saying the same thing:

Three feet are equivalent to one yard.	One yard is equivalent to 3 feet.
A yard is the same as 3 feet.	Three feet are in one yard.
There are 3 feet per yard.	A foot is one third of a yard.

Some examples of equivalence expressions between different units of time are:

There are 24 hours in one day.

A minute contains 60 seconds.

Sometimes you will see equivalence expressions presented as an equation:

$$5280 \text{ feet} = 1 \text{ mile}$$

 This book will not present equivalence expressions as equations since the units do not balance.

Before an equivalence expression can be used in a calculation, it must be restated as a conversion factor.

Conversion Factors

A conversion factor is a quantity that can be inserted into a calculation to convert one physical unit into another. It is an equivalence expression expressed as a rate. For example, consider the equivalence expression: There are 60 seconds per minute. The word 'per' means "for each one." This expression can be written as a conversion factor:

	60 (sec)/(1 min)
Which is shortened to:	60 (sec/min)
The reciprocal of this factor is:	1/60 (min/sec)
The reciprocal factor means:	1/60 of a minute for each one second

 Every conversion factor has a corresponding reciprocal factor. Notice that in a reciprocal factor both the value and the units are flipped.

Conversion Factor Rules

Here are important rules for conversion factors between two units:

- Conversion factors come in pairs, one being the reciprocal of the other.

- Conversion factors include a value followed by a ratio of units in parentheses.

- The value associated with the unit on the top of the ratio is the value of the factor.

- The value associated with the unit on the bottom of the ratio is the value 1.

For example:

The conversion factor 12 (in./ft) means that there are 12 inches in one foot.
The reciprocal factor 1/12 (ft/in.) means there is 1/12 of a foot in one inch.

Using Conversion Factors

Conversion factors multiply other terms so that like units divide out, leaving the desired result. To convert miles into feet, we use the fact that 5280 feet is equivalent to one mile. This gives the following reciprocal pair of conversion factors:

5280 (ft/mi)   and   1/5280 (mi/ft)

One of these factors is appropriate to convert miles into feet; the other converts feet into miles. Since we want to convert miles into feet, we select the factor that has miles on the bottom of the units ratio. Then miles will divide out leaving the desired result in feet. Thus, to convert 3 miles into feet, multiply by 5280 (ft/mi):

$$x(\text{ft}) = 3(\text{mi}) * 5280\left(\frac{\text{ft}}{\text{mi}}\right)$$

$$x(\text{ft}) = 3(\cancel{\text{mi}}) * 5280\left(\frac{\text{ft}}{\cancel{\text{mi}}}\right)$$

$$x(\text{ft}) = 15{,}840 \ (\text{ft}) \qquad\qquad \text{after the mile units divide out}$$

Note: In the example above the appropriate units of each term follow in parentheses. When two factors are multiplied, their units also multiply. Like units that cancel and are removed from the expression.

Conversion Examples

Here are several examples of unit conversion that multiply conversion factors from Table 4-3.3.

1. Convert 5 kilometers to miles using 0.6214 (mi/km).

$$x \ (\text{mi}) = 5 \ (\cancel{\text{km}}) * 0.6214 \left(\frac{\text{mi}}{\cancel{\text{km}}}\right) = 3.107 \ (\text{mi})$$

2. Convert 3 inches to centimeters using 2.540 (cm/in.).

$$x(\text{cm}) = 3(\cancel{\text{in.}}) * 2.54 \left(\frac{\text{cm}}{\cancel{\text{in.}}}\right) = 7.62 \ (\text{cm})$$

Table 4–3.1   Some Common Unit Abbreviations for U.S. Measures

Length		Time		Mass	
mi	mile	yr	year	lbm	pound mass
yd	yard	mo	month	slug	slug
ft	foot	wk	week		
in.	inch	h	hour		
		s	second		

Area		Volume		Weight	
sq in.	square inch	cu in.	cubic inch	lb	pound force
in.^{2}	square inch	in.^{3}	cubic inch		
sq yd	square yard	cu yd	cubic yard	**Temperature**	
yd^2	square yard	yd^3	cubic yard		
sq mi	square mile	gal	gallon	°F	degree Fahrenheit
mi^2	square mile	qt	quart		
ac	acre	pt	pint		

Table 4–3.2 Some Common Unit Abbreviations for Metric Measures

Length		Time		Mass	
km	kilometer	sec	second	kg	kilogram
m	meter	ms	millisecond	g	gram
cm	centimeter	μs	microsecond	mg	milligram
mm	millimeter	ns	nanosecond	μg	microgram
Area		**Volume**		**Weight**	
sq m	square meter	cu cm	cubic centimeter	N	newton
m^2	square meter	cm^3	cubic centimeter		
sq cm	square centimeter	L	liter		
cm^2	square centimeter	mL	milliliter		
sq mm	square millimeter				
mm^2	square millimeter				
Frequency		**Electricity**		**Temperature**	
Hz	hertz	V	volts (emf)	°C	degree Celsius
kHz	kilohertz	mv	millivolts		
MHz	megahertz	A	amps (current)		
GHz	gigahertz	ma	milliamps		
		Ω	ohms (resistance)		
		$m\Omega$	megaohms		
		W	watts		
		kW	kilowatts		

Table 4–3.3 Some Commonly Used Conversion Factors

Length	Time	Temperature
12 (in./ft)	60 (sec/min)	5/9 (°C/°F)
3 (ft/yd)	60 (min/hr)	
5280 (ft/mi)	24 (hr/day)	
1760 (yd/mi)	365.26 (day/yr)	
39.37 (in./m)	100 (yr/century)	
2.540 (cm/in.)		
0.6214 (mi/km)		
Area	**Volume**	**U.S. Liquid Volume**
144 ($in.^2/ft^2$)	1728 ($in.^3/ft^3$)	231 ($in.^3$/gal)
9 (ft^2/yd^2)	27 (ft^3/yd^3)	4 qt/gal
640 (ac/mi^2)	16.39 ($in.^3/cm^3$)	1.0566 (qt/L)
43,560 (ft^2/ac)	1000 (cm^3/L)	2 (pt/qt)
	61.02 ($in.^3$/L)	2 (cups/pt)
		16 (oz/pt)
Weight	**Mass**	**Mass to Weight***
16 (oz/lb)	1000 (g/kg)	2.205 (lb/kg)
2000 (lb/ton)	14.59 (kg/slug)	
4.448(lb/N)		

*At surface of earth

Note: Each conversion factor also has a reciprocal conversion factor (not shown).

3. Convert 4 liters to quarts using 1.0566 (qt/L).

$$x \text{ (qt)} = 4 \text{ (L)} * 1.0566 \left(\frac{\text{qt}}{\text{L}} \right) = 4.226 \text{ (qt)} \text{ (rounded to three places)}$$

4. Convert 30 days to hours using 24 (hr/day).

$$x \text{ (hr)} = 30 \text{ (day)} * 24 \left(\frac{\text{hr}}{\text{day}} \right) = 720 \text{ (hr)}$$

5. Convert 16 tons to pounds (lb) using 2000 (lb/ton).

$$x \text{ (lb)} = 16 \text{ (ton)} * 2000 \left(\frac{\text{lb}}{\text{ton}} \right) = 32,000 \text{ (lb)}$$

Here are some examples of unit conversions that require the reciprocal of the conversion factors from Table 4-3.3 to make the units cancel correctly. Notice that multiplying by a reciprocal factor is the same as dividing by the original factor.

1. Convert 5 miles to kilometers using 1/0.6214 (km/mi).

$$x \text{ (km)} = 5 \text{ (mi)} * \frac{1}{0.6214} \left(\frac{\text{km}}{\text{mi}} \right) = 8.046 \text{ (km)} \quad \text{(rounded to three places)}$$

2. Convert 3 centimeters to inches using 1/2.540 (in./cm).

$$x \text{ (in.)} = 3 \text{ (cm)} * \frac{1}{2.54} \left(\frac{\text{in.}}{\text{cm}} \right) = 1.181 \text{ (in.)} \quad \text{(rounded to three places)}$$

3. Convert 4 quarts to liters using 1/1.0566 (L/qt).

$$x \text{ (L)} = 4 \text{ (qt)} * \frac{1}{1.0566} \left(\frac{\text{L}}{\text{qt}} \right) = 3.786 \text{ (L)} \quad \text{(rounded to three places)}$$

4. Convert 30 hours to days using 1/24 (day/hr).

$$x \text{ (day)} = 30 \text{ (hr)} * \frac{1}{24} \left(\frac{\text{day}}{\text{hr}} \right) = 1.25 \text{ (day)}$$

5. Convert 16,000 pounds (lb) to tons using 1/2000 (lb/ton).

$$x \text{ (ton)} = 16,000 \text{ (lb)} * \frac{1}{2000} \left(\frac{\text{ton}}{\text{lb}} \right) = 8 \text{ (ton)}$$

Practice Problems
Use the given equivalence expressions to convert the following quantities into the units requested.

- Write the equivalence expression as a pair of reciprocal conversion factors.
- Let x equal the quantity you are trying to calculate (in the units requested).
- Then, multiply the given quantity by the appropriate conversion factor (either the given one or its reciprocal) so that the units cancel leaving the requested units.
- Your solution should show units being canceled as in the previous examples.
- Solutions that exceed two decimal places should be rounded to two places.

4-3.1 Express 10 feet in inches (use 1 foot is 12 inches).

4-3.2 Express 10 inches in feet (use 1 foot is 12 inches).

4-3.3 Express 10 meters in feet (use 1 meter is 3.28 feet).

4-3.4 Express 10 feet in meters (use 1 meter is 3.28 feet).

4-3.5 Express 10 miles in kilometers (use 1 kilometer is 0.6214 miles).

4-3.6 Express 10 kilometers in miles (use 1 kilometer is 0.6214 miles).

4-3.7 Express 10 kilograms in pounds (use 1 kilogram is 2.2 pounds).

4-3.8 Express 10 pounds in kilograms (use 1 kilogram is 2.2 pounds).

4-3.9 Express 200 Canadian$ in US$ (use 1 US$ is 1.33 Can$).

4-3.10 Express 200 US$ in Canadian$ (use 1 US$ is 1.33 Can$).

4-3.11 Express 16 quarts in gallons (use 1 gallon is 4 quarts).

4-3.12 Express 16 gallons in quarts (use 1 gallon is 4 quarts).

4-3.13 Express 15 gallons in liters (use 1 gallon is 3.79 liters).

4-3.14 Express 15 liters in gallons (use 1 gallon is 3.79 liters).

4-3.15 Express 10 square feet in square inches (use 1 square foot is 144 square inches).

4-3.16 Express 10 square inches in square feet (use 1 square foot is 144 square inches).

Intermediate Problems. Use the previous instructions to express the following quantities in the units requested. In these problems, two conversion factors will be required to make the units come out correctly.

4-3.17 Express 15 gallons per hour in liters per minute (use 1 hour is 60 minutes, 1 gallon is 3.79 liters).

4-3.18 Express 15 liters per minute in gallons per hour (use 1 hour is 60 minutes, 1 gallon is 3.79 liters).

4-3.19 Express 60 miles per hour in feet per second (use 1 hour is 3600 seconds, 1 mile is 5280 feet).

4-3.20 Express 60 feet per second in miles per hour (use 1 hour is 3600 seconds, 1 mile is 5280 feet).

Physical measurements often make use of standard prefixes that are appended to the beginning of a given base unit of measurement. Each prefix represents a given factor that multiplies the base unit and brings it into a particular range. For example, a *kilo*meter is 1000 meters and a *milli*watt is $\frac{1}{1000}$ of a watt. This is especially important for quantities that are very large or very small.

Two types of prefixes will be covered, decimal prefixes (based on powers of 10) and binary prefixes (based on powers of 2).

Decimal Prefixes

Table 4-4.1 presents some decimal prefixes that are based on powers of 10. Notice that most of these prefixes differ from one another by a factor of 10^3 or one thousand.

Table 4–4.1   Decimal Prefixes

Prefixes that differ by a factor of 1000 or $\frac{1}{1000}$:

Prefix	Symbol	Power of 10	Magnitude	
tera	T	10^{12}	1,000,000,000,000	1 trillion
giga	G	10^{9}	1,000,000,000	1 billion
mega	M	10^{6}	1,000,000	1 million
kilo	k	10^{3}	1,000	1 thousand
unity	none	10^{0}	1	one
milli	m	10^{-3}	1/1,000	1 thousandth
micro	μ	10^{-6}	1/1,000,000	1 millionth
nano	n	10^{-9}	1/1,000,000,000	1 billionth
pico	p	10^{-12}	1/1,000,000,000,000	1 trillionth

Prefixes that differ by a factor 10 or 1/10:

Prefix	Symbol	Power of 10	Magnitude	
hecto	h	10^{2}	100	1 hundred
deca	d	10^{1}	10	1 ten
centi	c	10^{-2}	1/100	1 hundredth
deci	d	10^{-1}	1/10	1 tenth

Beware: Some countries use different definitions for billion and trillion. In England, for example, one billion is 10^{12} and one trillion is 10^{9}, just the opposite of the U.S. definition. To avoid this confusion you will often hear international news reports refer to 10^{9} as a thousand-million and 10^{12} as a million-million.

Binary Prefixes

Additional prefixes, widely used in computing, are based on powers of 2. Hence they are called binary prefixes. Table 4-4.2 shows some of these binary prefixes.

Table 4–4.2   Binary Prefixes

Binary Prefix	Symbol	Power of 2	Decimal Value	Approximate Value
kilo	K	2^{10}	1,024	1 thousand
mega	M	2^{20}	1,048,576	1 million
giga	G	2^{30}	1,073,741,824	1 billion
tera	T	2^{40}	1,099,511,627,776	1 trillion

 Information technology is an emerging field, and you will often encounter binary prefixes having slightly different definitions from the pure binary definitions given in Table 4-4.2. Some of the most common variations follow.

Approximate Binary Prefixes

Because 1024 is approximately 1000, sometimes binary prefixes are based on a power of 10 that is approximately equal to the corresponding power of 2:

kilo	approximately	1,000	1 thousand (10^3)
mega	approximately	1,000,000	1 million (10^6)
giga	approximately	1,000,000,000	1 billion (10^9)
tera	approximately	1,000,000,000,000	1 trillion (10^{12})

Hybrid Binary Prefixes

At other times you will encounter binary prefixes that combine the binary and approximate definitions:

mega	1000 kilo	1,024,000
giga	1000 mega	1,024,000,000

Context Is Important. To add to the confusion, the binary prefixes use the same names as those employed for base 10 prefixes. Therefore, the *context* of the unit associated with the prefix must be used to determine its value. For example, terms such as *kilobyte, megabyte,* and *gigabyte* are used to measure the capacity of computer storage devices (disk drives and memory) and therefore have binary prefixes. On the other hand, *kilometer, kiloton, megawatt,* and *megaton* are terms that are not associated with computers and therefore imply the decimal prefix.

Application to Information Technology

Bits and Bytes

Computers store data by switching binary circuits to one of two states, either off or on. This off–on state is represented by a binary digit (a zero or a 1). For short, a binary digit is called a **bit.**

A collection of eight bits is called a **byte,** which can store exactly one character of data. Larger groupings of data are measured in kilobits and megabits as well as in kilobytes, and megabytes, and gigabytes. The meaning of any unit depends on which of the prefixes (defined earlier) is being employed.

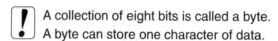

> A collection of eight bits is called a byte.
> A byte can store one character of data.

Binary Abbreviations

GB	gigabyte	MB	megabyte	KB	kilobyte

Binary Definitions

Again, be warned that the exact meaning of any of these terms, especially the megabyte, depends on where it is being used. The definition of a megabyte may be any of the following:

1 MB is 2^{20} or 1,048,576 bytes (the pure binary definition).

1 MB is 1,000 KB or 1,024,000 bytes (the hybrid definition).

1 MB is 1,000,000 bytes (the approximate definition).

How to Use the Prefix Tables

In theory, the factors given in Tables 4-4.1 and 4-4.2 can be applied to any base unit. However, in practice, prefixes are only applied to a few base units.

The decimal prefixes in Table 4-4.1 are usually applied to units of the metric system such as length in meters and time in seconds. The binary units given in Table 4-4.2 are usually applied to units of computer disk storage and memory (either bytes or bits).

To make use of a prefix given in either of the tables, you will first need to develop a reciprocal pair of conversion factors for that prefix. To do this, simply append the appropriate ratio of units to the factor given in the table. The conversion factor will have the form

$$\text{value} \left(\frac{\text{base unit}}{\text{prefix} + \text{base unit}} \right)$$

Here are some examples:

1. **Milliseconds**

   To use the value 10^{-3} associated with the prefix *milli* (m) in Table 4-4.1, with the base unit of seconds (sec), the ratio of units will be

   seconds/milliseconds or sec/ms

   Note: The base unit is alone on the top of the ratio, and the prefix modifies the base unit on the bottom.

   Combining the factor with the units gives the change factor:

   10^{-3} (sec/ms)   There is $\frac{1}{1000}$ of a second in each millisecond.

   And of course, there is a corresponding reciprocal change factor:

   10^{3} (ms/sec)   There are 1000 milliseconds in each second.

2. **Kilometers**

   To combine the value 10^{3} associated with the prefix *kilo* (k) in Table 4-4.1, with a base unit of meters (m), the ratio of units will be

   meters/kilometers or m/km

   Note: The base unit is alone on the top of the ratio, and the prefix modifies the base unit on the bottom of the ratio.

   Combining the factor with the units gives the change factor:

   10^{3} (m/km)   There are 1000 meters in each kilometer.

   And of course, there is a corresponding reciprocal change factor:

   10^{-3} (km/m)   There is $\frac{1}{1000}$ of a kilometer in each meter.

3. **Kilobytes**

   To combine the value 2^{10} representing the prefix *kilo* (K) in Table 4-4.2, with a base unit of bytes, the ratio of units will be

   bytes/kilobyte or bytes/KB

Note: The base unit is alone on the top of the ratio, and the prefix modifies the base unit on the bottom of the ratio.

Combining the factor given in Table 4-4.2 with the units gives the change factor:

2^{10} (bytes/KB)   There are 1024 bytes in each kilobyte.

And, of course there is a corresponding reciprocal change factor:

2^{-10} (KB/byte)   There is $\frac{1}{1024}$ of a kilobyte in each byte.

Note: The capital K indicates the binary prefix, while the lowercase k indicates the decimal prefix.

4. **Kilobits and megabits**
   Since there are 8 bits per byte:

$$8\left(\frac{\text{bits}}{\text{byte}}\right) = 8\left(\frac{\text{kilobits}}{\text{KB}}\right) = 8\left(\frac{\text{megabits}}{\text{MB}}\right)$$

Note: Since the prefix appears in both top and bottom of the ratio, it cancels, leaving the factor unchanged.

Changing Prefixes

Prefixes can be changed using the unit conversion method described in Section 4-2.

In prefix conversions only the prefix of the unit changes. The base unit remains unchanged. Thus, when milliseconds are converted to microseconds, both are forms of the base unit, seconds.

When the conversion is between one prefix unit and its base unit, one conversion factor is required.

▶ *Examples*

Converting milliseconds into seconds requires a conversion factor for seconds/millisecond.

Converting seconds into microseconds requires a conversion factor for microseconds/second.

When the conversion is between two prefixed units, two conversion factors are needed. Each factor relates a prefix unit to its base unit. When combined, the base unit will divide out.

▶ *Examples*

Converting milliseconds into microseconds requires a conversion factor for seconds/millisecond and a conversion factor for microseconds/second.

The process of prefix conversion is as follows:

- First, write a reciprocal pair of conversion factors for each prefix involved.

- Next, write an equation that begins with x (in units wanted) on the left side.

- Begin the right-hand side of the equation with the given value to be converted.

- Continue building the right-hand side of the equation by adding conversion factors, carefully selected from the list of reciprocal pairs prepared previously.

Select conversion factors that will let unwanted units cancel out, leaving only the units wanted.

Select only one factor from each reciprocal pair. Once a factor is selected, cross out its corresponding reciprocal factor and never use it again in this equation.

- Finally, carry out the multiplication. As you do, make sure that the units balance.

▶ Examples Converting Decimal Prefixes

1. Convert 3000 milliseconds (ms) to seconds (sec) (only one pair of conversion factors is needed).

$$1 \text{ ms is } 10^{-3} \text{ sec}$$

$$10^{-3} \left(\frac{\text{sec}}{\text{ms}} \right) \qquad 10^{3} \left(\frac{\text{ms}}{\text{sec}} \right)$$

$$x \text{ (sec)} = 3000 \text{ (ms)} * 10^{-3} \left(\frac{\text{sec}}{\text{ms}} \right)$$

$$= \frac{3000}{1000} \text{ (sec)}$$

$$= 3 \text{ (sec)}$$

2. Convert 100 milliseconds (ms) to microseconds (μs) (two pairs of conversion factors are needed).

$$1 \text{ ms is } 10^{-3} \text{ sec}$$

$$10^{-3} \left(\frac{\text{sec}}{\text{ms}} \right) \qquad 10^{3} \left(\frac{\text{ms}}{\text{sec}} \right)$$

$$1 \text{ μs is } 10^{-6} \text{ sec}$$

$$10^{-6} \left(\frac{\text{sec}}{\text{μs}} \right) \qquad 10^{6} \left(\frac{\text{μs}}{\text{sec}} \right)$$

$$x \text{ (μs)} =$$

$$100 \text{ ms} * 10^{-3} \left(\frac{\text{sec}}{\text{ms}} \right) * 10^{6} \left(\frac{\text{μs}}{\text{sec}} \right)$$

$$100 * 10^{3} \text{ (μs)}$$

$$100,000 \text{ (μs)}$$

3. Convert 3000 nanoseconds (ns) to microseconds (μs) (two pairs of conversion factors are needed).

$$1 \text{ ns is } 10^{-9} \text{ sec}$$

$$10^{-9} \left(\frac{\text{sec}}{\text{ns}} \right) \qquad 10^{9} \left(\frac{\text{ns}}{\text{sec}} \right)$$

$$1 \text{ μs is } 10^{-6} \text{ sec}$$

$$10^{-6} \left(\frac{\text{sec}}{\text{μs}} \right) \qquad 10^{6} \left(\frac{\text{μs}}{\text{sec}} \right)$$

$$x\,(\mu s) =$$

$$3000\text{ ns} * 10^{-9} \left(\frac{\text{sec}}{\text{ns}}\right) * 10^6 \left(\frac{\mu s}{\text{sec}}\right)$$

$$3000 * 10^{-3}\,(\mu s)$$

$$3\,(\mu s)$$

► Examples Converting Binary Prefixes

1. Convert 5000 bytes to kilobytes (KB) (only one pair of conversion factors is needed).

$$1 \text{ KB is } 2^{10} \text{ bytes or } 1024 \text{ bytes}$$

$$1024 \left(\frac{\text{bytes}}{\text{KB}}\right) \qquad \frac{1}{1024}\left(\frac{\text{KB}}{\text{byte}}\right)$$

$$x\,(\text{KB}) = 5000\,(\text{bytes}) * \frac{1}{1024}\left(\frac{\text{KB}}{\text{byte}}\right)$$

$$= \frac{5000}{1024}\,(\text{KB})$$

$$= 4.88\,(\text{KB})$$

2. Convert 2000 kilobytes (KB) to megabytes (MB) (two pairs of conversion factors are needed). (Use binary definition.)

$$1 \text{ MB is } 2^{20} \text{ bytes}$$

$$2^{20}\left(\frac{\text{bytes}}{\text{MB}}\right) \qquad 2^{-20}\left(\frac{\text{MB}}{\text{byte}}\right)$$

$$1 \text{ KB is } 2^{10} \text{ bytes}$$

$$2^{10}\left(\frac{\text{bytes}}{\text{KB}}\right) \qquad 2^{-10}\left(\frac{\text{KB}}{\text{byte}}\right)$$

$$x\,(\text{MB}) = 2000\text{ KB} * 2^{10}\left(\frac{\text{bytes}}{\text{KB}}\right) * 2^{-20}\left(\frac{\text{MB}}{\text{byte}}\right)$$

$$= 2000 * 2^{-10}\,(\text{MB}) = \frac{2000}{1024}\,(\text{MB})$$

$$= 1.95\,(\text{MB})$$

► Examples Converting Hybrid Prefixes

Following are some alternate definitions for megabyte and gigabyte that involve both a power of 2 and a power of 10:

1 megabyte is equivalent to 1000 kilobytes.

1 megabyte is equivalent to $1000 * 2^{10}$ bytes.

1 gigabyte is equivalent to 1000 megabytes.

1 gigabyte is equivalent to $1000 * 1000$ kilobytes.

1 gigabyte is equivalent to $1000 * 1000 * 2^{10}$ bytes.

> **!** These hybrid definitions are particularly confusing since mega and giga are defined in different ways in different situations.

1. Convert 1440 kilobytes to megabytes using 1 megabyte is 1000 kilobytes.

$$1 \text{ MB is } 1000 \text{ KB}$$

$$1000 \left(\frac{\text{KB}}{\text{MB}}\right) \quad \frac{1}{1000} \left(\frac{\text{MB}}{\text{KB}}\right)$$

$$x \text{ (MB)} = 1440 \text{ KB} * \frac{1}{1000} \left(\frac{\text{MB}}{\text{KB}}\right)$$

$$x \text{ (MB)} = 1.44 \text{ (MB)}$$

2. Convert 1.44 megabytes to bytes using 1 megabyte is 1000 kilobytes.

$$1 \text{ MB is } 1000 \text{ KB}$$

$$1000 \left(\frac{\text{KB}}{\text{MB}}\right) \quad \frac{1}{1000} \left(\frac{\text{MB}}{\text{KB}}\right)$$

$$1 \text{ KB is } 2^{10} \text{ bytes}$$

$$1024 \left(\frac{\text{bytes}}{\text{KB}}\right) \quad \frac{1}{1024} \left(\frac{\text{KB}}{\text{byte}}\right)$$

$$x \text{ (bytes)} = 1.44 \text{ (MB)} * 1000 \left(\frac{\text{KB}}{\text{MB}}\right) * 1024 \left(\frac{\text{bytes}}{\text{KB}}\right)$$

$$= 1.44 * 1000 * 1024 \text{ (bytes)}$$

$$= 1,474,560 \text{ (bytes)}$$

Note: The definition of a megabyte as 1000 kilobytes was employed by IBM to define the storage capacity of the $3\frac{1}{2}$-inch floppy disk. It is still the standard definition for the capacity of these disks.

Practice Problems
Convert the following quantities into the units requested according to these guidelines:

- Determine the reciprocal pairs of conversion factors from the prefix definitions in Table 4-4.1.

- Apply to appropriate conversion factor (select only one from each pair) in calculating the units requested.

- Solutions that exceed two decimal places should be rounded to two places.

Introductory Problems
4-4.1   Express 30 kilometers in meters.

4-4.2   Express 30 kilometers in centimeters.

4-4.3   Express 20,000 centimeters in meters.

4-4.4   Express 20,000 centimeters in kilometers.

4-4.5   Express 500 millimeters in meters.

4-4.6   Express 200 milliseconds in seconds.

4-4.7   Express 2000 nanoseconds in milliseconds.

4-4.8   Express 2000 nanoseconds in microseconds.

4-4.9 Express 3000 kilotons in megatons.

4-4.10 Express 20 megatons in kilotons.

Intermediate Problems

Convert the following quantities into the units requested using these guidelines:

- Use the equivalence expression given to write a pair of reciprocal conversion factors.

- Then apply the appropriate conversion factor (select only one from each pair) in calculating the units requested.

4-4.11 Express 2 kilobytes in bytes (use 1 kilobyte is 2^{10} bytes).

4-4.12 Express 1000 kilobytes in bytes (use 1 kilobyte is 2^{10} bytes).

4-4.13 Express 2 megabytes in bytes (use 1 megabyte is 2^{20} bytes).

4-4.14 Express 1.44 megabytes in bytes (use 1 megabyte is 2^{10} kilobytes and 1 kilobyte is 2^{10} bytes).

4-4.15 Express 1.2 gigabytes in bytes (use 1 gigabyte is 2^{30} bytes).

4-4.16 Express 0.25 kilobytes in bytes (use 1 kilobyte is approximately 1000 bytes).

4-4.17 Express 2 megabytes in bytes (use 1 megabyte is 1000 kilobytes and 1 kilobyte is 2^{10} bytes).

4-4.18 Express 1.44 megabytes in bytes (use 1 megabyte is 1000 kilobytes and 1 kilobyte is 2^{10} bytes).

4-4.19 Express 1.44 megabytes in bytes (use 1 megabyte is approximately 1,000,000 bytes).

4-4.20 Express 1.2 gigabytes in bytes (use 1 gigabyte is 10^{6} kilobytes and 1 kilobyte is 2^{10} bytes).

More Difficult Problems

Compare the following alternative definitions for binary prefixes a and b.

- Compute the difference between each pair of definitions as a percentage of the first.

- To do this, express both quantities in the same units. Then subtract b from a. Divide the difference by quantity a and multiply by 100.

- Round the result to one decimal place.

4-4.21 Compare: a) 1 kilobyte is 2^{10} bytes.
            b) 1 kilobyte is approximately 1000 bytes.

4-4.22 Compare: a) 1 megabyte is 2^{20} bytes.
            b) 1 megabyte is approximately 1,000,000 bytes.

4-4.23 Compare: a) 1 megabyte is 2^{20} bytes.
            b) 1 megabyte is 1000 kilobytes (use 1 kilobyte is 2^{10} bytes).

4-4.24 Compare: a) 1 gigabyte is 2^{30} bytes.
            b) 1 gigabyte is approximately 1,000,000,000 bytes.

4-4.25 Compare: a) 1 gigabyte is 2^{30} bytes.
            b) 1 gigabyte is 1000 megabytes (use 1 megabyte is 2^{20} bytes).

Alternative Definitions of Binary Prefixes

4-4.26 A PC with 48 megabytes of random access memory (RAM) has 50,331,684 bytes of memory. Which of the following definitions for a megabyte is used?
    a) 1 megabyte is approximately 1,000,000 bytes.
    b) 1 megabyte is 1000 kilobytes.
    c) 1 megabyte is 2^{20} bytes.

4-4.27 A 1.44-megabyte floppy disk has a total capacity of 1,474,560 bytes. Which of the following definitions for a megabyte is used?
   a) 1 megabyte is approximately 1,000,000 bytes.
   b) 1 megabyte is 1000 kilobytes.
   c) 1 megabyte is 2^{20} bytes.

4-4.28 A 1.192-gigabyte hard drive has a total capacity of 1,279,900,254 bytes. Which of the following definitions for a gigabyte is used?
   a) 1 gigabyte is approximately 1,000,000 kilobytes.
   b) 1 gigabyte is 1000 megabytes.
   c) 1 gigabyte is 2^{30} bytes.

The technique of multiplying factors to change units within an equation is a powerful problem-solving tool. When factors are multiplied, their units also multiply, forming a composite unit. When the same simple unit appears on both the top and the bottom of a composite unit ratio, they divide out and can be removed. In this way, the composite unit can be simplified. By carefully selecting which factors to multiply, you can make the result come out in any units you require.

Problems with Several Conversion Factors

Reciprocal Factors

At the heart of this process are reciprocal factors. Each factor available for the solution of a problem has a corresponding reciprocal factor. Reciprocal factors have both reciprocal values and reciprocal units. Only one of each reciprocal pair of factors will be used in a given problem.

 Since multiplication by a reciprocal of a quantity is the same as division by that quantity, all quantities will multiply. The decision of whether to multiply or divide is removed from the problem.

As you are learning to use this problem-solving tool, it is important to write down all the reciprocal pairs of factors that apply to a given problem. In this way all possible terms needed for the solution of the problem are at hand. Then the task is just to select from this list only those factors that will allow the units to divide out (cancel) and leave only the units needed for the solution.

After you become proficient at solving this type of problem, you may wish to save time by writing only one factor from each reciprocal pair. As you combine the factors, you can flip the factor in your mind to make any reciprocal as it is needed.

Composite Units Should Simplify

The terms must be selected skillfully so the units simplify to the desired result.

 The strategy is to select factors that when multiplied will allow composite units to simplify to the units of the result you want.

While some problems are simple unit conversions involving just two factors, other problems require many factors to be multiplied together. These factors may have simple units or composite units. They may include rates, conversion factors, and prefix conversion factors. The composite unit formed by the product should match the units of the result you are trying to find.

Procedural Approach to Problem Solving

A procedural approach that is useful when working problems with several conversion factors will be presented first. Then this procedure will be imbedded into the IPO worksheet. This worksheet will then be applied in several example problems.

1. **Examine the input** before you start.

   - Identify any additional information that is needed to solve the problem.

   - Carry out any preliminary calculations that will be needed.

   - Identify any equivalence expressions that involve unit conversions.

2. **Let the letter x** represent the quantity you are trying to find as output.

   - Wherever possible, make the units of x a ratio such as $/purchase or sec/file or mi/trip.

   - You may need to make up the unit on the bottom to fit the particular problem.

3. **Build a set of reciprocal factors.**

   - Translate the words of the problem statement into conversion factors:

   Each factor is a value followed by a ratio of units.

   The value represents the number of units on the top of the ratio that is equivalent to exactly *one* of the units on the bottom of the ratio.

   - Factors should be selected from the following sources:

   Values provided as input

   Values collected as additional information

   Values obtained from preliminary calculations

   - Factors come in reciprocal pairs.

   Make sure each factor includes units in parentheses.

4. **Build an equation** that relates *x* (on the left) to a set of factors (on the right).

   - Select factors from the reciprocal set in the following order: start with a factor that has the same units on the top as *x* does.

   - Next, select a factor that will cancel the unwanted unit in the first term.

   - Continue with other factors that allow unwanted units to cancel.

   - Finally, select a factor that has the same units on the bottom as *x* does.

   - Remember that *only one* of each pair of factors will be used in any problem.

   - The equation is complete after unwanted units are divided out and the units on the right balance the units on the left.

5. **Multiply the factors** on the right side of the equation to find the value of *x*.

   - Make sure the combined units on the right balance the units on the left.

6. **Check the result.**

   - Does the result seem reasonable?

   - Check reciprocal factors and their units.

   - Check calculation.

   - Check that units balance.

   - If possible, work the problem using an alternative solution. (For example, work the problem backwards or in a step-by-step fashion.)

Troubleshooting Tips

1. Check your reciprocal factors:

   Write down the reciprocal factors and check them carefully. Remember that incorrect reciprocal factors will lead to incorrect solutions.

   Make sure that the units of each factor are in a ratio.

   Make sure that each reciprocal factor has reciprocal units as well as reciprocal values.

   Make sure that each value represents the number of units on the top of the ratio that is equivalent to exactly one of the units on the bottom of the ratio.

2. As you build the equation:

   Make sure that each unit on the top of *x* appears once on the top of the right side.

Make sure each unit on the bottom of x appears once on the bottom of the right side.

Make sure that all other units (those not in x) appear on both the top and the bottom of the right side of the equation, thus canceling one another.

Make sure you do not use both reciprocal factors for a given conversion (once one is selected, the other is no longer needed).

Note: You do not have to consider when to multiply and when to divide because multiplying by the reciprocal of a factor is equivalent to dividing. Thus, all factors are multiplied.

3. If you have trouble balancing the units:

Check to see if some factor has been reversed (top to bottom) by accident.

If a needed factor is missing:

See if it can be calculated by combining other factors at hand.

See if it can be looked up in a reference. (Start with Tables 4-3.3, 4-4.1, and 4-4.2.)

4. After you have done the calculation:

Make sure the result is reasonable.

Try to check the result using an alternative approach.

5. Remember that evaluating a unit analysis equation involves two steps: balancing the units and then performing the calculation to find the value of x.

 Do not start the calculation until the units are balanced.

Selecting the Units of x

The value of what you are trying to find is usually represented by a single letter such as x, a, b, or c.

 Whenever possible, express the units you are trying to find in the form of a ratio or rate.

Let the unit on the bottom of the ratio be the whole that the unit on the top represents. For example, if you are trying to determine the number of miles traveled during an entire trip, let the unit be miles/trip rather than just miles. Similarly, if you are trying to find the number of megabytes on an entire disk, let the units be MB/disk rather than just MB. Remember that the unit on the bottom of a ratio is the whole, and represents exactly one of that quantity.

By selecting units x in this way, the factors on the right side of the equation will be clearer. To make the units balance, one of the factors on the right side must include that same unit in the bottom of its unit ratio as x does. Thus, for example,

$$x \left(\frac{\text{mi}}{\text{trip}} \right) = 25 \left(\frac{\text{mi}}{\text{gal}} \right) * 20 \left(\frac{\text{gal}}{\text{trip}} \right)$$

Or

$$x \left(\frac{\text{sec}}{1000 \text{ rec}} \right) = 100 \left(\frac{\text{sec}}{\text{rec}} \right) * 1000 \left(\frac{\text{rec}}{1000 \text{ rec}} \right)$$

If this last example seems a little artificial, try giving the term *1000 records* another name. Perhaps *1000 records* represent all the data on a file. Then

$$x \left(\frac{\text{sec}}{\text{file}} \right) = 100 \left(\frac{\text{sec}}{\text{rec}} \right) * 1000 \left(\frac{\text{rec}}{\text{file}} \right)$$

If this is not the case, try giving the term *1000 records* a generic name such as "group." Then

$$x \left(\frac{\text{sec}}{\text{group}} \right) = 100 \left(\frac{\text{sec}}{\text{ree}} \right) * 1000 \left(\frac{\text{ree}}{\text{group}} \right)$$

Unit Analysis Using the IPO Worksheet

The procedural steps for solving unit analysis problems, given at the beginning of this section, are now imbedded into the IPO worksheet introduced in Chapter 1.

IPO Worksheet for Unit Analysis

Problem Statement	The problem needs to be described in detail. Read the problem statement carefully.		
Output	Determine what you are trying to find and the units of that quantity.		
Input	**Examine the input** before you start. • Identify any additional information that is needed to solve the problem. • Carry out any preliminary calculations that will be needed. • Identify any equivalence expressions that involve unit conversions.		
Process	Notation	**Let the letter x represent the quantity you are trying to find.** • Whenever possible, make the units of x a ratio such as \$/purchase or sec/file or mi/trip. • You may need to make up the unit on the bottom to fit the particular problem.	

	Additional information	**Build a set of reciprocal factors.**
		• Translate equivalence expressions into conversion factors:
		    Each factor is a value followed by a ratio of units.
		    The value represents the number of units on the top of the ratio that is equivalent to exactly *one* of the units on the bottom.
		• Factors should be selected from the following sources:
		    Values provided as input
		    Values collected as additional information
		    Values obtained from preliminary calculations
		• Factors come in reciprocal pairs.
		    Make sure each factor includes units in parentheses.	

	Diagrams	Diagrams are not usually required to solve this type of problem.	

Process	Approach	**Build an equation** that relates x (on the left) to several factors (on the right).
		• Select a factor from the reciprocal set above that has the same units on the top as x does.
		• Next, select a factor that will cancel the unwanted unit in the first term.
		• Continue with other factors that allow unwanted units to cancel.
		• Finally, select a factor that has the same units on the bottom as x does.
		• Remember that *only one* of each pair of factors will be used in any problem.
		• The equation is complete after unwanted units are divided out and the units on the right balance the units on the left.
Solution		**Multiply the factors** on the right side of the equation to find the value of x.
		• Make sure the combined units on the right balance the units on the left.
Check		**Check the result.**
		• Does the result seem reasonable?
		• Check reciprocal factors and their units.
		• Check calculations.
		• Check that units balance.
		• If possible, work the problem using an alternative solution. (For example, work the problem backwards or in a step-by-step fashion.)

Examples Using the IPO Worksheet

Example Problem 1	A disk drive can read a record of data in 20 milliseconds (ms). If 1000 ms is 1 second, how many records can be read in 1 second?	
Output	Number of records read in one second	
Input	One record read in 20 ms	
Process	Notation	Let x = records read in one second.
	Additional Information	**Build a set of reciprocal factors:** $$20\left(\frac{ms}{rec}\right) \qquad \frac{1}{20}\left(\frac{rec}{ms}\right)$$ $$1000\left(\frac{ms}{sec}\right) \qquad \frac{1}{1000}\left(\frac{sec}{ms}\right)$$
	Diagram	None needed
	Approach	**Build an equation:** Start with what you are trying to find on the left. $$x\left(\frac{rec}{sec}\right) = \ldots$$ Select appropriate terms from the reciprocal factors to fill in the right side of the equation. The first term on the right should have the same units as on the top of x. $$x\left(\frac{rec}{sec}\right) = \frac{1}{20}\left(\frac{rec}{ms}\right)\ldots$$ The next term should cancel ms. $$x\left(\frac{rec}{sec}\right) = \frac{1}{20}\left(\frac{rec}{\cancel{ms}}\right) * 1000\left(\frac{\cancel{ms}}{sec}\right)$$ You are finished when the units on the left and right balance.
Solution	**Multiply the factors:** Multiply $\frac{1}{20} * (1000)$ to get 50. Therefore $x = \mathbf{50}\left(\frac{rec}{sec}\right)$	

Examples Using the IPO Worksheet (Continued)

Check	Does the result seem reasonable? Check reciprocal factors and their units. Check calculation. Check that units balance. Use an alternative approach to check the answer. In this case, work the problem backwards: $$\text{Given } x = 50 \left(\frac{rec}{sec}\right) \text{ confirm that there are } 20 \left(\frac{ms}{rec}\right)$$ $$y\left(\frac{ms}{rec}\right) = 1000 \left(\frac{ms}{see}\right) * \frac{1}{50}\left(\frac{see}{rec}\right)$$ $$y\left(\frac{ms}{rec}\right) = 20 \left(\frac{ms}{rec}\right), \text{ which agress with the original input.}$$

Example Problem 2	On your way home to the United States from a trip to Canada, you need to buy gasoline. Gasoline just north of the border in Canada costs 61 cents per liter. A service station a few miles south, across the border in the United States, sells gasoline for $1.31 per gallon. Is it less expensive to buy gas in Canada or in the United States? The currency exchange rate is $1.33 Canadian for each United States $.	
Output	Which costs less: Canadian or U.S. gasoline? (To compare, both quantities must be expressed in the same units. Use US$/gal to compare.)	
Input	Cost of gasoline in Canada is $0.61/L (in Canadian dollars). Cost of gasoline in the United States is $1.31/gal (in U.S. dollars).	
Process	Notation	Let x = cost of Canadian gasoline expressed in US$/gal.
	Additional Information	Information needed from Table 4-3.3: One liter is 1.05966 quarts. One gallon is equivalent to 4 quarts. **Build a set of reciprocal factors:** $$0.61\left(\frac{\text{Can\$}}{\text{L}}\right) \qquad \frac{1}{0.61}\left(\frac{\text{L}}{\text{Can\$}}\right)$$ $$1.33\left(\frac{\text{Can\$}}{\text{US\$}}\right) \qquad \frac{1}{1.33}\left(\frac{\text{US\$}}{\text{Can\$}}\right)$$ $$1.05966\left(\frac{\text{qt}}{\text{L}}\right) \qquad \frac{1}{1.05966}\left(\frac{\text{L}}{\text{qt}}\right)$$ $$4\left(\frac{\text{qt}}{\text{gal}}\right) \qquad \frac{1}{4}\left(\frac{\text{gal}}{\text{qt}}\right)$$
	Diagram	None needed

Process	Approach	
		Build an equation: Start with what you are trying to find on the left. $$x\left(\frac{\text{US\$}}{\text{gal}}\right) = \ldots$$ Select appropriate terms from the reciprocal factors to fill in the right side of the equation. The first term on the right should have the same units as on the top of x, US\$: $$x\left(\frac{\text{US\$}}{\text{gal}}\right) = \frac{1}{1.33}\left(\frac{\text{US\$}}{\text{Can\$}}\right) * \ldots$$ The next term should cancel Can\$. $$x\left(\frac{\text{US\$}}{\text{gal}}\right) = \frac{1}{1.33}\left(\frac{\text{US\$}}{\text{Can\$}}\right) * 0.61\left(\frac{\text{Can\$}}{\text{L}}\right) * \ldots$$ The next term should cancel liters: $$x\left(\frac{\text{US\$}}{\text{gal}}\right) = \frac{1}{1.33}\left(\frac{\text{US\$}}{\cancel{\text{Can\$}}}\right)$$ $$* 0.61\left(\frac{\cancel{\text{Can\$}}}{\cancel{\text{L}}}\right)$$ $$* \frac{1}{1.05966}\left(\frac{\cancel{\text{L}}}{\text{qt}}\right) * \ldots$$ The next term should cancel quarts: $$x\left(\frac{\text{US\$}}{\text{gal}}\right) = \frac{1}{1.33}\left(\frac{\text{US\$}}{\text{Can\$}}\right)$$ $$* 0.61\left(\frac{\text{Can\$}}{\cancel{\text{L}}}\right)$$ $$* \frac{1}{1.05966}\left(\frac{\cancel{\text{L}}}{\cancel{\text{qt}}}\right)$$ $$* 4\left(\frac{\cancel{\text{qt}}}{\text{gal}}\right)$$ You have finished when the units on the left and right balance.

Solution	**Multiply the factors:**
	Multiply: $\dfrac{1}{1.33} * (0.61) * \dfrac{1}{1.05966} * 4$ to get 1.7313.
	Therefore $x\left(\dfrac{\textbf{US\$}}{\textbf{gal}}\right) = \textbf{1.73}\left(\dfrac{\textbf{US\$}}{\textbf{gal}}\right)$ (rounded to two places).
	Compare cost of gasoline:
	U.S. gasoline at $1.31\left(\dfrac{\text{US\$}}{\text{gal}}\right)$
	Canadian gasoline at $1.73\left(\dfrac{\text{US\$}}{\text{gal}}\right)$
	Gasoline is less expensive in the United States.
Check	Does the solution seem reasonable? Make sure that the value you are trying to find has the correct units. Check each equation. Check calculation. Make sure units balance. Check using the following alternative, step-by-step approach.

Check the previous problem using the following alternative approach. In this case work the problem in steps, doing only one conversion in each step.

Alternative Check	1. Convert Can$ to US$:

$$a\left(\frac{US\$}{L}\right) = \frac{1}{1.33}\left(\frac{US\$}{\cancel{Can\$}}\right) * 0.61\left(\frac{\cancel{Can\$}}{L}\right)$$

$$= 0.459\left(\frac{US\$}{L}\right)$$

2. Convert liters to quarts:

$$b\left(\frac{US\$}{qt}\right) = 0.459\left(\frac{US\$}{\cancel{L}}\right) * \frac{1}{1.05966}\left(\frac{\cancel{L}}{qt}\right)$$

$$= 0.433\left(\frac{US\$}{qt}\right)$$

3. Convert quarts to gallons:

$$c\left(\frac{US\$}{gal}\right) = 0.433\left(\frac{US\$}{\cancel{qt}}\right) * 4\left(\frac{\cancel{qt}}{gal}\right) = 1.732\left(\frac{US\$}{gal}\right)$$

Round the result to two places: $\quad c = 1.73\left(\frac{US\$}{gal}\right)$

Compare cost of gasoline:

U.S. gasoline at $1.31 \left(\frac{US\$}{gal}\right)$

Canadian gasoline at $1.73 \left(\frac{US\$}{gal}\right)$

The answer agrees with the original solution.

Note: Gallons in the United States and Canada are defined differently. A U.S. gallon is 231 cubic inches, while the Canadian gallon, based on British imperial measure, is 277.42 cubic inches. This confusion is one reason Canada has changed from the British system of measure to the metric system, something the United States has yet to do.

Rounding Errors

In the step-by-step method, if rounding to the nearest cent is done at each step, the final result could be off by a cent or two. To avoid these rounding errors, the intermediate result after each step needs to be kept to at least three decimal places (the nearest mill). The final result, when rounded to two places (the nearest cent), should then be accurate.

When using a calculator to compute the combined solution, rounding should not take place until the final result. As each term is multiplied, the running product is kept in the calculator to its full accuracy (usually 10 places). Only at the end does the final value need to be rounded to the nearest cent.

Example Problem 3	Compute the area of a rectangular computer screen in square meters, given a height of 12 inches and a width of 16 inches.	
Output	Area of screen in square meters	
Input	Height of screen is 12 in. Width of screen is 16 in.	
Process	Notation	Let h = height of screen Let w = width of screen Let a = area of screen
	Additional Information	Area formula: $a = h * w$ Need conversion factor from Table 4-3.3 of 39.37 (in./m) **Build a set of reciprocal factors:** $$39.37 \left(\frac{\text{in.}}{\text{m}} \right) \qquad \frac{1}{39.37} \left(\frac{\text{m}}{\text{in.}} \right)$$
	Diagram	None needed
	Approach	**Build an equation:** Start with what you are trying to find on the left: $$a\,(\text{m}^2) = \ldots$$ Select appropriate terms from the reciprocal factors to fill in the right side of the equation. The first term on the right should have the same units as on the top of the left side (m^2). To do this you will need to square the factor with meters on top: $$a(\text{m}^2) = \frac{1}{39.37} \left(\frac{\text{m}}{\text{in.}} \right) * \frac{1}{39.37} \left(\frac{\text{m}}{\text{in.}} \right) * \ldots$$ The next term should cancel inches (use height): $$a(\text{m}^2) = \frac{1}{39.37} \left(\frac{\text{m}}{\text{in.}} \right)$$ $$* \frac{1}{39.37} \left(\frac{\text{m}}{\cancel{\text{in.}}} \right) * 12\,(\cancel{\text{in.}}) * \ldots$$ The next term should also cancel inches (use width): $$a(\text{m}^2) = \frac{1}{39.37} \left(\frac{\text{m}}{\cancel{\text{in.}}} \right) * \frac{1}{39.37} \left(\frac{\text{m}}{\cancel{\text{in.}}} \right)$$ $$* 12\,(\cancel{\text{in.}}) * 16\,(\cancel{\text{in.}})$$ You are finished when the units on the left and right balance.

Solution	**Multiply factors:** 1/39.37 * 1/39.37 * (12) * (16) to get 0.124387. Therefore: a (**m^2**) = **0.124 (m^2)** (rounded to three places)
Check	Check by converting height and width to meters and then computing area. Convert the height to meters: $h(\text{m}) = 12\,(\text{in.}) * \dfrac{1}{39.37}\left(\dfrac{\text{m}}{\text{in.}}\right) = 0.305\ (\text{m})$ Convert the width in meters: $w(\text{m}) = 16\,(\text{in.}) * \dfrac{1}{39.37}\left(\dfrac{\text{m}}{\text{in.}}\right) = 0.406\ (\text{m})$ Compute the area in m^2: $\quad a\ (\text{m}^2) = 0.305\ (\text{m}) * 0.406\ (\text{m}) = 0.124\ (\text{m}^2)$ The answer agrees with the previous solution.

Example Problem 4	How many kilobytes (KB) can be stored on a floppy disk when:
	Data is stored on two sides of the disk.
	Each side has 80 tracks (areas where data can be recorded).
	Each track is divided into 18 sectors (parts of a track).
	Each sector holds 512 bytes of information.
	Each kilobyte represents 1024 bytes (the binary unit).

Output	Capacity of disk in KB

Input	Data is stored on two sides of the disk.
	Each side has 80 tracks (areas where data can be recorded).
	Each track is divided into 18 sectors (parts of a track).
	Each sector holds 512 bytes of information.
	Each kilobyte represents 1024 bytes (the binary unit).

Process	Notation	Let x = the number of KB on the entire disk.
	Additional Information	No additional information is needed. Just write the equivalence information as reciprocal conversion factors.
		Build a set of reciprocal factors:
		$2\left(\dfrac{\text{sides}}{\text{disk}}\right)$ $\dfrac{1}{2}\left(\dfrac{\text{disk}}{\text{side}}\right)$
		$80\left(\dfrac{\text{tracks}}{\text{side}}\right)$ $\dfrac{1}{80}\left(\dfrac{\text{sides}}{\text{track}}\right)$
		$18\left(\dfrac{\text{sectors}}{\text{track}}\right)$ $\dfrac{1}{18}\left(\dfrac{\text{tracks}}{\text{sector}}\right)$
		$512\left(\dfrac{\text{bytes}}{\text{sector}}\right)$ $\dfrac{1}{512}\left(\dfrac{\text{sec}}{\text{byte}}\right)$
		$1024\left(\dfrac{\text{bytes}}{\text{KB}}\right)$ $\dfrac{1}{1024}\left(\dfrac{\text{KB}}{\text{byte}}\right)$
	Diagram	None needed

Process (continued)	Approach	**Build an equation:**

Start with what you are trying to find on the left.

$$x\left(\frac{\text{KB}}{\text{disk}}\right) = \ldots$$

Select appropriate terms from the reciprocal factors to fill in the right side of the equation.

The first term on the right should have the same units as on the top of the left side (KB):

$$x\left(\frac{\text{KB}}{\text{disk}}\right) = \frac{1}{1024}\left(\frac{\text{KB}}{\text{byte}}\right) * \ldots$$

The next term should cancel bytes:

$$x\left(\frac{\text{KB}}{\text{disk}}\right) = \frac{1}{1024}\left(\frac{\text{KB}}{\cancel{\text{byte}}}\right)$$
$$* 512\left(\frac{\cancel{\text{byte}}}{\text{sector}}\right) * \ldots$$

The next term should cancel sectors:

$$x\left(\frac{\text{KB}}{\text{disk}}\right) = \frac{1}{1024}\left(\frac{\text{KB}}{\cancel{\text{byte}}}\right)$$
$$* 512\left(\frac{\cancel{\text{byte}}}{\cancel{\text{sector}}}\right)$$
$$* 18\left(\frac{\cancel{\text{sector}}}{\text{track}}\right) * \ldots$$

The next term should cancel tracks:

$$x\left(\frac{\text{KB}}{\text{disk}}\right) = \frac{1}{1024}\left(\frac{\text{KB}}{\cancel{\text{byte}}}\right)$$
$$* 512\left(\frac{\cancel{\text{byte}}}{\cancel{\text{sector}}}\right)$$
$$* 18\left(\frac{\cancel{\text{sector}}}{\cancel{\text{track}}}\right)$$
$$* 80\left(\frac{\cancel{\text{track}}}{\text{side}}\right) * \ldots$$

	Approach	The next term should cancel sides:$$x\left(\frac{\text{KB}}{\text{disk}}\right) = \frac{1}{1024}\left(\frac{\text{KB}}{\cancel{\text{byte}}}\right)$$$$* \, 512\left(\frac{\cancel{\text{byte}}}{\cancel{\text{sector}}}\right)$$$$* \, 18\left(\frac{\cancel{\text{sector}}}{\cancel{\text{track}}}\right)$$$$* \, 80\left(\frac{\cancel{\text{track}}}{\cancel{\text{side}}}\right)$$$$* \, 2\left(\frac{\cancel{\text{side}}}{\text{disk}}\right)$$You have finished when the units on the left and right balance.
Solution		**Multiply the factors:**Multiply $\dfrac{1}{1024}$ * 512 * 18 * 80 * 2 to get 1440.Therefore: $x\left(\dfrac{\textbf{KB}}{\textbf{disk}}\right) = \textbf{1440}\left(\dfrac{\textbf{KB}}{\textbf{disk}}\right).$
Check		Does the result seem reasonable?Check reciprocal factors and their units.Check calculation.Check that units balance.Check using the following alternative, step-by-step approach.

Notice how this method makes perfectly clear which terms are multiplied and which terms are divided. No guesswork is needed here; the units make it clear!

Check the previous problem using the following alternative approach. In this case work the problem in steps, doing only one conversion in each step.

Alternative Check	Using a step-by-step approach, solve the problem again:

1. Compute the number of tracks per disk (a):

$$a\left(\frac{\text{tracks}}{\text{disk}}\right) = 80\left(\frac{\text{tracks}}{\text{side}}\right) * 2\left(\frac{\text{sides}}{\text{disk}}\right)$$

$$= 160\left(\frac{\text{tracks}}{\text{disk}}\right)$$

2. Use the information from step 1 to compute the number of sectors/disk (b):

$$b\left(\frac{\text{sectors}}{\text{disk}}\right) = 160\left(\frac{\text{tracks}}{\text{disk}}\right) * 18\left(\frac{\text{sectors}}{\text{track}}\right)$$

$$= 2880\left(\frac{\text{sectors}}{\text{disk}}\right)$$

3. Use the information from step 2 to compute the number of bytes per disk (c):

$$c\left(\frac{\text{bytes}}{\text{disk}}\right) = 2880\left(\frac{\text{sectors}}{\text{disk}}\right) * 512\left(\frac{\text{bytes}}{\text{sector}}\right)$$

$$= 1,474,560\left(\frac{\text{bytes}}{\text{disk}}\right)$$

4. Use the information from step 1 to compute the number of sectors/disk (d):

$$d\left(\frac{\text{KB}}{\text{disk}}\right) = \frac{1}{1024}\left(\frac{\text{KB}}{\text{byte}}\right) * 1,474,560\left(\frac{\text{bytes}}{\text{KB}}\right)$$

$$= 1440\left(\frac{\text{KB}}{\text{disk}}\right)$$

The answer agrees with the previous solution.

Example Problem 5	A disk drive is able to read a record of data in 20 milliseconds (ms). If each record contains 1000 characters of data, how many seconds will it take to read a data file of one million characters?	
Output	Number of seconds needed to read the entire data file	
Input	A record of data can be read in 20 milliseconds (ms). Each record contains 1000 characters of data. The file contains one million characters.	
Process	Notation	Let x = seconds to process entire file (sec/file).
	Additional Information	Recall that 1 millisecond is 1/1000 of a second. **Build a set of reciprocal factors:** $$10^6\left(\frac{\text{char}}{\text{file}}\right) \qquad 10^{-6}\left(\frac{\text{file}}{\text{char}}\right)$$ $$20\left(\frac{\text{ms}}{\text{rec}}\right) \qquad \frac{1}{20}\left(\frac{\text{rec}}{\text{ms}}\right)$$ $$10^3\left(\frac{\text{ms}}{\text{sec}}\right) \qquad 10^{-3}\left(\frac{\text{sec}}{\text{ms}}\right)$$ $$10^3\left(\frac{\text{char}}{\text{rec}}\right) \qquad 10^{-3}\left(\frac{\text{rec}}{\text{char}}\right)$$
	Diagram	None needed

Process	Approach	**Build an equation:** Start with what you are trying to find on the left.
		$$x\left(\frac{\text{sec}}{\text{file}}\right) = \ldots$$ Select appropriate terms from the reciprocal factors to fill in the right side of the equation. The first term on the right should have the same units as on the top of the left side (sec): $$x\left(\frac{\text{sec}}{\text{file}}\right) = 10^{-3}\left(\frac{\text{sec}}{\text{ms}}\right) * \ldots$$ The next term should cancel ms: $$x\left(\frac{\text{sec}}{\text{file}}\right) = 10^{-3}\left(\frac{\text{sec}}{\text{ms}}\right)$$ $$* 20\left(\frac{\text{ms}}{\text{rec}}\right)$$ The next term should cancel rec: $$x\left(\frac{\text{sec}}{\text{file}}\right) = 10^{-3}\left(\frac{\text{sec}}{\text{ms}}\right)$$ $$* 20\left(\frac{\text{ms}}{\text{rec}}\right)$$ $$* 10^{-3}\left(\frac{\text{rec}}{\text{char}}\right)$$ The next term should cancel char: $$x\left(\frac{\text{sec}}{\text{file}}\right) = 10^{-3}\left(\frac{\text{sec}}{\text{ms}}\right)$$ $$* 20\left(\frac{\text{ms}}{\text{rec}}\right)$$ $$* 10^{-3}\left(\frac{\text{rec}}{\text{char}}\right)$$ $$* 10^{6}\left(\frac{\text{char}}{\text{file}}\right)$$ You have finished when the units on the left and right balance.

Solution	**Multiply the factors:**
	Multiply $10^{-3} * 20 * 10^{-3} * 10^6$ to get 20.
	Therefore $x\left(\dfrac{\textbf{sec}}{\textbf{file}}\right) = 20\left(\dfrac{\textbf{sec}}{\textbf{file}}\right)$.
Check	Does the result seem reasonable?
	Check reciprocal factors and their units.
	Check calculation.
	Check that units balance.
	Check using the following alternative, step-by-step approach.

Alternative Check	Use the step-by-step method to solve the problem.

Use the step-by-step method to solve the problem.

1. Compute the number of records per file (a):

$$a\left(\frac{\text{rec}}{\text{file}}\right) = 10^{-3}\left(\frac{\text{rec}}{\cancel{\text{char}}}\right) * 10^{6}\left(\frac{\cancel{\text{char}}}{\text{file}}\right)$$

$$= 10^{3}\left(\frac{\text{rec}}{\text{file}}\right)$$

2. Use the information from step 1 to compute milliseconds per file (b)

$$b\left(\frac{\text{ms}}{\text{file}}\right) = * 20\left(\frac{\text{ms}}{\cancel{\text{rec}}}\right) * 10^{3}\left(\frac{\cancel{\text{rec}}}{\text{file}}\right)$$

$$= 20{,}000\left(\frac{\text{ms}}{\text{file}}\right)$$

3. Use the information from step 2 to compute seconds per file (c):

$$c\left(\frac{\text{sec}}{\text{file}}\right) = \frac{1}{1000}\left(\frac{\text{sec}}{\cancel{\text{ms}}}\right) * 20{,}000\left(\frac{\cancel{\text{ms}}}{\text{file}}\right)$$

$$= 20\left(\frac{\text{sec}}{\text{file}}\right)$$

Thus, the time to process the entire file is 20 seconds.
This agrees with the previous solution.

Additional Thoughts

Let the Units lead the Way. When you are not sure how to proceed, the units will show you the way. If you are uncertain whether to divide or multiply by a given term, the units will make it clear. Always select the appropriate conversion factor from the reciprocal pair of conversion factors, so unwanted units will cancel and the units desired for your solution are left.

Beware When Adding Units. Remember that canceling units only applies when quantities multiply and divide. When quantities are added and subtracted, canceling won't work. Additive terms must have the same units. You cannot add two lengths if one is in feet and the other is in meters. You must first convert one so that both terms have like units.

It makes no sense at all to add feet and square feet or to add hours and miles per hour. When you do so you are, as the old saying goes, comparing apples to oranges. Unless you are counting fruit, apples and oranges should not be added.

Nonsense Equations. In addition to having balanced units, equations must make physical sense. Don't include factors unless they relate to the problem at hand, *even if they make the units balance.* For example, imagine you are computing the average speed (miles/hour) traveled on a trip from San Francisco to Seattle. By multiplying the distance between New York and Chicago (750 miles) by the reciprocal of the overtime you put in at work last week (25 hours), you arrive at the following equation:

$$x\left(\frac{\text{mi}}{\text{hr}}\right) = 750\,(\text{mi}) * \frac{1}{25}\left(\frac{1}{\text{hr}}\right)$$

The units of this equation balance, but the value calculated has nothing to do with your trip!

Missing Factors. Unit analysis will not detect factors missing from an equation if those factors have no units. For example, suppose you are trying to compute the distance traveled by a car accelerating from a stoplight and you are provided the following information:

Acceleration of the car: $\quad a = 10 \left(\dfrac{ft}{sec^2} \right)$

Time to cover the distance: $\quad t = 10 \ (sec)$

Distance traveled: $\quad s = ? \ (ft)$

Formula that applies: $\quad s = \dfrac{1}{2} \, at^2$

Given this information you can work the problem, even without knowing much about physics.

$$s(ft) = \frac{1}{2} * 10 \left(\frac{ft}{sec^2} \right) * 10^2 (sec^2) = 500 \ (ft)$$

However, if you omit the factor of $\frac{1}{2}$, you will get an answer that is twice the correct value.

$$s(ft) = 10 \left(\frac{ft}{sec^2} \right) * 10^2 (sec^2) = 1000 \ (ft)$$

Because the factor $\frac{1}{2}$ has no units, the units in the equation still balance.

 Balancing units in an equation will not detect missing factors if those factors have no units.

Practice Problems

General Problems. Work the following problems using the procedural approach given at the beginning of the section.

4-5.1   If bulk candy costs $4.00 a pound, how much will 10 pounds of candy cost?

4-5.2   If bulk candy costs $4.00 a pound, how much will $\frac{1}{4}$ pound cost?

4-5.3   If bulk candy costs $4.00 a pound, how much will 10 ounces cost? (There are 16 ounces in a pound.)

4-5.4   If bulk candy costs $10.00 a kilogram, how much will 3 pounds cost? (There are 2.2 pounds in a kilogram.)

4-5.5   If bulk candy costs $4.00 a pound and 1 pound fills a box 5 inches by 3 inches by 2 inches, how much will it cost to fill a box 3 inches by 4 inches by 5 inches?

4-5.6   If bulk candy costs $10.00 a kilogram and 1 pound fills a box 5 inches by 3 inches by 2 inches, what is the weight of a box 3 inches by 4 inches by 5 inches that is full of candy?

4-5.7   A car averages 25 miles to the gallon. If gasoline costs $1.30 per gallon, how much does it cost to go 1000 miles?

4-5.8   A car averages 25 miles to the gallon. If a gallon of fuel costs $1.30, and you drive 1000 miles each month, how much does it cost to fuel the car for a year?

4-5.9   During a 250-mile trip a car uses fuel at a rate of 20 miles per gallon. At the start of the trip, $16 of fuel was pumped into the 15-gallon tank bringing the tank from a level of exactly $\frac{1}{4}$ full to exactly full. What is the cost of fuel for the trip?

4-5.10   Concrete delivered to a construction site costs $250 a cubic yard. What is the cost of the concrete needed to build a 6-inch-thick patio slab 20 feet by 10 feet?

4-5.11 The following problem was presented in Chapter 1. Try solving it using the unit conversion approach. If two men can paint two rooms in two hours, how long will it take one man to paint one room. Assume that both rooms are the same size.

Computer Problems. Solve these problems using the IPO worksheet on pages 139–140.

4-5.12 A computer can process a data record in 100 milliseconds. How many seconds will it take to process a file of 5000 data records?

4-5.13 A file on a computer hard drive consists of 5000 data records each 128 bytes long. The computer can process 1 kilobyte of data in 100 milliseconds. Use the binary definition of a kilobyte to find the number of seconds to process the entire file.

4-5.14 An older modem transfers data over a telephone line at a rate of 14,400 bits per second. How many seconds will it take to transfer a megabyte of data? Use the binary definition for megabyte and the fact that a byte is composed of eight bits.

4-5.15 A graphic image is being sent over the Internet. The image, consisting of 5 megabytes of information, is transferred along a data line at the rate of 1.5 megabits per second. How many seconds will the transfer take? Use the binary definition for megabyte and the fact that a byte is composed of eight bits.

4-5.16 What is the total storage capacity in kilobytes of a single-sided, single-density floppy disk used with original IBM PC? It had the following characteristics:

Heads:	1
Tracks per head:	40
Sectors per track:	8
Kilobytes per sector:	$\frac{1}{2}$

4-5.17 What is the total capacity in gigabytes, rounded to the nearest tenth of a gigabyte, for a computer hard drive with the following specifications:

Total heads:	16
Tracks per head:	6400
Sectors per track:	128
Bytes per sector:	512
Bytes per kilobyte:	1024
Kilobytes per megabyte:	1024
Megabytes per gigabyte:	1024

4-5.18 The CMOS setup of a computer auto detects an installed hard drive and provides the following information:

Heads:	16
Cylinders:	2484
Sectors:	63

Calculate the total capacity of the hard drive in gigabytes, rounded to the nearest tenth of a gigabyte. Note that the cylinder count represents the number of tracks per head, and the sector count represents the number of sectors per track. Use the definitions given in problem 4-5.17 for bytes per sector, bytes per kilobyte, kilobytes per megabyte, and megabytes per gigabyte.

4-5.19 A video monitor has a dot pitch of 0.28 mm or 28 hundredths of a millimeter. The dot pitch is a measure of the smallest dot that can be displayed on the screen. Another way of saying this is the dot pitch represents the pixel size at maximum resolution. If the maximum resolution of the monitor is 1024 by 768, and this represents the number of pixels along the horizontal and vertical sides of the screen, what are the dimensions of the display area of the screen in inches, to the nearest tenth of an inch? Use the fact that 1 millimeter is approximately $\frac{1}{25}$ inch.

4-5.20 A computer CPU can process 1 million machine language instructions per second. If a program includes 1000 lines of code and each line represents an average of 10 instructions, calculate the approximate amount of time in milliseconds for the CPU to process the entire program.

4-5.21 Once the desired track of data is located on a disk drive, data can be transferred to memory at a rate of 1.5 megabytes per second. How many milliseconds are needed to transfer 200 kilobytes of data? Define a megabyte as 1024 kilobytes.

Chapter Summary

Units

- **Units** provide a means to measure physical quantities such as time, length, and weight. The quantity measured is compared to a previously defined standard. This standard is called a unit of measurement.

- **Simple units** have a single unit of measurement. Examples: miles, feet, hours, and days.

- **Composite units** combine two or more units of measurement that multiply or divide. Examples: feet^2, yards^3, miles/gallon, feet/second, and pounds/inch^2.

- **Units cancel** (divide out) when the same unit occurs in both the numerator and denominator of a composite unit. Example: ~~gallon~~ * (miles/~~gallon~~)

- **Terms that add** or subtract must have the same units. Unless you are counting units of fruit, you can't add apples and oranges.

- **Units in an equation** must balance. After removing all units that divide out, the composite units on the left side must equal the composite units on the right side. Example: ~~gallon~~ * (mile/~~gallon~~) = mile.

- **Rates** are values with a ratio of units that relate two physical quantities. The value is the number of units on the top of the ratio for one of the units on the bottom. Examples: 30 feet/1 second, or 30 (feet/second); 20 mile/1 gallon or 20 (mile/gallon).

Conversions

- **Equivalence expressions** relate two measures of the same physical quantity in different units. Example: There are 60 seconds in 1 minute.

- **Conversion factors** are equivalence statements expressed as a rate. They include a value and a ratio of units and occur in reciprocal pairs. Example: 60 (sec/min) and its reciprocal 1/60 (min/sec).

- **Unit conversions** can be carried out by multiplying a given number of unit A by the appropriate conversion factor (relating unit B to one of unit A). In the calculation, unit A cancels (divides out). Example: 2 miles * 5280 (feet/mile) = 10,560 feet.

Prefixes

- **Decimal prefixes** are terms that modify a unit, making it larger or smaller by a power of **10.** Example: One **kilo**meter is 10^3 meters; one **milli**meter is 10^{-3} meter.

- **Binary prefixes** are terms that modify a unit, making it larger or smaller by a power of **2.** Example: One **kilo**byte is 2^{10} bytes or 1024 bytes, one **mega**byte is 2^{20} bytes or 1,048,576 bytes.

Working Problems with Several Conversion Factors

- Assign the quantity you are trying to calculate the value x with its appropriate units.

- Construct an equation that relates x on the left to a set of factors on the right. Wherever possible let the units of x be a rate (a ratio).

- Write down all the factors (and their units) that apply to the problem. For each factor, also write its reciprocal.

- Select the factors on the right side of the equation so that the units on both sides balance. Use reciprocal conversion factors to let like units divide out, (cancel).

- Get additional information as needed to make units in the equation balance.

- Types of terms that can occur in an equation include:

   Terms with simple units and composite units

   Rates, conversion factors, and prefix conversion factors

- Check all equations:

   Does the result seem reasonable?

   Check the equivalence information.

   Check the reciprocal factors.

   Check that the units in the equation balance.

   Finally, check the calculation.

Review Questions

1. Define the following terms: *unit, composite unit, rate, equivalence expression, conversion factor, prefix, decimal prefix, binary prefix.*
2. What constraints do units place on adding and subtracting terms?
3. How do composite units simplify when they are multiplied together?
4. What is the difference between weight and mass?
5. Of the two units in a conversion factor, which is associated with the value of the factor?
6. What quantity do the units in the denominator (bottom) of a conversion factor represent?
7. What arithmetic operation does taking the reciprocal represent?
8. What is the advantage of writing down all the reciprocal pairs of conversion factors?
9. How are prefixes and units related? Do prefixes have any particular units?
10. List the decimal prefixes from 10^{12} to 10^{-12} that differ by a factor of 1000.
11. List the decimal prefixes that differ by a factor of 100.
12. List the binary prefixes that are multiples of 1024.
13. Why do you need to be aware of alternate definitions for binary prefixes?
14. Write a procedure for approaching problems using units.
15. When a problem uses several conversion factors, which should be selected first?
16. Is it possible to add conversion factors?
17. When a problem with several conversion factors seems too difficult, what might help?
18. Explain why working a unit analysis problem involves solving two problems in tandem.

Problem Solutions

| Section 4-2 | 4-2.2 | seconds, hours, weeks, centuries | | |
| | 4-2.4 | miles/kilometer, hours/second | | |

Section 4-3	4-3.2	0.83 ft	4-3.4	3.05 m
	4-3.6	6.21 mi	4-3.8	4.54 kg
	4-3.10	266 Can$	4-3.12	64 qt
	4-3.14	3.96 gal	4-3.16	0.07 ft^2
	4-3.18	237.47 gal/hr	4-3.20	40.91 mi/hr

Section 4-4	4-4.2	3,000,000 cm	4-4.4	0.2 km
	4-4.6	0.2 sec	4-4.8	2 μs
	4-4.10	20,000 kilotons	4-4.12	1,024,000 bytes
	4-4.14	1,509,949 bytes (truncated)	4-4.16	250 bytes
	4-4.18	1,474,560 bytes	4-4.20	1,228,800,000 bytes
	4-4.22	4.6%	4-4.24	6.9%
	4-4.26	c)	4-4.28	c)

Section 4-5	4-5.2	$1.00	4-5.4	$13.64
	4-5.6	2 lb	4-5.8	$624.00
	4-5.10	$925.93	4-5.12	500 sec
	4-5.14	582.5 sec	4-5.16	160 KB
	4-5.18	1.2 GB	4-5.20	10 ms

CHAPTER

5

A Little Algebra

I am very well acquainted with matters mathematical. I understand equations, both simple and quadratical. About binomial theorem, I'm teeming with a lot o' news—with many cheerful facts about the square of the hypotenuse.
—W. S. Gilbert from *The Pirates of Penzance*

5-1 ■ Introduction

How to Use This Chapter

Is algebra really necessary? The answer is no . . . and also yes. If you are wondering whether you are likely to have a job in information technology in which you will use algebra as an everyday tool, the answer is probably no. If you are wondering whether learning algebra will sharpen your problem-solving skills and give you an understanding of symbolic notation that will prove useful in many other areas, including computer programming, then the answer is definitely yes! You don't need a lot of algebra, but some will be useful. Instead of the usual two courses in algebra, this book presents the essential parts of algebra in a single chapter.

The approach taken here will be to introduce some fundamental definitions followed by a discussion of how to work with equations. Once you understand the techniques of manipulating and solving equations, you will be able to put these skills to work solving real problems. It is essential to learn how to translate the words and phrases of a problem statement into algebraic notation. An English-to-algebra dictionary is presented to help you. A variety of problems are included, both as examples and for practice. Percents and proportions are covered using algebra to make them easier to understand.

Chapter 6 will introduce the idea of graphing and show how simple algebraic equations can be plotted on a rectangular coordinate system. Appendix B provides additional algebraic tools for those of you who want to go further.

Background

Algebra is an extension of arithmetic that uses letters to represent arbitrary numbers called variables. This branch of mathematics that deals with the general properties of numbers is broad and wide, and the material presented in this chapter is just a brief introduction. Nevertheless, this chapter will clarify two important aspects of algebra.

First, algebra is a powerful tool for generalizing. By using algebraic notation, ideas and concepts that would otherwise take many paragraphs (or even pages) to explain can be written with a handful of symbols. Those who learn to read and write these symbols are empowered with a new kind of literacy.

Second, algebra is a powerful tool for problem solving. Algebra provides a way to take a problem statement apart and express it in the symbolic notation of equations. Once the problem is expressed in equations, the order and precision of mathematics replace the ambiguity and confusion that are prevalent in thought and language. Algebra gives you step-by-step rules for manipulating and solving these equations. Once found, the result can be translated back into everyday language.

Chapter Objectives

In this chapter you will learn:

- About the components of an equation
- The rules for manipulating and transposing equations
- The step-by-step procedure for solving simple equations
- About linear (straight-line) equations with one or two variables
- To translate words into equations
- To use algebra as a tool to solve problems
- To solve percentage and proportion problems with algebra

Introductory Problem

Work this problem in teams of two:

You are mowing the lawn of a large house. The whole job consists of two parts: the front yard and the back yard. The front takes twice as long as the back. What fraction of the whole job is the back? Use the following notation:

Let w = time for whole job.

Let x = time for back lawn.

Use this notation to develop an algebraic equation and solve the problem.

Carefully check your result to make sure the solution satisfies the given conditions. Your instructor will provide the correct solution.

For those teams that complete the previous problem, try to solve this similar problem:

Suppose the whole job has three parts: front, back, and side. If the back takes twice as long as the side and the front takes twice as long as the back, what fraction of the whole job does the side represent? Notation: w = time for whole job; x = time for side lawn.

Components of an Equation

Algebraic equations are made up of algebraic expressions, which in turn are composed of algebraic variables, coefficients, and constants. Before the algebraic equation can be discussed, these components of the equation need to be defined.

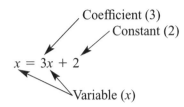

$$x = 3x + 2$$

Note: Algebra omits the multiplication symbol. Thus, 3*x* means 3 * *x*, and *xy* means *x* * *y*.

Algebraic Variables

Algebraic variables are letters such as *x, y,* and *z* used to represent unknown quantities having values yet to be determined. In fact, the words *variable* and *unknown* are often interchanged.

Sometimes variables can represent any value. In expressions such as

$$2x + 2$$

x can take on any value.

Sometimes a variable can take on any value except certain values. In the following example:

$$\frac{2}{(x - 1)}$$

x can take on any value except 1, since $(1 - 1)$ leads to division by zero.

Other times a variable can have only one value. For example, in the equation

$$2x = 6$$

3 is only value of *x* that will make the equation true.

Dependent and Independent Variables

When the value of one variable (*y*) in an equation depends on the value of another variable (*x*), then *x* is called the independent variable and *y* the dependent variable:

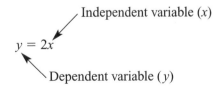

$$y = 2x$$

Constants

Algebraic quantities that do not change in relation to variables are called constants. Constants may be a specific value, for example:

$$x + 2$$

The constant 2 does not change as *x* takes on many values.

Other times constants are represented by a letter:

$$x^2 + y^2 = k$$

The letter k represents a constant that does not change with x and y.

The letters near the beginning of the alphabet such as $a, b, c \ldots k$ are usually reserved for constants, while the letters near the end of the alphabet such as x, y, z are usually reserved for unknowns. However, this is not a firm rule, and you will often encounter exceptions.

Coefficients

The value that multiplies an algebraic variable is called a coefficient. Sometimes a coefficient is a specific value:

$$2x + 3y$$

The coefficient of x is 2. The coefficient of y is 3.

Other times letters are used to represent coefficients:

$$ax^2 + bx + c$$

The coefficient of x^2 is a. The coefficient of x is b.

For the material presented in this book, coefficients are always constants.

Remember: In algebraic notation, adjacent terms imply multiplication. Thus bx means $b * x$.

Algebraic Expressions

Algebraic expressions are constructed from algebraic variables, coefficients, and constants. These parts are combined with arithmetic operators such as those for addition, subtraction, multiplication, division, exponents, and roots. There may be several terms or just a single term. A special case is a single constant term.

Here are a few examples of unrelated algebraic expressions with different numbers of terms:

Several Terms	*Two Terms*	*Single Term*	*Constant Term*
$x^3 + 2x^2 + 3x + 1$	$3 + x/y$	$4x$	2
$2x + y - 3$	$2x + 3y$	$3y$	3
$x^2 + y^2 + z^2$	$2xy + x$	$12xy$	12

In turn, expressions are used to form equations.

Equations

An equation is made up of two algebraic expressions that represent the same quantity. An equal sign separates the two expressions and signifies that the expressions are in balance. Algebraic expressions are made up of constants and algebraic variables "glued" together with arithmetic operators. The constants are specific numbers. The algebraic variables are letters that represent some number (or numbers) yet to be determined that will make the equation a true statement.

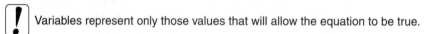 Variables represent only those values that will allow the equation to be true.

▶ Examples of algebraic equations:

$5 = 5$	Two constants are equivalent.
$x = 5$	The variable x must equal 5, and only 5, for the equation to be true.
$x = 2x - 3$	The variable x must be equal to the expression $2x - 3$. The only value of x that will satisfy this equation is 3. This is because $3 = (2 * 3) - 3$.

$x = y$	The variable x is equal to the variable y. There is no unique solution to this equation since there are an infinite number of $x = y$ pairs that will make it true.
$(x - 1)(x - 2) = 0$	The equation is true only when $x = 1$ or when $x = 2$. When $x = 1$, the first factor becomes zero. When $x = 2$, the second factor becomes zero. If either factor is zero, the product is zero.

 Every equation contains an equal sign.

Rules of Equations

The terms of an equation can be combined, separated, and moved from one side of the equal sign to the other in a variety of ways, as long as the rearranged expressions on both sides of the equal sign remain in balance.

An equation will remain in balance if the expressions on both sides are transformed by the same operation. Valid transformations include:

- **Adding** to the same quantity to both sides

- **Subtracting** the same quantity from both sides

- **Multiplying** both sides by the same quantity

- **Dividing** both sides by the same quantity (other than zero)

- **Raising** both sides to the same power

- **Taking** the square root of both sides

- **Exchanging** both sides with each other

The quantity used in a transformation may be a constant value, an algebraic variable, or another algebraic expression. Whatever they are, it is important to remember that this quantity must be applied to the *entire* expression on each side of the equation.

▶ Examples of transforming the equation $2x + 1 = 5$:

$2x + 1 + 2 = 5 + 2$	**Adding** 2 to both sides
$2x + 1 - 1 = 5 - 1$	**Subtracting** 1 from both sides
$2 * (2x + 1) = 2 * 5$	**Multiplying** both sides by 2
$(2x + 1) / 2 = 5 / 2$	**Dividing** both sides by 2
$(2x + 1)^2 = 5^2$	**Raising** both sides to the second power (squaring)
$\sqrt{2x + 1} = \sqrt{5}$	**Taking** the square root of both sides
$5 = 2x + 1$	**Exchanging** both sides

Notes: All of the previous equations are valid transpositions of the original equation. However, you must decide which transformations will help solve the problem and the correct order in which to perform them. Some transformations will lead you toward the solution and some will lead you away. To arrive at the solution you want, you must choose transformations skillfully.

A common mistake that some students make is to apply a transformation to only some of the terms in an equation. The transformation must apply to *all* terms, or else the result will be invalid. For example: If you intend to divide both sides of $2x + 1 = 5$ by 2, then $2x + 1 / 2 = 5 / 2$ gives an invalid result since the $2x$ term was not divided by 2. The proper transformation, using parentheses, is $(2x + 1) / 2 = 5 / 2$.

Multiplying Algebraic Expressions

To multiply two factors that are each an algebraic expression, let each additive term in the first expression multiply the entire second expression. Remember that according to the associative law of arithmetic (see Appendix A), parentheses can be removed from a set of additive terms. At the end, simplify by combining terms that have identical powers of the unknown. This is called combining like terms.

▶ **Some Simple Examples**

Notice the different notations used in the following examples

$(x)(x)$	is the same as	$x * x$
$x * x$	is the same as	x^2
$-2 * x$	is the same as	$-2x$
$(-2) * (-3)$	is the same as	$+(2 * 3)$

1. $(x)(x)$

The first expression is:	(x)
The second expression is:	(x)

In this case each factor has only one term. Multiply the factors.

This gives:	$x * x$
Replacing $x * x$ with x^2 gives the result:	x^2

2. $(x + 2)(x)$

First expression:	$(x + 2)$
Second expression:	(x)

Multiply each term in the first expression by the entire second expression.

This gives:	$(x * x) + (2 * x)$
Replacing $x * x$ with x^2 gives the result:	$x^2 + 2x$

3. $(x + 1)(x + 2)$

First expression:	$(x + 1)$
Second expression:	$(x + 2)$

Multiply each term of the first expression by the entire second expression.

This gives:	$x * (x + 2) + 1 * (x + 2)$
By the example above, this expands to:	$(x^2 + 2x) + (x + 2)$
Removing parentheses gives:	$x^2 + 2x + x + 2$
Combining like terms gives the result:	$x^2 + 3x + 2$

4. $(x - 2)(x - 3)$

First expression:	$(x - 2)$
Second expression:	$(x - 3)$

Multiply each term of the first expression by the entire second expression.

This gives:	$x * (x - 3) + (-2) * (x - 3)$
This expands to:	$(x^2 - 3x) + (-2x) + (-2) * (-3)$
Which is:	$(x^2 - 3x) + (-2x + 6)$
Removing parentheses:	$x^2 + -3x - 2x + 6$
Combining terms gives the result:	$x^2 - 5x + 6$

Note: Like terms are terms with identical powers of the unknowns.

Squaring a Sum

Now that you have seen how to multiply algebraic expressions, let's examine a special case that occurs frequently and is worth committing to memory. This is where the expression is the square of the sum of two terms.

$$(a + b)^2 = (a + b)(a + b) = a(a + b) + b(a + b) = a^2 + ab + ab + b^2 = a^2 + 2ab + b^2$$

This result can be confirmed using geometry by noting that increasing the side of a square with sides of a by an amount b increases the area a^2 by two rectangles of area ab and a square b^2.

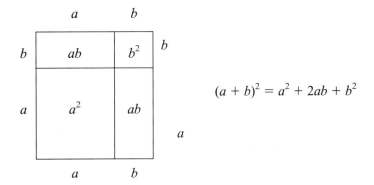

$$(a + b)^2 = a^2 + 2ab + b^2$$

▶ *Example*

Let $a = 6$ and $b = 2$. The sum of a and b is 8. The square of the sum is 64. The sum of the separate areas that make up the 8×8 square are

$$a^2 = 36 \qquad ab = 12 \qquad ab = 12 \qquad b^2 = 4$$

Summing the separate areas:

$$a^2 + 2ab + b^2 = 36 + 2 * 12 + 4 = 64$$

This confirms that $(a + b)^2 = a^2 + 2ab + b^2$ when $a = 6$ and $b = 2$.

Practice Problems

Multiply the following algebraic expressions and collect like terms.

5-2.1 $(2x)(3)$

5-2.2 $(2x)(x)$

5-2.3 $(2x + 1)(x + 1)$

5-2.4 $(2x - 1)(x + 1)$

5-2.5 Evaluate without a calculator: $\dfrac{1234567890}{(1234567891)^2 - (1234567890 * 1234567892)}$

Solving Equations

Solving an equation involves finding the values of the algebraic variables for which an equation is true. Depending on the equation, this process can range from extremely simple to extremely difficult. This section will focus on the solutions of the simplest equations in which the unknowns are raised to the first power. These first power equations are also known as linear equations because, as we will show in the next chapter, they graph as a straight line.

 This section shows how to apply the rules of equations to arrive at a solution in which the unknown appears alone on the left side of the equation and does not appear on the right side.

Solving Simple Equations with a Single Variable

An example of this type of equation is

$$6x + 3x = 2x + x + 24$$

To solve this equation, follow these four easy steps:

1. Combine the like terms on each side of the equation. (In this case combine x terms. In other cases you may also combine constant terms.)

$$9x = 3x + 24$$

2. Bring all the terms containing x to the left side and combine. (In this case subtract $3x$ from both sides.)

$$9x - 3x = 3x - 3x + 24 \qquad \text{or} \qquad 6x = 24$$

3. Then divide both sides by the coefficient of x to leave x alone on the left. (In this case divide both sides by 6.)

$$\left(\frac{6x}{6}\right) = \left(\frac{24}{6}\right) \qquad \text{or} \qquad x = 4 \quad \text{This is the solution.}$$

4. Check the solution by substituting the value found for x into the original equation. (In this case substitute 4 for x.)

$$6 * 4 + 3 * 4 = 2 * 4 + 4 + 24 \qquad \text{or} \qquad 36 = 36$$

Practice Problems

Find the value of x that makes these equations true.

5-2.6 $x + 2 = 4$

5-2.7 $2x + 1 = x + 2$

5-2.8 $x = 2x + 2 - 3x$

5-2.9 $2x + 3 = 4 - 4x - 7$

5-2.10 $3 - 3x + 2 = 2x - 3x - 4$

Ratio Equations

Another example of an equation with one unknown is one in which two ratios are equal. Ratio equations are in the form

$$\frac{a}{b} = \frac{c}{d}$$

Since the solution of this type of equation benefits from a technique called cross multiplication, it will be introduced first.

Cross Multiplication

Cross multiplication is a shortcut operation that applies to equations of two ratios. The bottom term on each side of the equation is multiplied by the opposite top term. Thus, when

$$\frac{a}{b} = \frac{c}{d}$$

then

$$ad = bc$$

Cross multiplication is really the result of multiplying both sides of the equation by the product of both denominators (the bottom part of each ratio):

$$(b * d) * \frac{a}{b} = \frac{c}{d} * (b * d)$$

which reduces to

$$ad = bc$$

The shortcut saves time, so use it.

Equations with *x* on One Side

To solve a ratio equation with unknown *x* on one side:

- Cross-multiply.

- Manipulate the equation so the unknown is alone on the left side of the equation and does not appear on the right side of the equation.

- The solution will be in the form: $x =$ some value.

▶ For example, consider the equation

$$\frac{3}{5} = \frac{6}{x}$$

To solve this equation, cross multiply to get

$$3x = 5 * 6$$

Divide both sides by 3 to leave *x* alone on the left side of the equation:

$$\frac{3x}{3} = 5 * \frac{6}{3}$$

Cancel out the 3s to give the solution:

$$x = 10$$

This type of equation applies to many problems. We will use it later to solve problems involving proportions and percentages.

Equations with *x* on Both Sides

When the variable *x* appears on both sides of the equation, the solution is a bit more complicated. Depending on the form of the equation, it may be that

- **Two specific values of *x*** satisfy the equation.

- **No value of *x*** satisfies the equation.

- **Only the value zero** satisfies the equation.

- **Any value of *x*, including zero,** satisfies the equation.

- **Any value except zero** satisfies the equation.

Remember: The solution will be in the form: $x =$ some value (or values).

Case 1: Variable x Appears on the Top of One Ratio and the Bottom of the Other Ratio.

Cross multiplication results in a term with x^2. Equations of this type, in which the highest power of the unknown is the second power of *x,* are called quadratic equations. The solution of a quadratic equation, in general, is beyond the scope of this text. However, in this special case the solution can be found easily.

Simply take the square root of both sides of the cross-multiplied equation.

▶ Example:

$$\frac{x}{2} = \frac{2}{x}$$
$$x * x = 2 * 2$$
$$x^2 = 4$$
$$\sqrt{x^2} = \sqrt{4}$$
$$x = 2$$

$x = 2$ is the positive, or principal, square root. There is also a negative square root that will solve the equation. Thus, there are two possible solutions to the equation: $x = 2$ and $x = -2$. **The full solution can be written as $x = \pm\, 2$.**

Note: When the signs of the constant terms are such that you are required to take the square root of a negative number, *there are no real numbers* that will satisfy the equation. The solution involves *imaginary numbers,* which are beyond the scope of this text.

Case 2: Variable x Appears on the Top of Both Sides.

There are two possibilities:

- When the equation solves to $x = x$, then the solution is **any value of x.**

  ▶ Example:

$$\frac{x}{2} = \frac{2x}{4}$$

$$4x = 4x$$

$$\frac{4x}{4} = \frac{4x}{4}$$

$$x = x$$

- When the equation solves to $x = 0$, the solution is **zero.**

  ▶ Example:

$$\frac{3x}{2} = \frac{2x}{4}$$

$$12x = 4x$$

$$12x - 4x = 0$$

$$8x = 0$$

$$x = \frac{0}{8}$$

$$x = 0$$

Case 3: Variable x Appears on the Bottom of Both Sides.

Again there are two possibilities:

- When the solution solves to $x = x$, then **any value except zero** will solve the equation.

  ▶ Example:

$$\frac{2}{x} = \frac{4}{2x}$$

$$4x = 4x$$

$$\frac{4x}{4} = \frac{4x}{4}$$

$$x = x$$

Note: Any value of x (except zero) will solve this equation. Zero is not allowed because division by zero in the original equation is not permitted.

- When the solution solves to $ax = bx$, then there is **no value of x** that will solve the equation.

▶ Example:

$$\frac{2}{x} = \frac{3}{x}$$

$$3x = 2x$$

$$3x - 2x = 0$$

$$x = 0 \qquad \text{However, this is not a valid solution.}$$

Note: The zero value solution that worked in case 2 doesn't work when x is on the bottom because division by zero is not allowed.

Practice Problems

Use cross multiplication to

- determine the value(s) of x that will make the equation true, or
- determine that any value of x will make the equation true, or
- determine that any value of x, except zero, will make the equation true, or
- determine that no value of x will make the equation true.

Introductory Problems

5-2.11	$x / 2 = 3 / 4$
5-2.12	$2 / 3 = 3x / 5$
5-2.13	$2 / x = 4 / 2$
5-2.14	$3 / 4 = 5 / 2x$
5-2.15	$-3 / x = -2 / 3$

Intermediate Problems

5-2.16	$x / 4 = 2x / 8$
5-2.17	$2x / 5 = x / 3$
5-2.18	$3 / x = 4 / 2x$
5-2.19	$4 / 3x = 8 / 6x$
5-2.20	$x / 2 = x / 3$

More Difficult Problems

5-2.21	$x / 2 = 2 / x$
5-2.22	$2 / 3x = 2x / 6$
5-2.23	$3x / 5 = 1 / 2x$
5-2.24	$x / 2 = -3 / x$
5-2.25	$-1 / x = x / 5$

Simple Equations with Two Variables

Another type of equation involves two unknowns, say x and y, both to the first power. For example:

$$y = 3x + 2$$

Unlike the equation with a single unknown, this equation does not have a single value of x that makes it true. Instead, there are many values of x for which the equation is true. For every value of x there is a corresponding value of y. In the next chapter it will be shown that the combination of two equations of this type are required to arrive at a solution with *single* values for x and y. The best we can hope for with only one equation of this type is to solve the equation for one unknown in terms of the other unknown.

▶ *Example*

Solve the following equation for x:

$$y = 4x - 1$$

Add 1 to both sides:

$$y + 1 = 4x - 1 + 1$$

Collecting terms gives

$$y + 1 = 4x$$

Flip the equation to get x on the left side:

$$4x = y + 1$$

Divide each side by 4 to leave x by itself on the left side:

$$x = \frac{1}{4}y + \frac{1}{4}$$

Thus, the solution has been found when x is by itself on the left side of the equation and no x terms are on the right side of the equation.

▶ *Another Example*

Temperatures measured in degrees Fahrenheit (F) can be computed from a temperature measured in degrees Celsius (C) by the familiar formula:

$$F = \frac{9}{5}C + 32$$

Solve this equation for C in terms of F by subtracting 32 from both sides:

$$F - 32 = \frac{9}{5}C + 32 - 32$$

Multiply both sides by 5 / 9 (reciprocal of 9 / 5):

$$\left(\frac{5}{9}\right)(F - 32) = \left(\frac{5}{9}\right)\left(\frac{9}{5}\right) * C$$

Combine terms:

$$\left(\frac{5}{9}\right)(F - 32) = C$$

Flip the equation:

$$C = \left(\frac{5}{9}\right)(F - 32)$$

The equation has been solved for C when C is alone on the left and there are no terms with C on the right.

 Note that $\dfrac{5}{9}$ multiplies (F − 32), not just F.

Practice Problems

Solve the following equations with one variable for an explicit value (or values) of the unknown.

5-2.26 $2x + 4x - 3 = 2x + 5$

5-2.27 $2y - y - 4 = 3y - 4$

5-2.28 $p + 3p - 1 = 2$

5-2.29 $3z - 7 = 5 + z$

5-2.30 $2p + \frac{1}{2}p + \frac{1}{4}p = 2$

5-2.31 $3v - 5 = 2v + 1$

5-2.32 $3 / x = 2$

5-2.33 $x / 3 = 1 + (x / 2)$

5-2.34 $2 / x = x / 2$

5-2.35 $3 / x = 2x / 3$

Solve the following equations for the variable indicated, in terms of the remaining variable(s).

5-2.36 $2y = x$ Solve for x.

5-2.37 $x + 3 = y$ Solve for x.

5-2.38 $3p + 1 = q$ Solve for p.

5-2.39 $2y + 3x = 5$ Solve for y.

5-2.40 $d = rt$ Solve for r.

5-2.41 $3y + 6x = 4$ Solve for y.

5-2.42 $2z + x = 2x + 3$ Solve for z.

5-2.43 $y = \frac{1}{2}x + 3$ Solve for x.

5-2.44 $z = \frac{1}{4}y + y + 2$ Solve for y.

5-2.45 $2p + 3v = 3p - 1$ Solve for p.

Now that you know how to solve equations by manipulating algebraic symbols, the next step is to learn how to write these equations. The equations are just mathematical notation for words. While mathematics is precise and unambiguous, words are often just the opposite: imprecise and ambiguous. So the process of translating words into equations is an important step in solving a problem and must be done carefully. Once done, the solution is more or less automatic because the rules of solving equations are precise.

Before starting this section, review Chapter 1 and make sure you understand the IPO method of problem solving. When solving any problem, it is expected that at this point you have done the following:

- Read and reread the problem and understand what the words mean
- Identified the output of the problem
- Identified the input to the problem
- Collected any additional information that is needed
- Decided on an appropriate algebraic notation (Let $x =$ etc.)
- Drawn a diagram (if needed)
- Outlined an approach
- Decided that algebraic equations are best for solving the problem

Only after doing all of this are you ready to translate the words into algebraic equations.

 As you write equations, remember to make sure the units balance.

Translating Words into Equations

Algebraic Notation

Before you can start the translation, you must assign a suitable letter for each term in the problem statement that does not have an explicit value. These become the unknowns.

In the problem "Find two consecutive integers that sum to 17," you can write

Let $x =$ an integer.

Let $y =$ the next consecutive integer.

In this case, two equations will be needed to solve for x and y.

In many cases it is possible to reduce the number of unknowns by writing one unknown in terms of another. For example, in the same problem ("Find two consecutive integers that sum to 17") you can write

Let $x =$ the first integer.

Let $x + 1 =$ the next consecutive integer.

With one unknown, only one equation is needed to solve for x.

Writing Equations

Next, the letters representing the unknowns need to be combined with other information in the problem statement to create one or more equations. To do this, you will need to translate the words and phrases in the problem statement into appropriate algebraic expressions that include algebraic variables and constant values "glued" together with arithmetic operators such as $+$, $-$, $*$, and $/$.

As you are practicing the translation process, you may find a dictionary helpful. The following is not a regular dictionary, but a special English-to-Algebra Dictionary.

English	Algebra	English Example	Algebraic Translation
Add	+ (addition)	Add 2 to 4	$2 + 4$
Away	− (subtract)	Take 2 away from 5	$5 - 2$
Be	=	Let x be 7	$x = 7$
By	multiplied by divided by reduced by increased by	3 multiplied by 2 is 6 6 divided by 2 is 3 3 reduced by 2 is 1 3 increased by 1 is 4	$3 * 2 = 6$ $6 / 2 = 3$ $3 - 2 = 1$ $3 + 1 = 4$
Decreased	− (subtract)	x decreased by 4	$x - 4$
Divide	/ or ÷ (divide)	Divide 6 by 2 to give 3	$6 / 2 = 3$
Divided by	/ or ÷ (divide)	6 divided by 2 is 3	$6 / 2 = 3$
Divided into	/ or ÷ (divide)	2 divided into 6 is 3	$6 / 2 = 3$
Exceeds	Def 1: − (subtract) Def 2: + (add)	5 exceeds 3 by 2 5 exceeds 3 by 2	$5 - 3 = 2$ $5 = 3 + 2$
Equals	=	x equals 2	$x = 2$
Fewer than	− (subtract)	Three fewer than x	$x - 3$
From	− (subtract)	3 from 5 is 2	$5 - 3 = 2$
From now (time)	+ (add)	Bill's age (x) 4 years from now will be 21	$x + 4 = 21$
Gives	= (equals)	3 * 4 gives 12	$3 * 4 = 12$
Greater than	+ (add)	5 is greater than 2 by 3	$5 = 2 + 3$
Half	$* \frac{1}{2}$ or divided by 2	Half of 10 is 5	$\frac{1}{2} * 10 = 5$ or $10 / 2 = 5$
Is	=	a is 5	$a = 5$
Less	− (subtract)	5 less 2 is 3	$5 - 2 = 3$
Less than	− (subtract)	5 less than 9 is 4	$9 - 5 = 4$
More than	+ (add)	5 more than 9 is 14	$5 + 9 = 14$
Multiply	*	3 multiplies 7 to give 21	$3 * 7 = 21$
Of	* (multiply)	$\frac{1}{2}$ of 10 is 5	$\frac{1}{2} * 10 = 5$
Product of	* (multiply)	The product of y and x	xy or $x * y$
Percent of	* (1 / 100)	3 percent	$3 * (1 / 100)$
Quotient	/ or ÷ (divide)	The quotient of 12 and 4	$12 / 4$
Results in	= (equals)	Adding 3 to 4 results in 7	$3 + 4 = 7$
Subtract	− (subtract)	Subtract 3 from 5	$5 - 3$
Take away	− (subtract)	Take 3 away from 4	$4 - 3$
Times	* (multiply)	3 times x	$3x$ or $3 * x$
Twice	* 2	Twice x	$2x$ or $2 * x$
Will be	= (will equal)	x will be 3	$x = 3$

► **Example Problems**

In the following examples we will first define the variables and then translate the problem statement into equations. The final step is to solve the equation and determine the value or values you are trying to find.

1. Find two consecutive integers that add to 5.

   Let x = the smallest integer.

   Let $x + 1$ = the next consecutive integer.

   Two integers sum to 5:

   $$x + (x + 1) = 5$$

Solve the equation:

$x + (x + 1) = 5$	The given equation
$2x + 1 = 5$	Collect terms.
$2x + 1 - 1 = 5 - 1$	Add -1 to both sides to remove $+1$ from the left side.
$2x = 4$	Collect terms.
$\dfrac{2x}{2} = \dfrac{4}{2}$	Divide both sides by 2 to leave x alone on the left side.
$x = 2$	The result is the smallest integer.
$2 + 1 = 3$	Adding 1 to the smallest integer gives the next consecutive integer.

Thus, the two consecutive numbers are 2 and 3.

Check: Make sure that *both* of the following conditions are satisfied:

1. The two numbers, 2 and 3, are consecutive integers.
2. The integers' sum is $2 + 3 = 5$.

2. Find two consecutive odd integers that sum to 8.

Let x = the smallest odd integer.

Let $x + 2$ = the next consecutive odd integer.

Two numbers that sum to 8 become

$$x + x + 2 = 8$$

Solve the equation:

$x + x + 2 = 8$	The given equation
$2x + 2 = 8$	Collect terms.
$2x + 2 - 2 = 8 - 2$	Add -2 to both sides to remove $+2$ from the left side.
$2x = 6$	Collect terms.
$\dfrac{2x}{2} = \dfrac{6}{2}$	Divide both sides by 2 to get x alone on the left side.
$x = 3$	The result is the smallest integer.
$3 + 2 = 5$	Adding 2 to the smallest integer gives the next consecutive odd integer.

The two consecutive odd numbers are 3 and 5.

Check: Make sure that *both* of the following conditions are satisfied:

1. The two numbers, 3 and 5, are consecutive odd integers.
2. The sum of the integers is $3 + 5 = 8$.

3. Find two numbers that sum to 40, where the second number is three times the first.

Let x = the first number.

Let $3x$ = the second number.

Two numbers sum to 40 becomes

$$x + 3x = 40$$

Solve the equation:

$x + 3x = 40$	The given equation
$4x = 40$	Collect terms.
$\dfrac{4x}{4} = \dfrac{40}{4}$	Divide both sides by 4 to get x alone on the left side.
$x = 10$	The result is the first number.
$3 * 10 = 30$	Multiplying the first number by 3 gives the second number.

The two numbers are 10 and 30.

Check: Make sure that *both* of the following conditions are satisfied:

1. The sum of the two numbers is $10 + 30 = 40$.

2. The second number is three times the first, since $3 * 10 = 30$.

4. Separate 33 into two parts so that one part is half the other.

Let x = one part.

Let $x / 2$ = the other part.

Separating 33 into two parts means that the two parts add to 33.

$$x + \frac{x}{2} = 33$$

Solve the equation:

$x + \dfrac{x}{2} = 33$	The given equation
$\dfrac{3x}{2} = 33$	Collect terms.
$\dfrac{3x}{2} * 2 = 33 * 2$	Multiply both sides by 2 to remove the $\dfrac{1}{2}$ on the left side.
$3x = 66$	Do the multiplication.
$\dfrac{3x}{3} = \dfrac{66}{3}$	Divide both sides by 3 to get x alone on the left side.
$x = 22$	The result is the first part.
$\dfrac{22}{2} = 11$	Divide the first part by 2 to get the second part.

The two parts that sum to 33 are 22 and 11.

Check: Make sure that *both* of the following conditions are satisfied:

1. Both parts combine to give 33 (22 + 11 = 33).
2. The first part is half the second part ($11 = \frac{1}{2} * 22$).

5. Find three consecutive integers that sum to 24.

Let x = the smallest integer.

Let $x + 1$ = the middle integer.

Let $x + 2$ = the largest integer.

Three consecutive integers, x, $(x + 1)$, and $(x + 2)$ sum to 24:

$$x + (x + 1) + (x + 2) = 24$$

Solve the equation:

$x + (x + 1) + (x + 2) = 24$	The given equation
$3x + 3 = 24$	Combine like terms.
$3x + 3 - 3 = 24 - 3$	Add -3 to both sides to remove $+3$ from the left side.
$3x = 21$	Collect terms.
$\dfrac{3x}{3} = \dfrac{21}{3}$	Divide both sides by 3 to get x alone on the left side.
$x = 7$	The result is the smallest integer.
$7 + 1 = 8$	The middle integer
$7 + 2 = 9$	The largest integer

The three consecutive integers are 7, 8, and 9.

Check: Make sure that *both* of the following conditions are satisfied:

1. The numbers, 7, 8, and 9, are consecutive integers.
2. The sum of the integers is 7 + 8 + 9 = 24.

Practice Problems
Translate the following problem statements into equations. Then solve the equations to find the requested result.

Introductory Problems
5-3.1 Find a number such that three times the number minus 6 is equal to 18.

5-3.2 Find four consecutive integers that sum to 34.

5-3.3 Find three consecutive integers such that the sum of the first two is 27 more than the third.

5-3.4 Separate 48 into two parts such that the first part is twice the second.

5-3.5 Separate 37 into two parts such that the first part is 5 less than the second part.

5-3.6 Separate 100 into three parts such that the second part is $\frac{2}{5}$ the first and the third is $\frac{3}{5}$ the first.

5-3.7 Diophantus of Alexandria, who is known as the father of algebra, lived in Greece about 250 A.D. We know how long he lived by this algebraic riddle that has been passed down over the centuries:

$\frac{1}{6}$ of his life was spent as a boy.

$\frac{1}{12}$ of his life was spent as a youth.

$\frac{1}{7}$ of his life later he married.

5 years after marriage a son was born.

The son lived only $\frac{1}{2}$ as long as his father did.

Diophantus died 4 years after his son's death.

Write an equation with one unknown and solve it to find how long Diophantus lived.

Hint: If you have trouble getting started, try using the IPO worksheet. Include a diagram that uses a time line to mark off the various life phases. From the information given, assign an algebraic expression to each phase of the time line. After doing this, it will be easier to write the equation.

Percents

Although you have probably learned to use percents before this course, it may be helpful to revisit them now that you know some algebra. As you will see, algebra makes solving percent problems easier. First, here is a quick review.

Percent means parts per hundred parts. A number expressed as a percent is followed by the percent sign (%). Percents have no physical units since the units of the parts and the units of the 100 parts cancel each other.

Percents give you a convenient way to compare values of different magnitudes that otherwise would be difficult to relate. Consider the following problem:

Suppose that of the 20,000 students at your college, 15,000 are state residents. In your math class 16 of 20 are state residents. Does your math class have a higher representation of state residents than the college as whole?

You cannot compare 16 and 15,000 directly. However, if each was expressed as a percentage of the whole, the comparison would be obvious.

The college has 15,000/20,000 * 100% or 75% residents.

The math class has 16/20 * 100% or 80% residents.

Therefore, the math class has a higher representation of resident students than does the college.

Problems Using Percents

It is easy to solve problems with percents if they can be expressed in the following form:

a is *b* percent of *c*

In this statement *a, b,* and *c* represent values yet to be determined. The statement can be rewritten as an equation by making the following substitutions:

- The word *is* can be replaced by the equal sign $=$.

- The word *percent* can be replaced by one-hundredth or $\frac{1}{100}$.

- The word *of* can be replaced by the multiplication sign *.

Then

$$a = \left(\frac{b}{100}\right) * c$$

If we are given any two of the quantities (*a, b, c*), the remaining unknown can be found by solving an equation with one unknown. There are three possibilities:

a is unknown and *b* and *c* are given.

b is unknown and *a* and *c* are given.

c is unknown and *a* and *b* are given.

▶ Example Problems
Here are examples of the three possible situations, renaming each unknown as *x:*

1. What is 40 percent of 25?

Start with the basic equation:	$a = (b/100) * c$ (*b* and *c* are known)
a, the unknown, becomes *x:*	$x = (40/100) * 25$
Which can be solved:	$x = 0.4 * 25 = 10$
Thus, **10** is 40 percent of 25.	

2. Ten is what percent of 25?

Start with the basic equation:	$a = (b/100) * c$ (*a* and *c* are known)
b, the unknown, becomes *x:*	$10 = (x/100) * 25$
Which can be solved:	$(10/25) = x/100$
	$x = 100 * (10/25) = 40$

Thus, 10 is **40 percent** of 25.

3. Ten is 40 percent of what?

Start with the basic equation:	$a = (b/100) * c$ (*a* and *b* are known)
c, the unknown, becomes *x:*	$10 = (40/100) * x$
Which can be solved:	$10 * 100/40 = x$
	$x = 100/4 = 25$

Thus, 10 is 40 percent of **25.**

Practice Problems
Answer the following percentage problems, rounding decimal answers to two places.

Introductory Problems
5-3.8   Ten is what percent of 25?

5-3.9   Ten is 15 percent of what number?

5-3.10 What is 70 percent of 90?

5-3.11 Thirty-five is what percent of 75?

5-3.12 Thirty-five is 70 percent of what number?

5-3.13 A restaurant meal costs $24.00. How much is a 15 percent tip?

Intermediate Problems
5-3.14 A restaurant meal costs $21.50. If you leave a 15 percent tip, what is the total cost?

5-3.15 You purchase a car for $9500. If the tax rate is 8 percent, how much is the tax?

5-3.16 You purchase a car for $11,500. What is the total cost if the tax rate is 6 percent?

5-3.17 The total cost, including tax, of a telephone is $23.64. The tax was $1.65. What percent is the tax rate?

5-3.18 What is the annual income of a $10,000 bond that yields 6 percent interest a year.

More Difficult Problems

5-3.19 The total cost of a car including tax at 8.6 percent is $12,500. What was the cost before tax?

5-3.20 A bond pays interest at 7 percent a year. The interest is taxed at a rate of 28 percent. What interest rate would a nontaxable bond need to pay to yield the same after-tax income as the taxable bond?

5-3.21 A nontaxable bond pays interest at 5 percent a year. What interest rate would a taxable bond need to pay to yield the same after-tax income as the nontaxable bond? The tax rate is 28 percent.

Percent Change

What about problems involving changes expressed as a percent? Suppose you have two values. To measure a change between them, you must decide which value you are starting from and which you are going to. The direction is important and may be either forward or backward in time.

Let the value you are starting *from* be called v_1 and the value you going *to* be called v_2. Then the change between them $(v_2 - v_1)$ is expressed as a percent of the starting value (v_1). You can compute this change using the percent formula given in the previous section (see the sidebar on page 184). However, it is faster to compute percent change using a ratio (v_2 / v_1) called a change factor. The value you are going *to* is always on the top of the ratio.

Finding percent change is a two-step process:

- First compute a change factor.

- Then convert the factor to a percent change.

To compute the change factor (f), divide the second value (v_2) by the first value (v_1).

$$f = \frac{v_2}{v_1} \qquad \text{except when } v_1 = 0$$

The value you are going to is on the top.

To compute the percent change (c), subtract 1 from the change factor (f) and multiply by 100.

$$c = (f - 1) * 100$$

Multiplying by 100 is the same as moving the decimal point two places to the right. The change between the two values will have the following effect on the factor:

If the change is an **increase**	then	$f > 1$
If the change is a **decrease**	then	$f < 1$
If there is **no change**	then	$f = 1$

Summary

- Let v_1 be the value you are starting from.

- Let v_2 be the value you are going to.

- The change goes from v_1 to v_2 (but is computed as $v_2 - v_1$).

- In some cases, that change may be backward in time.

- The change is expressed as a percent of the starting value, v_1.

For those of you who want to know where the percent change formula came from, here is confirmation of the formula from the original percent equation introduced in the previous section. Remember, this won't work when v_1 is zero.

Let the change between the first and second values be $v_2 - v_1$.

Then express the change $(v_2 - v_1)$ as a percent (c) of the first value (v_1).

Writing this statement as an equation:

$(v_2 - v_1) = (c / 100) * v_1$	From the percent equation
$(v_2 / v_1) - (v_1 / v_1) = (c / 100) * v_1 / v_1$	Divide all terms by v_1.
$[(v_2 / v_1) - 1] = (c / 100)$	Let $v_1 / v_1 = 1$.
$[(v_2 / v_1) - 1] * 100 = c$	Multiply both sides by 100.
$(f - 1) * 100 = c$	Replace v_2 / v_1 with the change factor f.
$c = (f - 1) * 100$	Exchange sides to get the formula.

▶ **Examples Going Forward in Time**

Sales of the XYZ Company increased *from* $50 million last year *to* $75 million this year. What is the percent change?

The change is from last year to this year, so v_1 = $50 million and v_2 = $75 million.

Change factor: $f = v_2 / v_1$ = $75 million/$50 million = 1.5

Percent change: $c = (f - 1) * 100 = (1.5 - 1) * 100 = 50\%$

The price this year is 50 percent greater than the price last year.
Suppose the company had a bad year and sales decreased *from* the $75 million reported last year *to* $50 million this year. Now what is the percent change?

The change is from last year to this year, so v_1 = $75 million and v_2 = $50 million.

Change factor: $f = v_2 / v_1$ = $50 million/$75 million = 0.667

Percent change: $c = (f - 1) * 100 = (0.667 - 1) * 100 = -33.33\%$

The price this year is 33.33 percent less than the price last year.

Notice that percent increases are positive and percent decreases are negative.

▶ **Examples Going Backward in Time**

It is possible to use the percent change formula to compute changes that go backward in time. In this case the first value v_1 will be measured at a time later than the second value v_2, the reverse of the previous examples. Just remember that the value the change starts from, or is compared to, is always the first value, v_1.

Consider the stock price of a share of XYZ Company. Suppose its price peaked in March at $50. Now in June, the price has dropped to $35. Find the change as a percent of the current value. (The same problem could have asked: "What is the change from June to March?")

The change is compared to June, so $v_1 = 35$ and $v_2 = 50$.

$$f = \frac{v_2}{v_1} = \frac{50}{35} = 1.43$$

$$c = (f - 1) * 100 = (1.42 - 1) * 100 = 42\%$$

The March stock price was 42 percent higher compared to the price in June.

Sales in May were $150 million and in June were $200. Compute the difference as a percent of the June value.

The change is compared to June, so $v_1 = 200$ and $v_2 = 150$.

$$f = \frac{v_2}{v_1} = \frac{150}{200} = 0.75$$

$$c = (f - 1) * 100 = (0.75 - 1) * 100 = -25\%$$

May sales were 25 percent lower compared to June sales.

Computing Percent Changes in Your Head

With a little practice you can convert change factors to percent changes in your head. Simply inspect the change factor, subtract 1, and multiply by 100.

▶ Examples of computing a percent change from a change factor, as you would do it in your head:

Increases:	*Factor*	*Reduced by 1*	*Times 100*	*Percent Change*
	$f = 1.1$	0.1	0.1 * 100	10%
	$f = 1.6$	0.6	0.6 * 100	60%
	$f = 2.5$	1.5	1.5 * 100	150%

Decreases:	*Factor*	*Reduced by 1*	*Times 100*	*Percent Change*
	$f = 0.9$	−0.1	−0.1 * 100	−10%
	$f = 0.7$	−0.3	−0.3 * 100	−30%
	$f = 0.3$	−0.7	−0.7 * 100	−70%
	$f = 0.1$	−0.9	−0.9 * 100	−90%

Practice Problems

Determine from the wording of the problem which value is v_1 and which is v_2. Compute percent change, rounding decimal answers to two places. Express your result in a sentence stating how much the second value is less than or greater than the first value. For example: "This year's sales are 50 percent less than last year's sales."

5-3.22 XYZ Company sales went from $12,000 to $14,000 a week.

5-3.23 XYZ Company sales went from $14,000 to $12,000 a week.

5-3.24 The price of XYZ Company stock went from $25 $\frac{1}{8}$ to $29 $\frac{3}{8}$.

5-3.25 The price of XYZ Company stock went from $29 $\frac{3}{8}$ to $25 $\frac{1}{8}$.

5-3.26 A TV is marked down from $300 to $200.

5-3.27 A TV is marked up from $200 to $300.

Use the data in the following table to answer the following problems. Determine from the wording of the problem which value is v_1 and which is v_2. Compute percent change, rounding decimal answers to two places. Express your result in a sentence stating how much the second value is less than or greater than the first value. For example: "August sales are 50 percent less than July sales."

Month	Sales ($ Million)
Jan	1.2
Feb	1.3
Mar	1.5
April	1.2
May	1.1
June	0.9

5-3.28 What is the percent change from January to February?

5-3.29 What is the percent change from January to March?

5-3.30 What is the percent change from March to April?

5-3.31 What is the percent change in sales between April and June in terms of June?

5-3.32 What would the July sales need to be to have a 20 percent increase from June?

5-3.33 What would the July sales need to be to have a 20 percent decrease from June?

5-3.34 What would the July sales need to be to have a 20 percent increase from January?

5-3.35 Compare May sales to June and express the change as a percent.

Proportions

Proportions are ratios used to show the relation of one quantity to another. They can be used to compare a part of something to its whole amount. Here are two examples:

- If 40 out of 50 people liked the movie they just saw, then the portion who liked it is $\frac{4}{5}$.

- If a right triangle has a height of 10 inches and a base of 20 inches, the proportion of height to base is $\frac{10}{20}$ or $\frac{1}{2}$. This is another way of saying that the height is $\frac{1}{2}$ of the base.

- In the same triangle the proportion of the base to the height is $\frac{20}{10}$ or 2. Thus, the base is twice the height.

 In each case, the number on the bottom of the ratio represents the whole and the number on the top represents part of that whole.

Alternate Notation
Sometimes proportions are written in an alternative notation using a colon in place of a ratio. The two notations are equivalent: 2:3 is the same as 2/3 or $\frac{2}{3}$.

Proportions Have No Units
In most cases, proportions are used to relate quantities with the same physical units. When this is true, the units on the top and bottom of the ratio cancel each other.

Equations Using Proportions
When quantities are in the same proportion, they can be related in an equation in which one ratio is equated to another ratio. If three of the four terms in the equation are known, it is a simple matter to solve the equation for the unknown quantity. Refer back to the section on ratio equations in Section 5-2 for details on how to solve this type of equation.

▶ *Example Problem*

Two computer monitors have rectangular screens of different but proportional sizes. That is, the ratios of height and width for both screens are the same. The larger screen has a height of 12 inches and a width of 16 inches. If the smaller screen has a height of 10 inches, what is its width?

Ratio of height to width of the larger screen is 12/16. Since both screens are proportional in size, the ratio of the smaller screen is 10/x, where x represents the unknown width in inches. The fact that both ratios are equal allows us to form an equation with one unknown, which can be solved easily.

$$\frac{12}{16} = \frac{10}{x}$$

$$x = 10 * \frac{16}{12}$$

$$x = 13.3 \text{ in.}$$

The problem could also have been solved using the ratio of width to height:

$$\frac{16}{12} = \frac{x}{10}$$

$$x = 10 * \frac{16}{12}$$

$$x = 13.3 \text{ in.}$$

Practice Problems

Introductory Problems

5-3.36 The length and width of a rectangle are in the proportion 3:1. If the width is 3 feet, what is the length?

5-3.37 The altitude and base of a right triangle are in the proportion 3:4. If the base is 20 feet, what is the altitude?

5-3.38 There are 12 men in a class of 30 college students. What is the proportion of men to women in the class?

5-3.39 The proportion of oil to vinegar in a salad dressing is 3:1. How much oil should be mixed with 2 ounces of vinegar?

Intermediate Problems

5-3.40 The proportion of oil to vinegar in a salad dressing is 3:1. How much vinegar and how much oil are needed to be combined to make a total of 8 ounces of dressing? (*Hint:* Express the amount of oil in terms of the total minus the amount of vinegar.)

5-3.41 Information stored on a hard drive consists of programs and data. The proportion of data storage space to program storage space is 10:1. How many megabytes (MB) of total storage space are needed to store a total of 22 megabytes of programs and data.

Chapter Summary

Components of Equations
Algebraic equations are made up of algebraic expressions, which in turn are composed of algebraic variables, coefficients, and constants.

Algebraic Equations
An equation is made up of two algebraic expressions that represent the same quantity. An equal sign separates the two expressions and signifies that the expressions are in balance with each other. Algebraic expressions are made up of constants and algebraic variables. Constants are specific numbers. Algebraic variables are letters that represent some number (or numbers) yet to be determined. An important characteristic of an algebraic variable in an equation is that **variables may represent only those values that will allow the equation to be true.**

Solving Equations
An equation will remain in balance if the expressions on both sides have the same operation performed on them. Valid operations include:

- **Adding** to the same quantity to both sides

- **Subtracting** the same quantity from both sides

- **Multiplying** both sides by the same quantity

- **Dividing** both sides by the same quantity (other than zero)

- **Raising** both sides to the same power

- **Taking** the square root of both sides

- **Exchanging** both sides with one another

An equation can be solved for one of its unknown terms by manipulating the equation so that the unknown term is by itself on the left side and all of the other terms (that do not involve that unknown) are on the right side. If there is a single unknown and it can be found to have a specific value (or values), the equation is said to have an exact solution. If more than one unknown occurs in an equation, it may only be possible to solve for one unknown in terms of the others.

Percents
Many problems involving percents can be expressed as

   a is b percent of c

Given any two of the values for a, b, and c, the third can be can be solved using the equation

$$a = \left(\frac{b}{100}\right) * c$$

Percent Change
Percent change involves the change of some quantity

- from a first value v_1

- to a second value v_2

The change is always *from* the first value (v_1) *to* the second value (v_2). The change is expressed as a percent of the first value (v_1) by first computing a change factor (f) that is the ratio of v_2 to v_1:

$$f = \frac{v_2}{v_1} \qquad \text{except when } v_1 = 0$$

and then computing the percent change using this factor:

$$c = (f - 1) * 100$$

Thus, to compute a percent change, just remember two simple steps:

1. First compute the change factor (f) as v_2/v_1.

2. Convert the change factor into a percent by subtracting 1 and multiplying the difference by 100.

Proportions

Proportion equations are in the form

$$\frac{a}{b} = \frac{c}{d}$$

Given any three of these terms you can solve for the fourth.

Review Questions

1. Define the following terms: *coefficient, constant, variable,* and *equation.*
2. When an equation has two variables, which variable is independent and which is dependent? 165
3. What operations can be performed to an equation that will permit it to remain in balance? 167
4. What does it mean to solve an equation? 169
5. What is a linear equation?
6. In your own words, write a procedure to solve a linear equation with a single variable.
7. Is there an explicit solution to a single linear equation with two variables?
8. Describe how to solve a linear equation with two variables for the independent variable.
9. What are the steps involved in translating word problems into equations? 176
10. What is the English statement that can be used to describe many percent problems? Percent of
11. What equation results when you translate the percent statement stated in question 10? 1/100
12. What are the two equations involved with percent changes? Define terms in the equations. $(v_2 - v_1)$ (v_2/v_1)
13. What is the rule for determining if a change factor represents an increase or a decrease? 183
14. Which of the two values in a change factor appears in the numerator of the ratio? the 2nd value
15. In percent change, the change going from v_1 to v_2 is expressed as a percent of which value? v_1, the starting value
16. What is the difference between a percent and a proportion?

Problem Solutions

Section 5-2

5-2.2 $2x^2$		5-2.4 $2x^2 + x - 1$
5-2.6 $x = 2$		5-2.8 $x = 1$
5-2.10 $x = 4\frac{1}{2}$ or 4.5		5-2.12 $1\frac{1}{9}$
5-2.14 $3\frac{1}{3}$		5-2.16 Any value
5-2.18 No value of x works.		5-2.20 $x = 0$
5-2.22 $x = \pm\sqrt{2}$		5-2.24 No solution with real numbers

	5-2.26 $x = 2$	5-2.28 $p = \frac{3}{4}$	
	5-2.30 $p = 8/11$	5-2.32 $x = 1\frac{1}{2}$	
	5-2.34 $x = \pm 2$	5-2.36 $x = 2y$	
	5-2.38 $p = (\frac{1}{3})(q - 1)$	5-2.40 $r = d/t$	
	5-2.42 $z = (\frac{1}{2})(x + 3)$	5-2.44 $y = (\frac{4}{5})(z - 2)$	
Section 5-3	5-3.2 7, 8, 9, 10	5-3.4 32, 16	
	5-3.6 50, 20, 30	5-3.8 40%	
	5-3.10 63	5-3.12 50	
	5-3.14 $24.73	5-3.16 $12,190	
	5-3.18 $600	5-3.20 5.04%	
	5-3.22 16.67% (increase)	5-3.24 16.92% (increase)	
	5-3.26 −33.33% (decrease)	5-3.28 8.33% (increase)	
	5-3.30 −20% (decrease)	5-3.32 $1.08 million	
	5-3.34 $1.44 million (increase)	5-3.36 9 ft	
	5-3.38 2:3	5-3.40 2 oz vinegar, 6 oz oil	

Graphing

A picture is worth a thousand words.

6-1 ■ Introduction

How to Use This Chapter

Everyone working in information technology needs to know a little about graphs. The field overflows with all kinds of data. In fact, the old-fashioned name for the field is *data processing*. Data in its original "raw" form is very hard to digest. To make sense of it, it must be displayed in a way that is easier to understand. Graphing data is one of the best ways to accomplish this.

Spreadsheet software is likely to be the first place you will be exposed to graphing data. Although spreadsheets refer to graphs as charts, the two are really the same thing. This book will not attempt to teach you how to use spreadsheet software. That is best done elsewhere. But a few of the different graphing (charting) techniques employed by spreadsheets will be covered from two points of view that are often overlooked by most texts on using spreadsheets. They are:

- What types of data are best suited to be displayed for each type of chart

- What is going on behind the screen as the software automatically creates a chart

Another important aspect of graphing, which is a continuation of the previous chapter on algebra, is the graphing of equations. Graphing of the linear (straight-line) equation will be covered thoroughly. You will see how two linear equations, each with two unknowns, can be solved graphically for an explicit value of each unknown. Seeing algebra through your "graphic glasses" will help you appreciate better what was going on in the previous chapter.

Background

The pictorial presentation of data provides a way to see at a glance how the various data points are related to one another. There are many different techniques for presenting data graphically. The next section will focus on spreadsheet charts. Here, a series of data is displayed in pie charts, bar charts, line charts, or *x-y* charts. The section that follows this will show how to graph data generated by algebraic equations.

The concluding section will cover two special topics concerning graphs. First, a technique will be covered for computing the shortest distance between two points on a graph using the famous Pythagorean theorem. The chapter concludes with a brief introduction to the coordinate system used for displaying graphics on a computer screen, something important for programmers who need to display more than text in their program output.

Chapter Objectives

In this chapter you will learn:

- The basic types of graphs used by spreadsheet software, how they each work, and when to use each

- How to use a rectangular coordinate system to graph equations

- That a linear equation with two variables (equations with all variables raised to the first power) will plot a straight line on a rectangular coordinate system

- That linear equations with two variables have a specific slope (a measure of the orientation of the straight-line plot). The more vertical the orientation of line, the larger is the absolute value of the slope

- That linear equations with two variables have a specific y-intercept (the value of y where the straight line crosses the y-axis)

- That a linear equation written in a special form, called the slope-intercept form, allows you to know the slope and y-intercept of the equation directly, without further calculation

- How to compute the slope of a liner equation, given two points along its line

- How to compute the shortest distance between two points on a graph

- How computer graphics coordinates work

Introductory Problem

Use a piece of graph paper to display the following data as a bar chart.

<u>*XYZ Company Earnings ($ Millions)*</u>

Qtr1	$215
Qtr2	$175
Qtr3	$300
Qtr4	$415

To draw the bar chart, you should accomplish the following tasks:

1. Determine a scale.

   - Depending on the size of the small divisions on your graph paper, let a suitable number of divisions represent an appropriate number of million dollars.

   - Pick a scale so the largest bar will fill about half of the page. This will leave room for labels and titles.

2. Draw the axes.

   - Leave appropriate margins.

   - Draw the vertical axis as a line along the left side of the page.

- Mark the increments of the scale along the vertical axis
  (You need to decide what increments are appropriate.)

- Draw the horizontal axis as a line along the bottom of the page.

3. Create labels and titles.

   - Put the units of the scale beside the vertical axis: $ Million.

   - Put the quarter number beneath the horizontal axis: Qtr1, Qtr2, etc.

   - Put a title at the top of the graph: XYZ Company Quarterly Earnings.

4. Draw the bars.

   - Let each vertical bar represent the earnings for one quarter.

   - The length of each bar should be proportional to the value it represents, according to the scale on the left.

Spreadsheet programs provide a variety of ways to graphically display your numeric data. This section will introduce the four most fundamental types of graphs, or charts as they are more commonly called.

- Pie chart

- Bar chart

- Line chart

- *x-y* chart

Many other types of charts, which are variations of these four, are also available. Each type of chart provides its own unique way of displaying the data. Depending on the type of data you have and what aspect of the data you wish to bring out, one type of chart will often offer advantages over the others.

Data Series

Data to be graphed or charted come in a series. A series of data is a collection of related observations of some quantity. All of the data in a series must be collected in a consistent manner and measured in the *same physical units.*

One of the most typical ways to collect data is by making observations of some quantity at different times. Such a collection is called a *time series.* Examples of time series are the daily circulation figures of a newspaper or the monthly sales totals of a company.

Pie Charts

Pie charts are a great way to show parts of a whole. The whole pie represents the sum of all of the data items, and each piece of pie represents one data item. The size of each piece, determined by the angle of its wedge, is proportional to the share of the whole for that data value.

Pie charts work best when there are just a few data items, so the chart is not cluttered with too many slices. For example, if you have sales data for the four quarters of the year, the amount of sales for each quarter is represented by a slice of pie. The size of each slice is proportional to a share of the annual sales for that quarter.

Consider the following data:

Quarterly Sales for XYZ Company ($Millions)

Qtr1	$100
Qtr2	$150
Qtr3	$200
Qtr4	$400
Year	$850

Spreadsheet software will display this data in a pie chart, as shown in Figure 6.1.

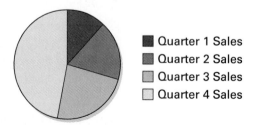

Figure 6.1 *XYZ Company Sales*

The pie chart in Figure 6.1 divides the full 360 degrees of the circular pie into four angles, each representing the sales for one quarter. The angle of each piece of pie is proportional to the share of the annual sales for one quarter.

Here is how you compute the size (angle) of each piece of pie:

For any quarter:

- Proportion on circle: Angle / 360

- Proportion of sales: Quarter Sales / Annual Sales

- Both are the same: Angle / 360 = Quarter Sales / Annual Sales

First quarter: $(\text{Angle}_1 / 360) = (100 / 850)$
$\text{Angle}_1 = 360 * 100 / 850$
$\text{Angle}_1 = 42.35$ degrees
This is 11.8 percent of the whole pie.

Second quarter: $(\text{Angle}_2 / 360) = (150 / 850)$
$\text{Angle}_2 = 63.52$ degrees
This is 17.6 percent of the whole pie.

Third quarter: $(\text{Angle}_3 / 360) = (200 / 850)$
$\text{Angle}_3 = 84.7$ degrees
This is 23.5 percent of the the whole pie.

Fourth quarter: $(\text{Angle}_4 / 360) = (400 / 850)$
$\text{Angle}_4 = 169.4$ degrees
This is 47.1 percent of the whole pie.

Of course the spreadsheet software does all of these calculations for you automatically.

Bar Charts

Bar charts are good for comparing the various items in a data series to one another. They are also good for showing how data varies from one observation in the series to another.

The data is represented by a set of bars, each of which represents one data item. The bars can be displayed either horizontally (in rows) or vertically (in columns). The length of each bar is proportional to the value of one data item.

To accomplish this, a scale needs to be assigned that is consistent with the data to be displayed. This scale is like a ruler that has been shrunk to fit on the side of your chart. Each unit on the ruler is proportional to a unit of your data. For example, 1 inch on the ruler might represent a number of dollars or days, or whatever units your data is measured in. The ruler (scale) must be long enough to fit the largest data item and perhaps a little more so things are not too crowded.

A bar chart of the XYZ Company quarterly sales data given earlier is shown in Figure 6.2.

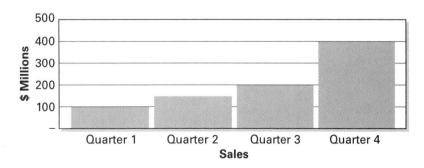

Figure 6.2 *XYZ Company Sales*

Determining the Scale

The maximum value on the scale of the bar chart in Figure 6.2 needs to be at least $400 million, the largest data value. It is a good idea to make the maximum a little larger so there is some room between the longest bar and the top of the scale. A maximum of $800 million would be too large since all the bars would then be in the lower half of the chart. In this case, $500 million does nicely. Next, it is necessary to divide the maximum scale amount into parts. In this case five parts, each representing $100 million, works best.

The spreadsheet software will do all of this for you automatically. But sometimes you may want to adjust the scale yourself. The scale on most spreadsheet charts can be adjusted manually if you do not like the one the software assigns.

Line Charts

The line chart is an alternate way to display data suitable for a bar chart. However, a dot is placed in the center of the top of each bar, the dots are connected by lines, and the bars are removed. The scaling is the same as for the bar chart. As in bar charts, the data in line charts needs to be a series of single data items recorded in the same units.

A line chart of the XYZ Company quarterly sales data given earlier is shown in Figure 6.3.

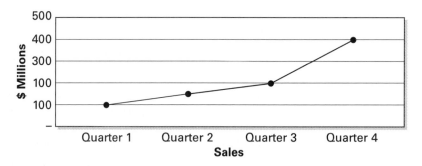

Figure 6.3 *XYZ Company Sales*

x-y Charts

Although they are similar to line charts, *x-y* charts allow *corresponding pairs* of data observations to be plotted. The first value in a data pair is called the *x*-value and the other is called the *y*-value. The chart has two scales, called axes. The horizontal axis (the *x*-axis) is used to plot *x*-values. The vertical axis (the *y*-axis) is used to plot *y*-values.

When the data on the horizontal scale is in equal increments (for example, the numbered quarters of the year) the *x-y* chart looks like a line chart. However, the special feature of the *x-y* chart is that the *x*-values plotted on the horizontal scale need not be in equal increments. Rather, they can be values that fluctuate relative to a corresponding *y*-value.

As shown in Figure 6.3, the data for each pair of values represents one point on the chart. This point is at the intersection of two imaginary lines, one vertical and the other horizontal. The vertical line passes through the *x*-value on the *x*-axis. The horizontal line passes through the *y*-value on the *y*-axis. This arrangement is called a *rectangular coordinate system*. Don't be too concerned if it is not yet perfectly clear since it will be described in detail in the next section.

The following data pairs represent the number of widgets that can be packed into boxes of different sizes. The boxes are cubes with all sides of equal length. The box size represents the length in inches of one side of the cube.

Box Size (inches)	Number of Widgets
6	20
10	100
12	175
16	400

Letting box size be the *x*-values and number of widgets be the *y*-values, the *x-y* chart would look like Figure 6.4.

Figure 6.4   *Widget Packing*

It takes both a box size and a corresponding number of widgets to mark each of the four dots on the chart. Vertically, a point is positioned directly above the value on the box size scale that corresponds to the number of boxes. Horizontally, a point is positioned to the right of the value on the widgets scale that corresponds to the number of widgets. The intersection of the vertical path and the horizontal path marks the location of the point.

For example, to locate the third pair of data items (12, 175) on the chart:

- Draw a vertical line passing through 12 on the box size scale.

- Draw a horizontal line passing through 175 on the widgets scale.

The intersection of these two lines mark the position of the third point in the data series.

This section will focus on a technique that best suits displaying the data generated by an equation. But it is also suitable for plotting data from a table when the data comes in corresponding pairs. This technique is called a *graph*. Of the spreadsheet charts that were introduced in the previous section, the *x-y* chart is the best for this purpose.

Rectangular Coordinate System

In order to graph an equation, you need some sort of grid that relates distances in different directions. This grid is called a coordinate system. Mathematicians employ many different types of coordinate systems, usually related to some geometric shape such as the rectangle, circle, or ellipse. The simplest of these, and the one most useful for the types of graphs you are likely to encounter, is the *rectangular coordinate system.*

The rectangular coordinate system builds on the number line (see Appendix A), which is a one-dimensional coordinate system. So, if you have only one variable to plot, the number line will do. But when you have two variables, such as x and y, that come in corresponding pairs, you need a two-dimensional coordinate system.

The two-dimensional rectangular coordinate system employs two number lines at right angles to each other. One is horizontal and one is vertical. The most common arrangement is as follows:

- The horizontal number line is called the x-axis (usually with $+x$ to the right and $-x$ to the left).

- The vertical number line is called the y-axis (usually with $+y$ up and $-y$ down).

The intersection of the x-axis and the y-axis is called the origin (usually at $x = 0, y = 0$).

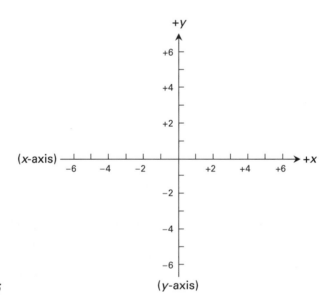

Figure 6.5

When you draw a rectangular coordinate system, be sure to include all of its components:

- Two axes drawn at right angles to each other

- Arrows indicating the positive direction of each axis

- Labels indicating the variable (usually x and y) assigned to each axis
- A suitable scale labeled for each axis (usually the same for both)

Plotting Data Points

Every point on a graph represents a specific value of x and a specific value of y.

- The value of x is determined by the position of the point along the x-axis.
- The value of y is determined by the position of the point along the y-axis.

Points on the graph may be designated by a pair of numbers written within parentheses (x, y).

- The x-value comes first.
- The y-value comes second.

Figure 6.6 shows some examples of points plotted on a rectangular coordinate system.

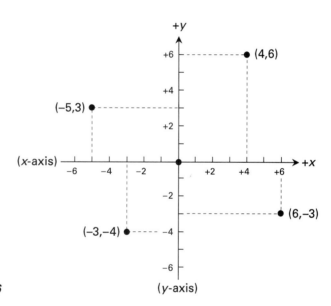

Figure 6.6

Notice that the origin marks the point $(0,0)$.

Practice Problems

Use graph paper to draw a rectangular coordinate system and plot the following points.

6-3.1 $x = 5$, $y = 6$	6-3.7 $(2, -2)$
6-3.2 $x = 2$, $y = 3$	6-3.8 $(-4, -3)$
6-3.3 $x = 2$, $y = 0$	6-3.9 $(2, 0)$
6-3.4 $x = 0$, $y = 4$	6-3.10 $(-4, -2)$
6-3.5 $x = 0$, $y = 0$	6-3.11 $(0, 1)$
6-3.6 $x = -2$, $y = 2$	6-3.12 $(1, 0)$

Plotting a Series of Data Points

Suppose you have several pairs of (x, y) data points that are related. Each of the pairs can be plotted as a single point on the graph. Then, because the data is a series of related points, the dots can be connected to make a graph.

Consider the following series of related *x-y* data points:

x	y
-3	9
-2	4
-1	1
0	0
1	1
2	4
3	9

The graph of the data is shown in Figure 6.7.

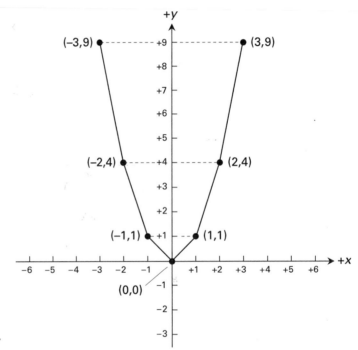

Figure 6.7

Note: The data points of this series were derived from the equation $y = x^2$.

Plotting an Equation

A rectangular coordinate system is ideal for plotting equations with two variables such as *x* and *y*. However, keep in mind that the coordinates can represent any pair of variables designated with letters (such as *p* and *q* or *r* and *s*) or names (such as *time* and *distance* or *sales amount* and *year*).

Before you plot an equation, you must first derive a suitable number of data points from it. To do this, make a table containing pairs of data points. For the independent variable (the one on the right of the equation) write down a few values. Then compute the corresponding value for the dependent variable (the one on the left of the equation). Of course you can't plot all the points defined by the equation, so select a few points of interest, usually near the origin.

For example, plot the equation $y = 2x + 1$. First make a table of data points by computing y at selected values of x near the origin.

A Few Positive Values of x			A Few Negative Values of x	
x	y		x	y
0	1		-1	-1
1	3		-2	-3
2	5		-3	-5
3	7			

The equation has many other x-y points between these points and at points far away from the origin. But we will be able to see the "shape" of the data by plotting only these few points (see Figure 6.8).

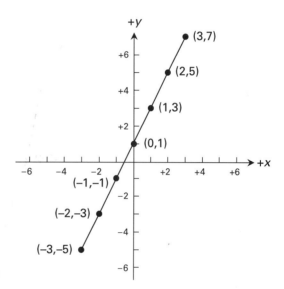

Figure 6.8

 Notice that the equation $y = 2x + 1$ plots points that lie on a straight line.

Two Points Define a Line

You don't need an entire table of data to plot a straight-line equation. You only need two points. Plot the graph of $y = 3x + 3$ using only two points. Although any two points will do, it is often easiest to use:

- A first point when $x = 0$.

- A second point when $y = 0$.

To complete coordinates at each point, you will need to substitute zero for the value on one axis to find the corresponding point on the other axis:

- At $x = 0$ the equation becomes $y = 0 + 3$ or $y = 3$. Coordinates are $(0, 3)$.

- At $y = 0$ the equation becomes $0 = 3x + 3$ or $x = -1$. Coordinates are $(-1, 0)$.

Figure 6.9 shows the straight line between the two points.

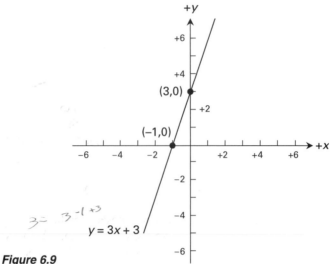

Figure 6.9

Note: For equations that pass through the origin (0, 0), $x = 0$ and $y = 0$ is not sufficient to define the line. These equations require an additional point away from the origin to define the line.

Checking the Graph. It is always important to check your results. When using just two points to define a straight-line graph, it is important to plot one or two additional points to confirm that the first two points were done correctly. All the points must lie along the same straight line.

Practice Problems
Plot the following equations using only the two data points. First determine the two (x, y) points, then plot and label the points on a coordinate system, and finally draw the equation line between the points. Try one point where $x = 0$ and another point where $y = 0$. If these points coincide, pick another (x, y) point.

6-3.13 $y = 3x + 2$

6-3.14 $y = -2x + 1$

6-3.15 $y = -4x - 3$

6-3.16 $y = 4x + 2$

6-3.17 $y = \frac{1}{2}x + \frac{3}{4}$

Plot the following equations that pass through the origin using any two points.

6-3.18 $y = 2x$

6-3.19 $y = \frac{1}{2}x$

6-3.20 $y = 3x$

Use graph paper to draw a rectangular coordinate system and plot the following data.

6-3.21

x	y
0	3
1	0
2	-3
3	-6
4	-9

6-3.22

x	y
-2	-7
0	-3
2	1
4	5
6	9

The Slope-Intercept Form of the Linear Equation

Linear equations have all unknowns raised to the first power. No unknowns are squared, and no unknowns are raised to powers higher than 1.

Linear equations with two variables can be written in what is called the slope-intercept form:

$$y = mx + b$$

This form of the equation can represent any linear equation (with the exception of a vertical line) by providing specific values for m and b terms. You have seen equations like this before. The equation used as an example in the previous section, $y = 2x + 1$, is a linear equation. When graphed, its points plot a straight line.

 All equations of this form plot a straight line on a rectangular coordinate system.

The coefficient of x (represented by m) is called the *slope* of the equation. The constant b is called the y-*intercept.*

 The advantage of writing a linear equation in this form is that the slope and y-intercept can be read directly from the equation without further calculation.

The y-Intercept

The y-intercept is the point at which the line of an equation crosses the y-axis. This occurs when $x = 0$. By setting x to zero in the equation $y = mx + b$, the x term in the equation drops out, leaving

$$y = b \text{ (when } x = 0)$$

The value represented by b is called the y-intercept.

Slope

The slope of a linear equation is a measure of how much the line deviates from the horizontal. A horizontal line has a slope of zero. A line at 45 degrees above the positive x-axis has a slope of 1. The more vertical a line becomes, the larger its slope. However the slope of a vertical line is undefined since, as we will see later, it involves division by zero. Defined mathematically

$$\text{slope} = \frac{\text{rise}}{\text{run}}$$

where the rise is the change of the value of y between any two points (here called points 1 and 2):

$$\text{rise} = y_2 - y_1$$

The run is defined as the change in the value of x between those same two points:

$$\text{run} = x_2 - x_1$$

Then, slope = rise / run can be expressed in terms of x and y at any two points:

$$\text{slope} = \frac{(y_2 - y_1)}{(x_2 - x_1)}$$

Check the slope of the equation $y = 3x + 3$ using the points $(0, 3)$ and $(-1, 0)$ in Figure 6.10.

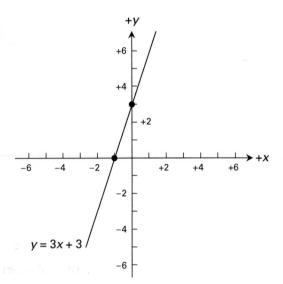

Figure 6.10

Start with the slope equation: slope $= (y_2 - y_1) / (x_2 - x_1)$
Substitute values: slope $= (0 - 3) / (-1 - 0)$
Compute slope: slope $= -3 / -1$
    slope $= 3$

Notes: The computed slope is the same as the coefficient of x in the slope-intercept form equation.

- Any other two points could have been used to find the slope. These two were the most convenient.

- The order of the points is not important, as long as you are consistent. Try computing the slope using $(-1, 0)$ as point 1 and $(0, 3)$ as point 2.

Equations with Different Slopes

Changing the slope of a linear equation will cause its straight-line graph to rotate while keeping its y-intercept fixed. The equations of the graphs in Figure 6.11 have the same y-intercept but different slopes. In this case the y-intercept is zero. Thus, the slope-intercept form of these equations are all

$$y = mx$$

Notice that as the slope (m) is assigned different values, the resulting straight-line graphs rotate, while retaining the same y-intercept of zero.
    Notice that:

- Positive slopes *move up* as they move right.

- Negative slopes *move down* as they move right.

- A slope of $+1$ is a line at 45 degrees above the positive x-axis.

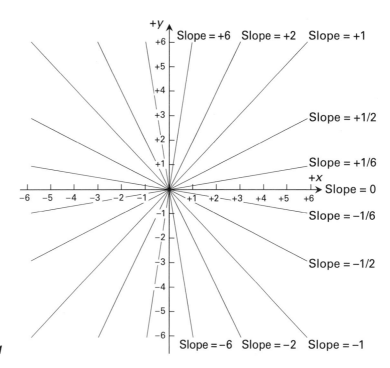

Figure 6.11

- A slope of zero is a horizontal line.

- A slope of -1 is a line at 45 degrees below the positive x-axis.

- As slopes become greater than $+1$, the line becomes closer to being vertical.

- As slopes become less than $+1$, the line becomes closer to being horizontal.

- Perpendicular lines have slopes that are negative reciprocals of each other. (For example, a line with the slope of $+6$ is at right angles to a line with a slope of $-1/6$. Vertical and horizontal lines are perpendicular, even though a slope of zero has no reciprocal.)

Equations with Different *y*-Intercepts

Changing the y-intercept of a linear equation will cause its straight-line graph to translate (move up or down) on the y-axis, while keeping its rotational orientation.

The equations in the graphs in Figure 6.12 all have the same slope but different y-intercepts. In this case the slope is $+1$. The slope-intercept form of these equations is

$$y = (+1)\, x + b$$

As the y-intercept (b) is assigned different values, the resulting straight-line graphs move up and down on the coordinate system, while retaining their rotational orientation.

Note: Line graphs of linear equations with the same slope plot as *parallel* lines.

Special Case: Horizontal Slope

When the slope of a linear equation is zero, the form of the equation simplifies from

$$y = mx + b$$

to

$$y = b$$

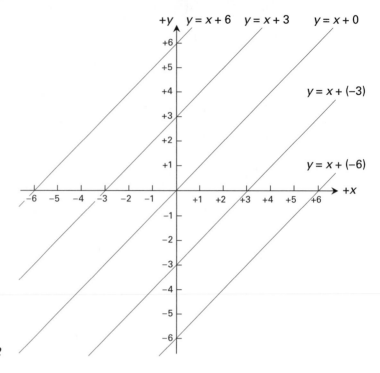

Figure 6.12

This is an equation of a horizontal line. The value of b (the y-intercept) determines the height above (or below) the x-axis where the line lies. Not only does the line intercept the y-axis at this value, but the equation implies that y has this value for all values of x.

> **!** Equations of the form $y = b$ will graph as horizontal lines.

Examples of equations that graph as horizontal lines (see Figure 6.13) are

$$y = 5 \qquad y = 2 \qquad y = -2$$

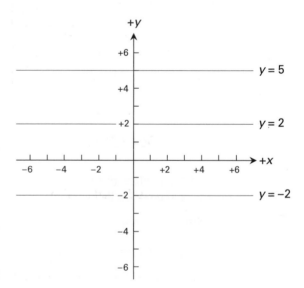

Figure 6.13

Also, the x-axis is the line

$$y = 0$$

Note: These "horizontal" equations imply that y has the same specific value for all values of x.

Special Case: Vertical Slope

A linear equation with a vertical slope cannot be written in the slope-intercept form ($y = mx + b$) since the run is zero and this implies an infinite slope (m). The way around this is to write the equation in a general form that does not involve the slope directly:

$$(c_1)y + (c_2)x + c_3 = 0$$

When the value of the coefficient of y (represented by c_1), is zero, the equation becomes

$$x = -\frac{c_3}{c_2}$$

The degenerative case in which $c_2 = 0$ can be disregarded since, with both c_1 and c_2 equal to zero, there would be no line to plot, vertical or otherwise. You don't have to know the exact values of c_2 and c_3 to see that this vertical line equation will be in the form

$$x = \text{some constant}$$

Since this equation does not depend on y (i.e., it is the same for all values of y), it will graph as a vertical line. The constant represents the point at which the line crosses the x-axis (the x-intercept).

Examples of equations that graph as vertical lines (see Figure 6.14) are

$$x = -3 \qquad x = 4$$

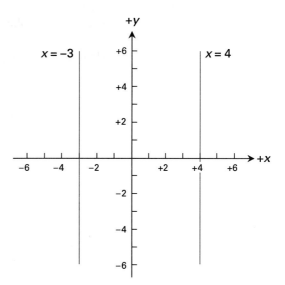

Figure 6.14

Also, the y-axis is the vertical line

$$x = 0$$

Note: These "vertical" equations imply that x has the same specific value for all values of y.

Tips for Determining the Slope and *y*-Intercept

Case 1: The Equation is Already in the Slope-Intercept Form.

The slope and y-intercept of a linear equation with two variables can be easily determined if the equation is written in the slope-intercept form:

$$y = mx + b$$

In this form, the slope and y-intercept can be read directly from the equation. The slope will be the coefficient of the x term, and the y-intercept will be the constant term. Here are a few examples:

$y = 4x + 2$	or $y = +4x + 2$	slope $= +4$	y-intercept $= +2$
$y = x - 3$	or $y = +1x - 3$	slope $= +1$	y-intercept $= -3$
$y = -x + 1$	or $y = -1x + 1$	slope $= -1$	y-intercept $= +1$
$y = 2$	or $y = 0x + 2$	slope $= 0$	y-intercept $= +2$

Case 2: The Equation is Not in the Slope-Intercept Form.

You cannot read the slope and y-intercept directly from a linear equation with two variables that is not in the slope-intercept form. The equation must first be translated into the slope-intercept form. Here are a few examples:

$x = 2y + 1$	$2y = x - 1$	$y = \frac{1}{2}x - \frac{1}{2}$	slope $= +\frac{1}{2}$	y-intercept $= -\frac{1}{2}$
$3x = 4y - 8$	$4y = 3x + 8$	$y = \frac{3}{4}x + 2$	slope $= +\frac{3}{4}$	y-intercept $= +2$

Case 3: The Equation is Not Yet Known.

In cases in which there is a straight-line plot of some data but no equation, the slope can be calculated using the differences between the x-y values at any two points as follows:

$$\text{slope} = \frac{\text{rise}}{\text{run}} = \frac{(y_2 - y_1)}{(x_2 - x_1)}$$

The y-intercept can be read directly from the graph as the value of y where the line crosses the y-axis (where $x = 0$). Then, with both the slope (m) and y-intercept (b) known, the equation of the line can be written in the form

$$y = mx + b$$

Case 4: Vertical Lines

Vertical lines are a special case because they have equations that *cannot* be put in slope-intercept form. Equations of vertical lines have no y-intercept, and because of division by zero, the slope is undefined. Here are a few examples of equations of vertical lines:

$x = 0$	no y-intercept	and	slope undefined
$x = 2$	no y-intercept	and	slope undefined
$x = -2$	no y-intercept	and	slope undefined

 Summary of Properties of the Slope

- A horizontal line has a slope of zero.

- Positive slopes run from left to right as y values get larger.

- Negative slopes run from left to right as y values get smaller.

- When the line runs at 45 degrees above the positive x-axis, the slope is $+1$.

- When the line runs at 45 degrees below the positive x-axis, the slope is -1.

- Lines with slopes greater than $+1$ or less than -1 are steeper than 45 degrees.

- As a line approaches the vertical, its slope becomes a very large positive or negative number.

- The equation $y = mx + b$ can never have a vertical slope because of division by zero.

Practice Problems

Put the following equations into the slope-intercept form ($y = mx + b$) and read the value of the slope and y-intercept.

6-3.23 $y = 3x + 2$ 6-3.24 $y = 2x + 1$

6-3.25 $y = x$ 6-3.26 $y = 5$

6-3.27 $x = y - 1$ 6-3.28 $y = -3x - 2$

6-3.29 $2y = 6x + 4$ 6-3.30 $x = y + 1$

6-3.31 $2x = 3y + 4$ 6-3.32 $2y + x + 1 = y - x + 2$

Solving Two Linear Equations

Solution by Graphing

A single linear equation cannot be solved for x and y explicitly. It can only be solved for y in terms of x (or x in terms of y). In the section on graphing equations it was shown that any of the x-y points that lie along a straight line will satisfy the equation defined by that line. So how can we solve a linear equation for a specific value of x and a specific value of y?

The answer is that two linear equations are required—not any two equations, but two "independent" equations, that is, two equations that have different slopes and will intersect when plotted. The intersection is the key thing. At the intersection there is a single x, y point that satisfies both equations. No other points will do this.

▶ **Example Problem**

$$y = 2x + 1$$

$$y = -x + 4$$

What values of x and y will satisfy both equations? Let's solve this problem by graphing both equations and finding the point of intersection (see Figure 6.15).

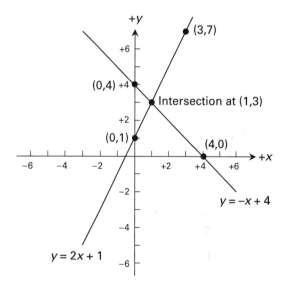

Figure 6.15

The point of intersection is at $(1, 3)$. The values $x = 1$ and $y = 3$ will satisfy both equations.

Note: The solution of two linear equations can only be found if the two graphs intersect. Two parallel lines (equations having the same slope) will have no solution.

Solution by Algebra

The equations in the previous example give a whole number solution. This makes reading the coordinates of the intersection easy. In general, solutions will not always be whole numbers.

Therefore, in many cases the graphing method will find only an *approximate* solution to equations. For an exact solution you need to use algebra.

Appendix B presents a detailed method for using algebra to solve two linear equations with two unknowns. Here is a preview, using the previous example:

$$y = 2x + 1$$
$$y = -x + 4$$

The solution of these equations is the explicit value of x and y that will satisfy *both* equations at the same time. Because the value of y in the solution will be the same for both equations, we can combine the equations to form a new equation with a single variable:

$$2x + 1 = -x + 4$$

This new equation is easily solved for x using the methods presented in Chapter 5.

$$3x + 1 = 4 \qquad \text{after adding } -x \text{ to both sides}$$
$$3x = 3 \qquad \text{after subtracting 1 from both sides}$$
$$x = 1 \qquad \text{after dividing both sides by 3}$$

To get the y part of the solution, substitute $x = 1$ into either original equation, in this case, the first.

$$y = 2x + 1$$
$$y = 2(1) + 1$$
$$y = 3$$

Thus, the solution is

$$x = 1 \text{ and } y = 3$$

To check the solution, substitute the value of $x = 1$ into the other original equation:

$$y = -x + 4$$
$$y = -(1) + 4$$
$$y = 3$$

Thus, the solution checks and also agrees with the solution found by graphing.

Practice Problems

Solve the following pairs of linear equations by graphing. Confirm the solution using algebra.

6-3.33	$y = 2x + 1$	and	$y = 3x + 2$
6-3.34	$y = 3x - 2$	and	$y = x$
6-3.35	$y = 1$	and	$y = 2x + 2$
6-3.36	$y = 4x + 2$	and	$y = 2x - 2$
6-3.37	$2y = 4x - 2$	and	$3y = 9x - 6$

The Shortest Distance Between Two Points

The old saying goes, The shortest distance between two points is a straight line. In this section you will learn how to calculate this straight-line distance between any two points on a graph. Because two points plotted on a rectangular coordinate system define a right triangle, we will first introduce the right triangle and show how the lengths of its three sides are related.

The Right Triangle

A right triangle is a three-sided figure that has two sides perpendicular to each other. Perpendicular sides have an angle between them that is exactly a quarter of a circle. The angle that is a quarter of a circle is more commonly called a *right angle.*

The longest side of a right triangle (the side opposite the right angle) is called the *hypotenuse.* Consider the right triangle in Figure 6.16 with lengths of the three sides represented by a, b, and c. The longest side, c, is the hypotenuse; b is the base; and a is the altitude.

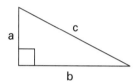

Figure 6.16

Pythagorean Theorem

The ancient Greeks discovered a special relationship between the lengths of the sides of a right triangle. This relationship is called the Pythagorean theorem[1]. The Pythagorean theorem states that for any right triangle

 The square of the hypotenuse (the longest side) is equal to the sum of the squares of the two remaining sides.

The Pythagorean theorem can be expressed as an algebraic equation in terms of the lengths of the three sides:

$$c^2 = a^2 + b^2$$

Since the Pythagorean equation has second-power terms, it is not linear. However, it can be easily solved when two of the three sides are given by taking the square root of both sides of the equation. (Remember, taking the square root was one of the rules of equations.)

► For example, if a = 3 feet and b = 4 feet, then c can be found as follows:

1. Substitute these values into the equation: $c^2 = 3^2 + 4^2$
2. Evaluate the squares: $c^2 = 9 + 16$
3. Add terms: $c^2 = 25$
4. Take the square root of both sides: $\sqrt{c^2} = \sqrt{25}$
5. This simplifies to: $c = 5$ ft

The negative square root is discarded because is has no meaning for this problem. Be aware that most right triangles will not have the length of all three sides come out as whole numbers. Even if two sides are whole numbers, the third side is likely to involve a radical.

---

[1]Mathematicians in China and India discovered this relationship before the Greeks. However, the Greeks and Pythagoras are commonly given credit today.

▶ Here is an example involving a radical:

a = 1 foot and b = 1 foot

1. Substitute these values into the equation: $c^2 = 1^2 + 1^2$
2. Evaluate the squares: $c^2 = 1 + 1$
3. Add terms: $c^2 = 2$
4. Take the square root of both sides: $\sqrt{c^2} = \sqrt{2}$
5. This simplifies to: $c = \sqrt{2}$ ft

The length $\sqrt{2}$ ft is approximately 1.414 ft (rounded to three decimal places).

▶ The following example solves for a side other than the hypotenuse:

Suppose the hypotenuse, c, has a length of 5 feet and another side, a, has a length of 4 feet. Find the length of the remaining side, b.

1. Substitute these values into the equation: $5^2 = 4^2 + b^2$
2. Evaluate the squares: $25 = 16 + b^2$
3. Transpose the equation (subtract 16 from both sides): $25 - 16 = 16 - 16 + b^2$
4. Combine terms and flip the equation: $b^2 = 9$
5. Take the square root of both sides: $\sqrt{b^2} = \sqrt{9}$
6. This simplifies to: $b = 3$ ft

Notice that this is the same problem as the first example, but with a different unknown.

▶ Here is a final example:

Find the length of a square (a figure with four equal sides) with a diagonal of 2 feet.

In this problem the diagonal divides the square into two right triangles, and the diagonal is the hypotenuse of each triangle. Thus, c = 2 feet. The shorter sides are equal to each other and will be called x.

1. Substitute these values into the equation: $2^2 = x^2 + x^2$
2. Evaluate the square and combine terms: $4 = 2x^2$
3. Divide both sides by 2 and flip the equation: $x^2 = 2$
4. Take the square root of both sides: $\sqrt{x^2} = \sqrt{2}$
5. This simplifies to: $x = \sqrt{2}$
6. Evaluating the square root approximately: $x = 1.414$ (ft)

Shortest Distance on a Graph

Plotting a graph on a rectangular coordinate system makes it easy to find the shortest distance between any two points.

Suppose the two points have coordinates (x_1, y_1) and (x_2, y_2). Then a right triangle can be drawn so that the difference between the x values $(x_2, -x_1)$ represents the length of the base of the triangle and the difference between the y values $(y_2 - y_1)$ represents the altitude of the triangle.

The hypotenuse of the right triangle is the distance between the points. Because the hypotenuse is a straight line, it is the shortest distance between these points.

The shortest distance can be calculated by substituting the known values of (x_1, y_1) and (x_2, y_2) into the Pythagorean equation and solving for the distance (d) between the points as follows:

$$d^2 = (x_2 - x_1)^2 + (y_2 - y_1)^2$$

or

$$d = \sqrt{(x_2 - x_1)^2 + (y_2 - y_1)^2}$$

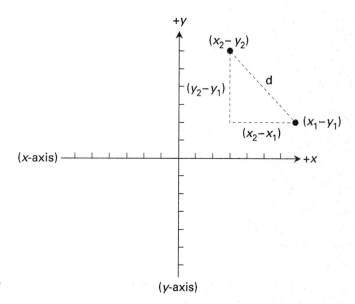

Figure 6.17

(x-axis) (y-axis)

▶ **Example with positive differences:**

What is the straight-line distance between the points (2, 1) and (6, 4)?

1. Substitute the values into the equation: $d^2 = (6 - 2)^2 + (4 - 1)^2$
2. Evaluate the differences: $d^2 = 4^2 + 3^2$
3. Evaluate the squares: $d^2 = 16 + 9$
4. Combine terms: $d^2 = 25$
5. Take the square root of both sides: $d = \sqrt{25} = 5$

Notice that the 5, 4, 3 right triangle has popped up again.

▶ **Example with negative differences:**

The same formula works when you let (6, 4) be the first point and (2, 1) be the second point. Even though the differences $(x_2 - x_1)$ and $(y_2 - y_1)$ are now negative, the square of these negative values is the same positive value as in the previous example.

1. Substitute the values into the equation: $d^2 = (2 - 6)^2 + (1 - 4)^2$
2. Evaluate the differences: $d^2 = (-4)^2 + (-3)^2$
3. Evaluate the squares: $d^2 = 16 + 9$
4. Combine terms: $d^2 = 25$
5. Take the square root of both sides: $d = \sqrt{25} = 5$

Notice that the quantity d is the same regardless of the order of the two points.

▶ **Example with negative coordinates:**

The same formula will work when points involve negative coordinates.

Suppose the points from the previous example are on the negative sides of both the x-axis and the y-axis. Then the coordinates are (−6, −4) and (−2, −1).

Applying the formula but being careful to handle the negative signs correctly gives

1. Substitute the values into the equation: $d^2 = (-2 - (-6))^2 + (-1 - (-4))^2$
2. Evaluate the differences: $d^2 = (-2 + 6)^2 + (-1 + 4)^2$
   $d^2 = (4)^2 + (3)^2$
3. Evaluate the squares: $d^2 = 16 + 9$
4. Combine terms: $d^2 = 25$
5. Take the square root of both sides: $d = \sqrt{25} = 5$

Notice that the length of the hypotenuse is unchanged by moving the points, as long as they are moved in such a way that length of the sides of the right triangle are not changed.

▶ **Example with vertical distances:**

When one point is directly above the other, the triangle collapses into a vertical line. This special case can still be handled by the Pythagorean theorem.

The same formula will work when points lie along a vertical line.

Compute the distance between points $(3, 2)$ and $(3, 5)$.

Applying the formula gives

1. Substitute the values into the equation:
2. Evaluate the differences:

3. Evaluate the squares:
4. Take the square root of both sides:

$$d^2 = (3 - 3)^2 + (5 - 2)^2$$
$$d^2 = (0)^2 + (3)^2$$
$$d^2 = (3)^2$$
$$d^2 = 9$$
$$d = \sqrt{9} = 3$$

Note: In this special case the distance can more simply be computed as the absolute value of the difference between the two *y*-values.

$$5 - 2 = 3$$

▶ **Example with horizontal distances:**

When one point is directly alongside the other point, the triangle collapses into a horizontal line. This special case can still be handled by the Pythagorean theorem.

The same formula will work when points lie along a horizontal line.

Compute the distance between points $(5, 2)$ and $(1, 2)$.

Applying the formula gives

1. Substitute the values into the equation:
2. Evaluate the differences:

3. Evaluate the squares:
4. Take the square root of both sides:

$$d^2 = (5 - 1)^2 + (2 - 2)^2$$
$$d^2 = (4)^2 + (0)^2$$
$$d^2 = (4)^2$$
$$d^2 = 16$$
$$d = \sqrt{16} = 4$$

Note: In this special case the distance can more simply be computed as the absolute value of the difference between the two *x*-values.

$$5 - 1 = 4$$

Practice Problems

Use the Pythagorean equation to compute the distance between the following pairs of (x, y) points. Assume that the units of the coordinate system represent feet and evaluate square roots to two decimal places when necessary.

6-4.1 $(0, 0)$ and $(1, 1)$

6-4.2 $(0, 0)$ and $(3, 4)$

6-4.3 $(1, 1)$ and $(2, 2)$

6-4.4 $(6, 7)$ and $(3, 1)$

6-4.5 $(-2, 2)$ and $(3, -3)$

Computer Graphing Coordinates

Computer graphing uses a two-dimensional rectangular coordinate system. However, it is slightly different from the system introduced earlier in this chapter. First, the negative part of each axis is chopped off. This is because computer screen graphics allow no negative screen locations. Second, the vertical y-axis is flipped around so that positive y is down. The origin (0, 0) is in the upper left-hand corner of the computer screen. The result looks like this for the point (600, 400):

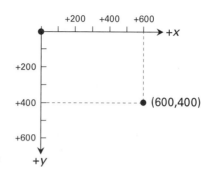

Figure 6.18

The increments along the x-axis and the y-axis are measured in pixels (picture elements). In turn, the pixel size matches the screen resolution you select. The video adapter card in each computer determines which resolutions are possible. Here are a few resolutions supported by a typical SVGA card: 640 by 480, 800 by 600, 1024 by 768. With more video memory, even higher resolutions are possible. Notice that all of these resolutions are in the ratio 4:3, the same width to height ratio as used for traditional analog TV screens. The new HD (high definition) TV, on the other hand, has a screen aspect ratio of 16:9. As this is written, it remains to be seen if computers will follow suit.

Each resolution is defined in terms of a maximum x-value followed by a maximum y-value. The point that corresponds to these maximum values is at the lower right-hand corner of the screen. Thus, when the 800 by 600 resolution is in effect, the coordinates of a point at the lower right corner of the screen will be (800, 600). This point is 800 pixels along the x-axis and 600 pixels along the y-axis.

Distance between Two Points

For a given screen resolution, the distance between any two points on the screen depends on the size of the pixel. Pixels can be thought of as circular dots packed in a matrix of square boxes. The diameter of a pixel is the same length as the side of one of the matrix boxes. Thus, the distance between two points along the x-axis is the number of pixels between the points multiplied by the diameter of a pixel. The same is true for two points on the y-axis.

At first, the distance between two points along some sloping line may seem more confusing since the number of pixels along a sloping line makes no sense. But the sloping length on the screen can be determined in the same way as the shortest distance on a graph in the previous section.

1. Determine the length along the x-axis (the base of a right triangle) by multiplying the number of pixels along the x-axis by the diameter of a pixel.

2. Determine the length along the y-axis (the altitude of a right triangle) by multiplying the number of pixels along the y-axis by the diameter of a pixel.

3. Use the Pythagorean theorem to find the sloping distance (the hypotenuse of the right triangle).

As a shortcut, you can even calculate the hypotenuse in pixels and convert the result into some other length unit at the end. Even though the number of pixels along the hypotenuse makes no sense, converting that nonsense number to a length (by multiplying by the diameter of a pixel) gives the same result as if you did the conversion of the base and altitude lengths as shown previously.

Practice Problems

6-4.6 Compute the height of a 15-inch computer screen if the screen size represents the diagonal measurement and the width of the screen is 12 inches.

In the following problems, compute the distance in millimeters between the following *x, y* screen coordinates measured in pixels. Assume that a pixel is a dot with a diameter of 0.28 millimeters.

6-4.7 (0, 0) and (0, 1000)

6-4.8 (0, 100) and (1000, 0)

6-4.9 (0, 0) and (1000, 1000)

6-4.10 (100, 100) and (200, 200)

In the following problems, compute the radius in *inches* of the following circles drawn on a computer screen. Two points, with coordinates measured in pixels, define each circle. The first point marks the center of the circle and the second marks a point on the circumference. The radius is the distance between the two points. Assume that a pixel is a dot with a diameter of 0.28 millimeters.

6-4.11 (500, 500) and (100, 100)

6-4.12 (650, 350) and (350, 650)

6-4.13 (200, 600) and (100, 200)

6-4.14 (400, 100) and (600, 300)

Chapter Summary

Spreadsheet Graphing

Pie charts: Display a series of data as slices of pie. Best for comparing each data item as a proportion of the sum of all data items.

Bar charts: Display a series of data as a set of bars. Best for comparing data items to one another or showing how data varies over time.

Line charts: Display a series of equally spaced data points connected by lines in a way very similar to the bar chart.

x-y charts: Display a series of x-y data pairs on a rectangular system. Best to show how the x-value of each pair varies in relation to the corresponding y-value.

Graphing Equations

In order to display data from an equation in a graph, you need some sort of grid that relates distances in different directions. This grid is called a coordinate system. Mathematicians employ many different types of coordinate systems, usually related to some geometric shape such as the rectangle, circle, or ellipse. The simplest of these and the one most useful for the type of graphs you are likely to encounter, is called the *rectangular coordinate system.*

The two-dimensional *rectangular coordinate system* employs two number lines at right angles to each other. One is horizontal and one is vertical. The most common arrangement is the following:

- The horizontal number line is called the x-axis (usually with $+x$ to the right).

- The vertical number line is called the y-axis (usually with $+y$ up, $-y$ down).

Every point on a graph marks a specific value of x and a specific value of y.

- The value of x is determined by the position of the point along the x-axis.

- The value of y is determined by the position of the point along the y-axis.

Only two points are needed to define a line. When plotting the graph of $y = 3x + 3$ using only two points, the easiest two points to find are $x = 0$ and $y = 0$. Substitute the zero values to find the coordinates for each point:

- At $x = 0$ the equation becomes $y = 0 + 3$ or $y = 3$.   point $(0, 3)$

- At $y = 0$ the equation becomes $0 = 3x + 3 = -1$ or $x = -1$.   point $(-1, 0)$

Draw the straight line between the two points.

Linear Equations

Linear equations with two variables can be written in the slope-intercept form

$$y = mx + b$$

All linear equations plot a straight line on a rectangular coordinate system.

- The coefficient of x (represented by m) is called the slope of the equation.

- The constant b is called the y-intercept.

The slope can be expressed in terms of x and y at two points, 1 and 2:

$$\text{slope} = \frac{(y_2 - y_1)}{(x_2 - x_1)}$$

Special cases of the slope:

- When the equation plots a *horizontal* line, the slope is zero.
- When the equation plots a line 45 degrees *above* the horizontal, the slope is 1.
- When the equation plots a line 45 degrees *below* the horizontal, the slope is -1.
- When the equation plots a *vertical* line, the slope is undefined.

Shortest Distance between Two Points

The shortest distance between two points (x_1, y_1) and (x_2, y_2) on a rectangular coordinate system can be found by using the equation derived from the Pythagorean theorem:

$$d = \sqrt{(x_2 - x_1)^2 + (y_2 - y_1)^2}$$

Review Questions

1. Which type of spreadsheet chart will best display the following data:
   a) Daily production values for an entire year
   b) Total monthly sales for each of five salespersons
   c) Amounts for each of seven budget categories that make up the total budget
   d) Temperature of a vat, measured each minute over several hours
2. Within a pie chart what component is proportional to the data being displayed?
3. Describe how you would determine the horizontal and vertical scales for a bar chart needed to plot monthly revenue figures that ranged from $2.1 million to $16 million.
4. How is a rectangular coordinate system related to the number line?
5. How does a linear equation with two variables graph on a rectangular coordinate system?
6. What is the minimum number of points needed to plot a straight-line equation?
7. Define the following terms: *linear equation, slope,* and *y-intercept.*
8. For each of the following descriptions of straight-line equations, tell whether the slope is positive, negative, equal to 1, equal to zero, or undefined.
   a) A horizontal line
   b) A vertical line
   c) A line sloping up from left to right
   d) A line sloping up from left to right that is halfway between vertical and horizontal
   e) A line sloping down from left to right
9. When each of the following equations is graphed on a rectangular coordinate system, is it horizontal, vertical, along the x-axis, or along the y-axis?
   a) $x = 0$ b) $y = 0$ c) $x = 1$ d) $y = 1$
10. When two lines are perpendicular, how are their slopes related?
11. How can a graph be used to solve two linear equations with two unknowns by finding explicit values of x and y that satisfy both equations?
12. What formula finds the hypotenuse of a right triangle, given the base and altitude?
13. What formula finds the base of a right triangle, given the hypotenuse and altitude?
14. What formula finds the altitude of a right triangle, given the hypotenuse and base?
15. What formula finds the distance between two x-y points on a graph?
16. How do computer screen coordinates differ from the traditional rectangular coordinate system?

Problem Solutions

Section 6-3

6-3.2 to 6-3.12

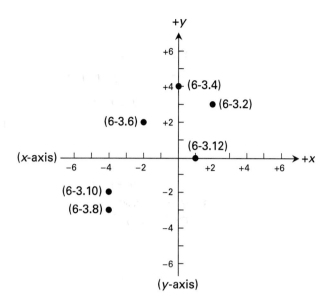

6-3.14

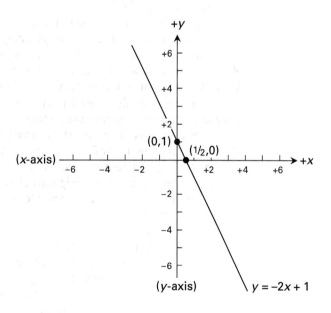

6-3.16

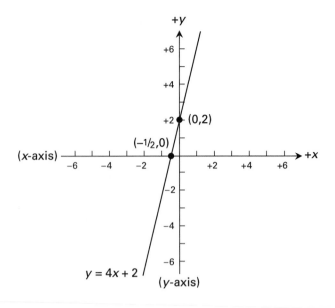

$y = 4x + 2$

6-3.18

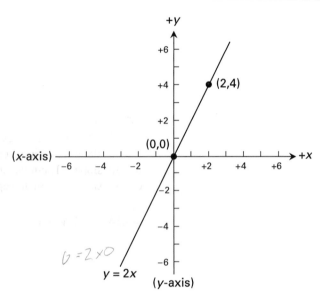

$y = 2x$

6-3.20

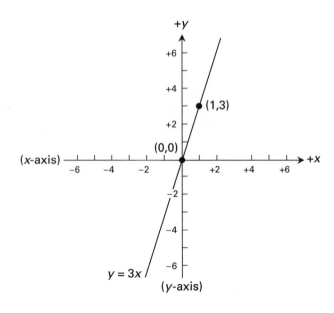

$y = 3x$

6-3.22

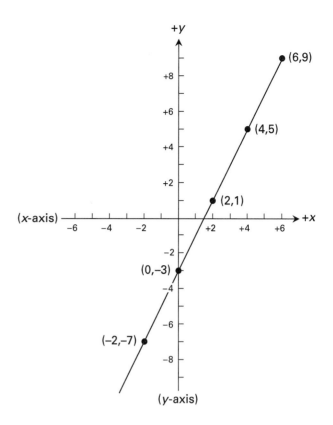

6-3.24 slope $+2$, y-intercept $+1$
6-3.26 slope 0, y-intercept $+5$
6-3.28 slope -3, y-intercept -2
6-3.30 slope $+1$, y-intercept -1
6-3.32 slope -2, y-intercept $+1$

6-3.34 $x = 1$, $y = 1$

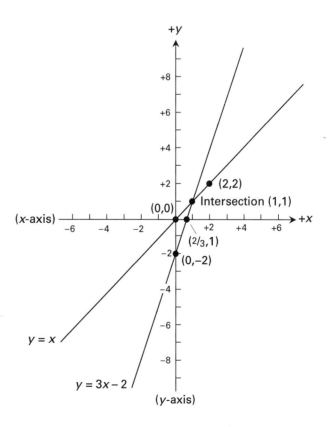

6-3.36 $x = -2, y = -6$

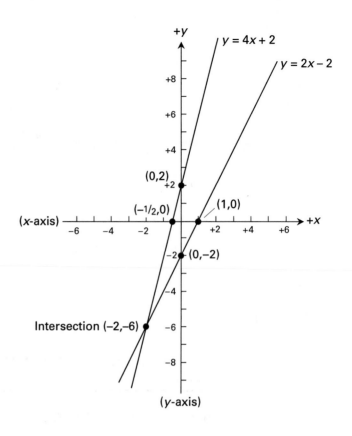

Computer Programming Concepts

Mr. Natural sez: "Always use the right tool for the job."
—R. Crumb, Zap Comix

7-1 ■ Introduction

How to Use This Chapter

The concluding chapters in this book are about computer programming. The material is in no way intended to be a course in programming, but rather is an introduction to the topic that will prepare you for your first course in programming.

In the past, programming was at the center of any college curriculum related to information technology. Today, it is more on the periphery since most information technology careers do not involve programming directly. Those students wanting to become programmers must take a series of programming courses. So, many of you might ask, Why not avoid programming altogether?

The answer is that programming skills are transferable to many other areas of information technology besides that of the software developer. Even when learning to use the most basic office applications such as word processing and spreadsheets, programming concepts play a role. Macro languages available in both of these applications are simplified programming languages available to customize your work. Conditional functions are available to carry out some processing only when some condition is true. In addition, scripting languages are at work behind the scene of any page on the World Wide Web.

Macro languages, conditional functions, and scripting languages all draw on the concepts presented in this chapter. So even if you never become a computer programmer, you must understand programming concepts to get ahead in the field of information technology.

No Computer Is Necessary

This introductory material may be used without a computer. Your mind plays the role of the computer as you step through each part of a process. Paper and pencil can be used in place of computer memory when it is necessary to store the result from one operation for later use in another operation.

This chapter will focus on basic programming concepts, the building blocks of any simple program. The chapters that follow will expand on these concepts.

Background

Programming involves algorithms to specify the step-by-step procedure needed to accomplish some task. Algorithms are much like recipes. Each step must be explained completely and unambiguously. Our English language is often ambiguous and therefore not best suited for writing instructions that will be read by a computer.

Computers are very unintelligent machines. It is software that gives them the ability to simulate intelligence. The machine itself can "understand" only a very crude and restrictive language called *machine language*. Because machine language is in a special binary code, it is almost unreadable by humans. Therefore, translators (called compilers) have been built to enable human programmers to write instructions that are more like the English language. BASIC, C, and Java are examples of computer programming languages that use English-like instructions.

Although computers are able to "understand" a few key words and phrases, at present the hardware and software are not yet capable of taking instructions in free-flowing conversational language with all of its ambiguities. Even when this happens, and it surely will happen someday, a layer of software instructions will always be needed beneath the surface to interpret the conversational instructions and translate them into more structured instructions.

This chapter will introduce the basic concepts of such a computer language. To do this, a greatly simplified language will be presented. This language is very much like the BASIC computer language, stripped of much of its complexity to make it easier to understand. Yet even in this simplified form, you will appreciate how structured the instructions in a computer program must be. Hopefully, this introduction will pave the way for the successful completion of your first programming course.

Chapter Objectives

In this chapter you will learn:

- Some beginning definitions involved with computer programming:

    Computer, computer programs, and computer programming language

    Constants and variables

    Arithmetic and conditional expressions

- A simplified computer language with just five statements that accomplish all the fundamental tasks of any programming language:

INPUT	To bring data from outside the program to memory
OUTPUT	To send data from memory to outside the program
LET	To assign data from one memory location to another
IF-THEN-ELSE	To do processing conditionally
LOOP	To repeat some processing

- How to combine these five statements into simple programs

- How a program can be modified to make it more functional

Introductory Problem

This introductory problem is your opportunity to see where we are headed and to "play" with a few basic issues before you tackle the details of computer programming.

The problem:

Write a step-by-step procedure called an algorithm that describes how to **enjoy a cup of tea.**

Write your algorithm using a technique called *structured English.* (Algorithms will be covered in detail in Section 9-2.)

Structured English is a method for describing a step-by-step process in outline form, using indentation to show different levels of detail.

Here are two examples:

▶ Example 1:
    Start your day
        Wake up
        Get out of bed
        Drag a comb across your head

▶ Example 2:
    Brush your teeth
        Get ready
            Get equipment
                Get out brush
                Get out toothpaste
            Put some toothpaste on toothbrush
                Take cap off toothpaste
                Squeeze paste on brush
                Put cap back on toothpaste
            Wet brush
        Brush teeth
            Repeat until teeth are clean
                Brush top
                    Brush upper right
                    Brush upper left
                Brush bottom
                    Brush lower right
                    Brush lower left
            End repeat
            Rinse out
        Clean up
            Wash brush
            Store brush
            Store toothpaste

Your structured English for the tea algorithm should have several levels:

Level 0:	Begin your outline with the top-level process naming the entire procedure (no indentation).
Level 1:	Indented one tab stop beneath level 0 should be the major steps involved in carrying out the top-level process.
Level 2:	Indented an additional tab stop beneath *each* level 1 statement should be the detailed steps telling how to carry out the previous level 1 statement. For example:

    Enjoy a Cup of Tea
        Heat the water
            All the steps needed to explain how to heat the water
        Brew the tea
            All the steps needed to explain how to brew the tea
        Drink the tea
            All the steps needed to explain how to drink the tea

Other levels: Similarly, other instructions may be indented at levels beneath level 2 when additional detail is needed. Level 3 statements may be used to explain how to accomplish a level 2 instruction, level 4 instructions may be used to explain how to accomplish a level 3 instruction, and so on.

Assume that the following equipment and supplies are at hand and ready for use:

Box of tea bags	Electric range (for heat)
Kitchen sink (for water)	Teakettle (for heating water)
Teapot (for brewing tea)	Teacup (for drinking tea)

Try to make your algorithm clear and *without errors.* Include such processes as

Opening and closing the box of tea bags

Turning the range and the sink on and off

Filling the kettle with cold water until full

Removing and replacing the tops on the kettle and teapot

A step (or group of steps) may be repeated until some condition becomes true. For example:

```
Repeat until cup is empty
      Take a sip of tea
      Enjoy the moment
      Wait a while
End repeat
```

By skillfully using the "repeat" feature, try to think of a way to include the feature of refilling your cup to be able to drink several cups of tea until you are full or the teapot is empty. Be sure to test your algorithm.

What Is a Computer?

A computer is an electronic device that accepts data as input, processes that data into information, and delivers the information as output. The input, processing, and output are under the control of a set of instructions called a program. Both the data and the program reside in a storage area called memory.

Programs and their data can be easily changed so the computer can be applied to a variety of problems. This concept of the stored program makes the computer the most versatile problem-solving machine yet devised.

Computer Programs

A computer program is a set of instructions that controls the operation of a computer. Programs fall into two principal types: systems programs and application programs. *Systems programs* control the operation of the most primitive aspects of the computer hardware and allow application programs (defined next) and their data to be loaded into memory and run. *Application programs* are run under the control of the operating system to apply the computer to a specific task such as spreadsheets or word processing.

Programming Language

A computer can only run programs written in a special code consisting of zeros and 1s called machine language. While this binary code presents no difficulty for computers to process, it is very difficult for humans to understand. Therefore, other computer languages are available for human programmers. Before these higher-level languages can be run on a computer, their English-like statements must be translated into machine language. A special program called a compiler is available to do this translation.

Many higher-level languages are popular today. A few examples are:

Visual Basic   Java   C   C++   COBOL

While each of these languages has special features that make it unique, at the core they are all quite similar. Some essential features shared by all languages include:

- A means to input data

- A means to output processed information

- A means to store data until needed

- A means to do arithmetic

- A means to evaluate logic (statements that are either true or false)

Computer Variables

Computer programming languages use variables to store data. These variables are named *locations* in the computer's memory that can store a result from one operation until it is needed by another operation.

A memory location can store only one piece of data at a time. It is possible to change the contents of a computer variable by replacing old data with new data. The computer can be instructed to do this very quickly so the data appear to vary over time. However, at any given instant, only one piece of data can be stored in one variable.

Name vs. Content

It is important to distinguish the name of a computer variable from its content. As described here, a box represents a location in memory where the contents of a variable are stored.

```
┌─────────┐
│         │          A given memory location for a variable
└─────────┘
```

Beneath the box is the name of the variable. This name can be a letter, a word, or a phrase. It is useful to let the variable name describe the data that will be stored in the variable.

```
┌─────────┐
│         │          A variable named age
└─────────┘
    age
```

Inside the box, the currently stored data is shown.

```
┌─────────┐
│   27    │          The number 27 stored in the variable age
└─────────┘
    age
```

Computer variables can also store text data.

```
┌─────────┐
│   Jim   │          The text "Jim" stored in a variable called name
└─────────┘
   name
```

Because any text can be stored in the variable called *name*, confusion can arise when the content and the name of a variable are the same. Remember: the *name* refers to a storage location, while the *content* is the actual data stored at that location.

```
┌─────────┐
│  Name   │          The text "Name" stored in a variable called name
└─────────┘
   name
```

Note: Real computer languages allow you to declare the type of data that can be stored in a variable, restricting it to only text, only integers, or only decimals, for example. To keep things simple, this book will make no such restrictions. However, data items that are text must be have quotes around them.

Algebraic vs. Computer Variables

Algebraic variables and computer variables share the word *variable* in their name. They are similar in that they both use a symbolic notation to represent quantities. However, the similarity is only skin deep and there are fundamental differences between the two types of variables. The following table highlights some of these differences.

Algebraic Variables	Computer Variables
Named with a single letter	Named with a letter, a word, or a phrase
May represent one or many values	Stores only one value at a time
Values represented are numbers	May be used to store either numbers or text

Naming Conventions

Algebraic variables are usually named with a single letter of the alphabet making them easy to write and manipulate. The names of computer variables may be named with a letter, but also with words and phrases. While an algebraic expression may have one, two, or at most a handful of variables, a computer program may have many dozens or hundreds of variables. Thus, it is important to name computer variables with meaningful names.

Spaces are not allowed in the names of computer variables. Different techniques are used to avoid spaces in variable names and still make them readable. For example, a variable that will store the last name of a student might be named any one of the following:

STUDENT_LAST_NAME

Student_Last_Name

StudentLastName

studentlastname

student_last_name

This text will use a convention in which the variable names use only lowercase letters. This is because the key words in a statement like INPUT and OUTPUT (which will be introduced in the next section) will be written using all capital letters. This is not a requirement, but rather a stylistic consideration for clarity. Thus, of the several variable names in the previous example, the last one,

student_last_name

is the one that would be selected for programs written in this text.

Arithmetic and Logic

Arithmetic Expressions

Arithmetic expressions, also called numeric expressions, evaluate to a specific number. Arithmetic expressions may include:

Constants, such as 5, or 3.1459, or −6.5

Variables, such as age or labor_cost

Arithmetic operators

^	For exponentiation
*	For multiplication
/	For division
+	For addition
−	For subtraction

Parentheses for grouping various calculations

Here are some examples of arithmetic expressions:

current_year − birth_year + 1   cost * (1 + tax_rate)   length * width

Note: Within a computer program, arithmetic expressions cannot occur in isolation. They must be used in conjunction with language statements. This will be shown in the next section.

Order of Precedence

Within an expression calculations proceed in a predetermined order. The various parts of an expression are evaluated in the following order:

1. Parentheses   Innermost first, then left to right

2. Exponentiation   Left to right

3. Multiplication **or** Division   Left to right

4. Addition **or** Subtraction   Left to right

Using this order of precedence, complex calculations done on a computer will always yield a single result. Although complex calculations do not confuse the computer, humans writing these expressions often are confused. **To avoid confusion, make extensive use of parentheses.**

Arithmetic expressions may also include a variety of arithmetic functions. These functions may be built into the language (intrinsic) or may be custom designed by the programmer. Each function takes a given input, performs a predetermined operation, and returns a result. The result is made available as the other parts of an arithmetic expression are evaluated. Most languages have built-in functions to take a square root, determine an absolute value, truncate (or round) a decimal, convert a number into an integer, or do trigonometry. Other functions are available to manipulate text.

To keep things simple, this book will not include functions in the discussion that follows.

Conditional Expressions

The ability to perform *logic* is the thing that distinguishes a computer from a simple calculator. This is done within a computer program by using conditional expressions to direct processing, one way or another, depending on a true/false outcome.

Conditional expressions, also called logical expressions, evaluate to either *true* or *false*. No other outcome is possible. In their simplest form they include arithmetic expressions (like the ones in the previous section) linked by relational operators (like the ones that follow).

Here are the six relational operators that may be used in a conditional expression:

$=$	Equal to
$>$	Greater than
$<$	Less than
$<>$	Not equal to
$<=$	Less than or equal to
$>=$	Greater than or equal to

Examples of Conditional Expressions

The logical outcome of the following expressions depend on the value assigned to the variable x.

$x = 5$	*True* for values of x equal to 5, *false* for all other values
$x > 5$	*True* for all values of x greater than 5, *false* for all other values
$x < 2$	*True* for values of x less than 2, *false* for all other values
$x <> 3$	*True* for values of x not equal to 3, *false* for all other values
$x <= 3$	*True* for values of x less than or equal to 3, *false* for all other values
$x >= 0$	*True* for values of x greater than or equal to 0, *false* for all other values

Like arithmetic expressions, conditional expressions also have an order of precedence that determines how they are evaluated. The details of this and many more features of computer logic will be covered in the next chapter. For now, simple conditional expressions like those already mentioned will be used in the programming language introduced in the following section.

Most programming languages have many dozens of different programming statements that give the programmer a great deal of flexibility when designing a program. This section will introduce you to programming with an extremely simple language consisting of only a few statements. This simple programming language, sometimes referred to as *pseudocode,* is not intended to run on a computer. Rather, it will allow you to understand how a computer program works before being faced with the complexity of the real thing.

This simple programming language uses five instructions, given by the following key words:

INPUT	Assigns a value to a variable with data from outside the program
OUTPUT	Delivers the results of processing to outside the program
LET	Assigns a value to a variable or stores the result of a calculation
IF-THEN-ELSE	Processes other statements depending on a logical condition
LOOP	Repeats some processing depending on a logical condition

At first glance these five types of instructions may seem so simple that they can only be used to build simple programs. But this is not the case. These five are the only types of instructions that are needed to build programs of immense complexity.

INPUT Statement

The INPUT statement allows the data from the outside world to be brought into the program as it is running. The data is assigned to variables where it is available for processing. The key thing about input is that it allows the same program to produce different output depending on the data provided.

Thus, a program designed to produce a sales report will produce a January sales report when January sales data are provided, will produce a February sales report when February sales data are provided, and so on. None of the programming instructions need to change in order to produce the monthly sales report for a different month. Only the data change.

 For the same processing, different input will produce different output.

▶ Examples:

INPUT name	Inputs data to the variable called *name*
INPUT age	Inputs data to the variable called *age*
INPUT name, age	Inputs data to variables called *name* and *age*

For clarity, key words are in capital letters and variables are in lowercase.

Data Lists

In real programming languages, input can come from a variety of sources such as a file, a database, or the keyboard. In this simple programming language, input will come from a list of data. This data, either numeric or text, will be in a *data list* kept separate from the program instructions.

Each time an INPUT statement is encountered within a program, the variable (or variables) listed after the key word is filled with data from the list. The program will automatically keep track of the data values that have been used previously so that each data item is used only once and in the order provided.

▶ *Example 1*

Data List: 5, 7, 9

INPUT n   Assigns the value 5 to the variable *n*.

INPUT n	Assigns the value 7 to the variable *n*.
INPUT n	Assigns the value 9 to the variable *n*.

▶ *Example 2*

Data List:	"Bill", 21, "Sue", 19, "Jim", 23
INPUT name, age	The variable *name* is assigned "Bill" and *age* is assigned 21.
INPUT name, age	The variable *name* is assigned "Sue" and *age* is assigned 19.
INPUT name, age	The variable *name* is assigned "Jim" and *age* is assigned 23.

> **Note:** Any data item that is text (i. e., "Bill", "Sue", "Jim") must be enclosed within quotes. Also, in these examples several INPUT statements were used. Later you will see how a single INPUT statement placed inside a loop can accomplish the same thing.

Practice Problems

Use the INPUT statement to accomplish the following tasks:

7-3.1   Input values to variable *x* and variable *y* using two INPUT statements.

7-3.2   Input values to the variables *x* and *y* using a single INPUT statement.

7-3.3   Input a value into a variable named *number_on_hand*.

7-3.4   Input text into a variable named *car_manufacturer.*

7-3.5   Input data into the variables *car_manufacturer* and *number_on_hand* using a single INPUT statement.

7-3.6   Input values to the following three variables using a single INPUT statement: *item_number, number_on_hand,* and *number_ordered.*

7-3.7   Write a sample data list for the statement: INPUT car_manufacturer.

7-3.8   Write a sample data list for the statement: INPUT vehicle_model.

7-3.9   Write a sample data list for the statement: INPUT vehicle_model, number_on_hand.

7-3.10 Write a sample data list for the statement: INPUT name, age, height. (Give height in inches.)

OUTPUT Statement

The OUTPUT statement allows information to be sent from memory (where it is invisible) to some output device where it can be read or saved for later use. In the real programming languages many output devices are available. Output may be directed to the printer, the screen, or a database on your hard drive. Output to your hard drive from one program may become input to some other program.

In this simple programming language output will be written only to the screen. The book will sometimes show this as a box (representing the screen) with data written inside. Output may be numeric, text, or a mixture of both.

▶ *Examples*

OUTPUT "The sales amount is"   *A message (given in quotes)*

```
The sales amount is
```

OUTPUT sales_amount   *A value from a variable, for example, 22.50*

```
22.50
```

OUTPUT "The sales amount is ", sales_amount   *Both message and value*

```
The sales amount is 22.50
```

Practice Problems

Use the OUTPUT statement to accomplish the following tasks:

7-3.11 Output the contents of the variable *average.*

7-3.12 Output the message "The average is".

7-3.13 Combine the preceding two problems into a single statement.

7-3.14 Use three statements to output the contents of variables *num1, num2,* and *num3.*

7-3.15 Use one statement to output the contents of variables *num1, num2,* and *num3.*

What will the output for the following statements look like when the variable *name* is "Bill" and the variable *age* is "31"?

7-3.16 OUTPUT name

7-3.17 OUTPUT name, age

7-3.18 OUTPUT " The age of ", name, " is ", age

7-3.19 OUTPUT age, " is the age of ", name

7-3.20 OUTPUT name, " has reached the age of ", age

LET Statement

The LET statement allows you to assign data (numeric or text) to a given variable. Numeric values may come from several sources:

- A constant specified in the assignment statement

- A value copied from another variable

- A value calculated from an arithmetic expression including both variables and constants

▶ *Examples*

LET name = "Bill"	Assigns "Bill" to the variable *name*
LET sales = 5	Assigns the constant 5 to the variable *sales*
LET year = previous_year	Assigns a copy of the contents of the variable *previous_year* to the variable *year*
LET average = (value1 + value2)/2	Assigns a computed value to the variable *average*

Notes:

- The content of the variable on the **left of the equal sign** is overwritten.

- The content of any variables on the **right of the equal sign** remains unchanged (unless a variable on the left also appears on the right).

- The INPUT and the LET statements are the only ways to assign a value to a variable. The INPUT statement assigns values from data *outside* the program. The LET statement assigns values from *inside* the program.

- When the data to be assigned is text, it must have quotes around it.

Practice Problems

Use the LET statement to assign data to variables as directed.

7-3.21 Assign the value 21 to the variable *age.*

7-3.22 Assign the text "Bill" to the variable *name.*

7-3.23 Assume that variable *num1* contains the value 33 and variable *num2* contains the value 44. What values will each variable contain after the following statement is executed: LET num2 = num1 + 1?

7-3.24 Compute the sum of the values stored in variables *num1, num2,* and *num3* and assign the result the variable *total.*

7-3.25 Compute the average of the three values stored in variables *num1, num2,* and *num3* using the fact that the sum of the three values is stored in the variable *total.*

7-3.26 Assign the contents of the variable *age* to the variable *previous_age.*

7-3.27 Add 1 to the variable *previous_age* and store the result in the variable *age.*

7-3.28 Add 1 to the contents of the variable *age* and store the result in the variable *age.*

7-3.29 Use three LET statements to swap the contents of variables *num1* and *num2.*

7-3.30 What will display as a result of the following instructions?

LET message = "The answer to life, the universe, and everything is"

LET answer = 42

OUTPUT message, answer

IF-THEN-ELSE Statement

This statement allows processing to occur conditionally. Suppose you are writing a program for a driving license application and you want to issue a warning if an age seems too large. If the age is >100, a warning is issued and processing continues. If the age is < = 100, processing continues.

If age > 100 THEN OUTPUT "Warning: Age may be invalid. Please check."

A statement that does nothing (under certain conditions) may at first seem a little troublesome. But as you see, it is often very useful to control the processing in exactly this way.

Now suppose you want the computer to determine if it is OK to sell beer to a young person. You can program the computer to display a message depending on the age of the person. Assume the age has been previously assigned to the variable *age.* The conditional expression *age* > = 21 can be used to direct this processing:

IF age > = 21 THEN OUTPUT "OK to sell beer."

In this case, when the variable *age* has a value less than 21, the statement does nothing. When the variable *age* contains a value of 21 or older, the message "OK to sell beer" is displayed as output. Another version of the IF statement allows the same instruction to be written on several lines.

IF age > = 21 THEN
        OUTPUT "OK to sell beer."
END IF

This second form of the IF statement also allows you to do more than one thing when *age* > = 21:

IF age > = 21 THEN
        LET sell = "Yes"
        LET sell_code = 1
        OUTPUT "OK to sell beer."
END IF

Suppose you want two messages to appear: one when it is OK to sell beer and another when it is not. Another version of the IF statement provides for that:

```
IF age > = 21 THEN
        LET sell = "Yes"
        LET sell_code = 1
        OUTPUT "OK to sell beer."
ELSE
        LET sell = "No"
        LET sell_code = 0
        OUTPUT "Don't sell beer."
END IF
```

Note: The final example is much more functional than the preceding ones because it covers all the bases and leaves nothing unstated regardless of the true/false outcome.

Exactly the same processing as in the previous example will occur in the following example by reversing the conditional expression and at the same time flipping the two blocks of statements:

```
IF age < 21 THEN
        LET sell = "No"
        LET sell_code = 0
        OUTPUT "Don't sell beer."
ELSE
        LET sell = "Yes"
        LET sell_code = 1
        OUTPUT "OK to sell beer."
END IF
```

In general the IF-THEN-ELSE statement has the following key words and components:

IF	Marks the beginning of the entire statement. A conditional expression follows immediately on the same line.
THEN	Follows the conditional expression (on the same line) and marks the beginning of the *true* block of statements. The *true* block of statements follows on as many lines as needed.
ELSE	Marks the beginning of the *false* block of statements. The *false* block of statements follows on as many lines as needed.
END IF	Marks the end of the entire statement.

Note: The *true* and *false* blocks of statements each may include one or more statements, or be left empty. Each block may contain any of the five kinds of statements, including other IF statements.

▶ *Example*

Compute a four-digit year from a two-digit year between 1970 and 2069.

```
INPUT yr                          Input a two-digit year from a data list
IF yr > = 70 THEN
        LET year = 1900 + yr      Executed only when the condition is true
ELSE
        LET year = 2000 + yr      Executed only when the condition is false
END IF
```

Notice how the statements inside the IF-THEN-ELSE instructions are indented.

Note: For years before 1970 and after 2069, this statement will not work.

Practice Problems
Apply the rules for the IF statement to accomplish the following processing.

Introductory Problems
7-3.31 Use a single-line IF statement to assign the value zero to the variable *count* when the value of *count* exceeds 99.

7-3.32 Use a multiple-line IF statement to assign the value zero to the variable *count* when the value of *count* exceeds 99.

7-3.33 Use a single-line IF statement to output the message "Count exceeds 99" when the value of the variable *count* is greater than 99.

7-3.34 Use a multiple-line IF statement to output the message "Count exceeds 99" when the value of the variable *count* is greater than 99.

Intermediate Problems
7-3.35 Depending on the value previously assigned to the variable *x*, use an IF-THEN-ELSE statement to assign the appropriate text below to the variable *answer*.
"The value of *x* is less than zero."
"The value of *x* is greater than or equal to zero."

7-3.36 Write a multiple-line IF statement that outputs the message "Yes" when the variable *x* equals 1, and the message "No" when the variable *x* doesn't equal 1.

7-3.37 Write an IF statement that assigns the text "False" to the variable *answer* when variable $x = 0$; otherwise, assign "True". Write an equivalent statement that has the same outcome but uses the opposite condition, $x <> 0$.

7-3.38 Write an IF statement that assigns the text "No" to the variable *answer* when the variable *age* is 65 or greater; otherwise, assign "Yes". Write an equivalent statement using the opposite condition; $age < 65$.

More Difficult Problems
7-3.39 In the following problems, the variable *color_code* will contain only integer values. There are three valid codes: 1 = Red, 2 = Blue, 3 = Yellow. All other codes are invalid.
a) Write a single-line IF statement that outputs the message "Invalid Code" when *color_code* is less than 1.
b) Write a single-line IF statement that outputs the message "Invalid Code" when *color_code* is greater than 3.
c) Write three single-line IF statements that output the name of the corresponding color to each valid *color_code*.
d) Combine the previous statements into a multiple-line IF statement containing other multiple-line IF statements but no single-line IF statements.

7-3.40 Use three separate single-line IF statements to assign the appropriate text to the variable *answer* depending on the value of the variable *x*. The three possible texts are:
"The value of *x* is less than zero."
"The value of *x* is equal to zero."
"The value of *x* is greater than zero."
Then combine the three into an IF statement with another IF statement inside.

LOOP Statement

A LOOP statement allows a block of statements to be repeated while some condition is true. Key words mark the different parts of the loop statement:

DO WHILE   Marks the beginning of the loop followed by the condition.
The block of statements to be repeated comes next.
LOOP   Marks the bottom of the statement block.

► **Example**

The following program repeats a block of three statements while the values of the variable *input* are not equal to 99. The three things done in the block statements are:

Input one value into variable *n*.

Compute the square of *n* and store the result in variable *s*.

Output the value of the variable *s*.

Assume the initial value of zero has been assigned to variable *n* and variable *s* before the loop begins.

```
DO WHILE n < > 99
        INPUT n
        LET s = n * n
        OUTPUT s
LOOP
```

The condition is evaluated at the top of the loop. Thus, if the condition is false at the beginning, the statement block is skipped entirely. To make sure the loop is entered at least once, the variable *n* could be set to some initial value (not shown) such as zero. This process is called *initialization* and is discussed later in the chapter.

Notes:

- Indention of the statements inside the loop makes the program more readable.

- To avoid an *infinite loop* (a loop that never terminates), there must be a statement inside the loop that can change the logical outcome of the condition. In this case the statement "INPUT n" serves that purpose. The condition will be true for all values of n except $n = 99$. Thus, when the condition is evaluated for $n = 99$, the condition is no longer true and the loop terminates.

- The design of this loop permits the terminating value (99) to be squared and sent to output, just like any other data item. Later in the chapter you will see how to design a loop that avoids this.

Practice Problems

Introductory Problems
Apply the looping rules to accomplish the specified processing. Assume that the value of the variable *n* is equal to zero as the loop begins.

7-3.41 What condition will allow a DO WHILE loop to repeat while the variable *n* is zero?

7-3.42 What condition will allow a DO WHILE loop to repeat while the variable *n* is greater than zero?

7-3.43 What condition will cause a DO WHILE loop to terminate before it starts? (Remember, the value of *n* is zero before the loop begins.)

7-3.44 Build a DO WHILE loop that repeats the output "Once more around the loop" and provides no way to stop. (This so-called "infinite loop" is one programmers avoid.)

7-3.45 Modify the infinite loop built in problem 7-3.44 to make it terminate after one time through.

Intermediate Problems
Again, assume the value of the variable *n* is equal to zero as the loop begins.

7-3.46 Build a DO WHILE loop that will output values of the variable *n* after they have been inputted. Repeat to loop until the value 99 is encountered.

7-3.47 Use an IF statement to modify problem 7-3.46 to exclude the value 99 from the output. Can you think of a better way to do this that does not use an IF statement?

7-3.48 Design a program to output the counting numbers from 1 to 5. Use two variables, *count* and *previous_count,* inside a DO WHILE loop. The variable *count* represents the counting number to be outputted the current time around the loop. The variable *previous_count* represents the counting number outputted the previous time around the loop. Assume the values of *count* and *previous_count* are both 1 before the loop starts.

More Difficult Problem

7-3.49 What output will result from running the following program with one loop nested inside another?

```
LET sum = 0
LET j = 1
DO WHILE j < 3
        LET k = 1
        DO WHILE k < 3
                LET product = j * k
                OUTPUT product
                LET sum = sum + product
                LET k = k + 1
        LOOP
        LET j = j + 1
LOOP
OUTPUT "The sum of the products is", sum
```

What would the final sum be if both conditional statements were changed to $<\,=3$?

Variations on LOOP Statements

There are several variations on the basic loop statement. (Again, we assume the variable n has a value of zero when the loop starts.)

Case 1: DO WHILE vs. DO UNTIL
DO WHILE:

As you have seen in the introduction to the LOOP statement, the key word WHILE directs the loop to:

- Repeat as long as the condition is true.

- Stop once the condition becomes false.

▶ Example:

```
DO WHILE n <> 99
        INPUT n
        LET s = n * n
        OUTPUT s
LOOP
```

Note: This code repeats a block of statements WHILE the variable *n* is not equal to 99. The condition "*n* <> 99" is tested each time at the *top* of the loop. Should *n* be set to 99 before the loop begins, all the statements inside the loop will be skipped entirely.

DO UNTIL

If the key word UNTIL is substituted for WHILE, then the loop will:

- Repeat as long as the condition is false.

- Stop once the condition becomes true.

▶ Example:

```
DO UNTIL n = 99
        INPUT n
        LET s = n * n
        OUTPUT s
LOOP
```

Note: This code repeats the statement block UNTIL the variable n equals 99. The condition "$n = 99$" is tested each time at the *top* of the loop. The inner-loop block of statements will be skipped entirely should the variable n be set to 99 before the loop begins.

General Note: The two preceding examples accomplish exactly the same thing. Changing WHILE to UNTIL allowed the condition "$n <> 99$" in the first example to be stated in a more positive way as "$n = 99$" in the second example.

Case 2: Top vs. Bottom

If the condition (along with WHILE or UNTIL) is moved from the *top* of the loop to the *bottom,* the conditional statement is evaluated at the bottom of the loop. In this case the statement block will always be executed at least once.

LOOP WHILE Example:

```
DO
        INPUT n
        LET s = n * n
        OUTPUT s
LOOP WHILE n <> 99
```

Note: This code repeats the statement block WHILE the variable n is not equal to 99. The condition "$n <> 99$" is tested each time at the *bottom* of the loop. Regardless of the value of n when the loop begins, the statements inside the loop will be executed at least once.

LOOP UNTIL Example:

```
DO
        INPUT n
        LET s = n * n
        OUTPUT s
LOOP UNTIL n = 99
```

Note: This code repeats the statement block UNTIL the variable n equals to 99. The condition "$n = 99$" is tested each time at the *bottom* of the loop, so the inner-loop statements will be executed at least once.

General Notes:

- Again, the two preceding examples accomplish exactly the same thing. Changing WHILE to UNTIL permits the condition "$n <> 99$" to be stated in a more positive way as "$n = 99$."

- The difference between case 1 and case 2 is the place where the condition is tested. In the first two examples it is tested at the top of the loop. In the latter two examples it is tested at the bottom of the loop. The consequence of this is that, depending on the value of variable n as the loop begins, case 1 permits the loop to be skipped entirely, while case 2 does not.

Practice Problems

Use the loop variation given in problems 7-3.50 to 7-3.53 to write a program that inputs values into variables *num1* and *num2,* stores their difference in variable *diff,* and outputs the result. End the loop when the *diff* becomes zero. Assume that the value of variable *diff* is 99 before the loop starts.

7-3.50 DO WHILE

7-3.51 DO UNTIL

7-3.52 LOOP WHILE

7-3.53 LOOP UNTIL

The simple programming language introduced in the preceding section will now be employed to write simple programs. This section will show how minor changes in the statements of a program produce significant changes in its operation. A method of checking the operation of programs that does not require a computer will also be introduced.

Recall the five statements of our simple programming language:

- INPUT
- OUTPUT
- LET
- IF-THEN-ELSE
- LOOP

A typical program usually has these three features:

- A means to **input** data
- A means to **process** that data into new information
- A means to **output** that information

Input, process, and output are usually not three distinct components that operate independently. More often they are integrated so that they occur repeatedly throughout the program. For example, to produce a report of the sums of several pairs of numbers, do the following:

1. Input, process, and output the first sum.
2. Input, process, and output the second sum.
3. Input, process, and output the next sum.
4. Continue until all sums have been completed.

This will be demonstrated in many of the examples that follow in this and the next section.

Important Assignment Tasks

Initializing

Whenever a variable is part of an expression, it must have a value that has been previously assigned. Otherwise, the expression cannot be evaluated and an error will often occur. This is true for conditional expressions and arithmetic expressions. It is true for any variable to the right of the equal sign in an assignment statement. To avoid problems, a variable may be assigned an initial or temporary value. This temporary value is used to get things started and prevent errors before the "real" data are assigned. This process is called initializing the variable.

> **Note:** Initialization is only needed when a variable is used in an expression *before* input or other data have been assigned. Initialization is not needed in cases in which a variable is used in an expression only *after* input or other data have been assigned.

Initialization is always accomplished by a LET statement. For example:

LET n = 0

Once initialized, the value stored in *n* can be used as needed to get the program started. Consider this example:

LET n = 0	Initializes the variable *n*
DO WHILE n < 99	Loops while the condition *n < 99* is true
INPUT n	Assigns a value to the variable *n* from a data list
OUTPUT n	Prints out the values stored in the variable *n*
LOOP	Returns to the top of the loop

Notes:

- The preceding example will continue to print out values read from a data list as long as the values remain below 99. When a value of 99 of above is encountered, the program stops.

- The first data item is not assigned by the INPUT statement until *after* the logical condition "*n < 99*" in the DO WHILE statement is tested for the first time. Thus, for the condition test to succeed, it is necessary to assign an initial value to the variable *n before* the loop begins.

- Although the value zero was used in the initialization statement, the program would have worked the same had *n* been initialized to *any* value less than 99. The value used depends on the situation. In some situations the program will only work when a specific initialization value is used.

- In this case, the initial value of *n* is only used to get things started. The value zero will not show up in the output.

- Some programming languages automatically initialize variables, relieving the programmer of this task. However, it is always best to *explicitly* initialize variables.

Practice Problems

7-4.1 When is initialization needed? When is it not needed?

7-4.2 Write a statement to initialize the variable *count* to the value 1.

7-4.3 Write a statement to initialize the variable *sum* to the value 0 (zero).

7-4.4 In the example program preceding these problems, what initialization statement would cause the program to produce no output?

7-4.5 Modify the example program preceding these problems so that an assignment statement would not be needed to initialize the variable *n*.

Counting

A counting statement can be constructed by assigning a numeric value to a variable that is one more than the current value of that variable. To count using the variable *n,* the following assignment statement can be written:

LET n = n + 1

 The expression $n = n + 1$ should not be confused with an algebraic equation. As an assignment statement, it has a very different meaning.

The computer interprets counting statements in the following order:

- Add 1 to the *current* value of the variable *n*.

- Assign the result as a *new* value of the variable *n*.

Here is a simple program followed by a diagram that demonstrates how counting works:

LET *n* = 0
LET *n* = *n* + 1
LET *n* = *n* + 1
LET *n* = *n* + 1

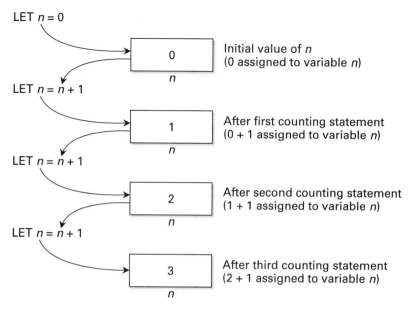

LET $n = 0$ 0   Initial value of n
(0 assigned to variable n)

n

LET $n = n + 1$ 1   After first counting statement
(0 + 1 assigned to variable n)

n

LET $n = n + 1$ 2   After second counting statement
(1 + 1 assigned to variable n)

n

LET $n = n + 1$ 3   After third counting statement
(2 + 1 assigned to variable n)

n

Figure 7.1

Figure 7.1 shows that, each time the counting statement is executed, the value of the variable n is incremented by 1. This is accomplished by adding 1 to the current value of the variable n. The result of the addition is assigned to the variable n, replacing the *current* value with a *new* value that is 1 more.

Counting statements are often placed inside loops to count the number of times the statements inside the loop are executed.

```
LET n = 1
DO WHILE n <= 5
        OUTPUT n            The output will be: 1 2 3 4 5
        LET n = n + 1       Increase n by one each time through the loop.
LOOP
```

In this program the statements in the loop will execute five times. Notice that the variable n is assigned an initial value of 1 before the loop begins. This enables n to be used in the conditional statement "$n <= 5$" at the top of the loop as well as in the arithmetic expression "$n + 1$" inside the loop. In many languages, not initializing a variable will lead to an error that stops the program prematurely. Therefore, always initialize.

Note:

- The last statement inside the loop sets the variable n to a new value. Immediately after this statement, the condition is tested at the top of the loop. After the fifth time around the loop, the condition tests false and the loop stops.

- It is best if the statement that will cause the loop condition to change is the last statement inside the loop. In this way, no processing takes place between the time the variable is changed and when it is tested in the loop condition.

Suppose you want to count five numbers starting at 6. The program to do this is as follows:

```
LET n = 6
DO WHILE n <= 10
        OUTPUT n          The output will be: 6 7 8 9 10
        LET n = n + 1
LOOP
```

Practice Problems

7-4.6 Describe the two-step process that takes place as the computer executes a counting statement.

7-4.7 Write an assignment statement for counting with the variable *count.*

7-4.8 Write an assignment statement for counting with the variable *number.*

7-4.9 Modify the example program immediately preceding these practice problems to count from 0 to 10.

7-4.10 Modify the example program immediately preceding these practice problems to count from 10 to 21.

7-4.11 Modify the example program immediately preceding these practice problems to count from −10 to +10.

Incrementing

Counting is a special case of the more general technique of incrementing. While counting changes a variable by 1, incrementing adds (or subtracts) any constant value you may like.

The following example will output five consecutive *even* numbers, starting with 2:

```
LET n = 2                  Initialize n to the first value to be outputted.
DO WHILE n < = 10          The loop condition begins as true, since 2 < = 10.
        OUTPUT n           Output the current value of n (2, 4, 6, 8, 10).
        LET n = n + 2      Increment n by 2 each time through the loop.
LOOP                       This statement marks the bottom of the loop.
```

There are several ways to design an incrementing program. The previous version is especially well designed because the sequence of numbers that are outputted is easily changed. To modify the program to give another sequence:

- Set the value in the initialization statement to the starting value.

- Set the relational operator to match the direction of change: < = for incrementing up, > = for incrementing down.

- Set the value in the conditional statement to the ending value.

- Set the value in the incrementing statement to the difference between the terms.

For example, to produce the sequence 200, 175, 150, 100, 75, 50, 25, 0, the program becomes:

```
LET n = 200                Starting value is 200
DO WHILE n > = 0           Incrementing down needs > = and the ending value is zero
        OUTPUT n
        LET n = n - 25     Difference between terms is −25
LOOP
```

Note: By avoiding the relational test of exactly equal, and using < = or > = instead, the possibility of an infinite loop is avoided. Depending on the incrementing value used, the program will terminate even when the output value jumps over the ending value.

Practice Problems

7-4.12 What is the difference between counting and incrementing?

7-4.13 Write a statement to increment variable n in the following sequence: 10, 20, 30, 40, etc.

7-4.14 Write a statement to increment variable n in the following sequence: 10, 9, 8, 7, etc.

7-4.15 Write a statement to increment variable n in the following sequence: 40, 30, 20, 10, etc.

7-4.16 Modify the example program immediately preceding these practice problems to produce the following output: 15, 10, 5, 0, -5, -10, -15.

7-4.17 Modify the example program immediately preceding these practice problems to produce the following output: -8, -5, -2, 1, 4, 7, 10.

Accumulating

The LET statement can be used to *accumulate* a set of values from one variable into another variable. Consider the accumulation statement:

    LET sum = sum + n

Notice that an accumulation statement is similar to an incrementing statement, except that the incrementing amount is now a variable (in this case n) rather than a constant.

The variable *sum* was selected to describe the data that it will store. The choice of the variable name does not affect its function, but rather makes it easier to understand. It could have been named *total* or even *mickey_mouse,* although the later would just serve to make things confusing.

The previous accumulation statement is interpreted by the computer in the following order:

- Add the current value of n to the **current** value of *sum.*

- Store the result as the **new** value of *sum.*

Here is a simple program followed by a diagram that demonstrates how accumulation works:

    LET sum = 0
            LET n = 1
    LET sum = sum + n
            LET n = 2
    LET sum = sum + n
            LET n = 3
    LET sum = sum + n

Figure 7.2 shows how the values 1, 2, and 3 are assigned in turn to the variable n. After each assignment the current value of n is accumulated into the variable *sum.* The first accumulation statement assigns the value 1 to the variable *sum.* This value is computed by adding the initial value of *sum* (zero) to the current value of n (1), to give 1. The second accumulation statement assigns the value 3 to the variable *sum.* This value was computed by adding 1 and 2. The third accumulation statement assigns the 6 to the variable of *sum.* This value is computed by adding $3 + 3$.

Accumulation statements are often placed inside loops so that each time they are encountered, a new value (n in this case) will be added to the accumulation variable (*sum* in this case). Thus, accumulation statements are usually used in conjunction with a statement that changes the value of the variable to be added (n in this case).

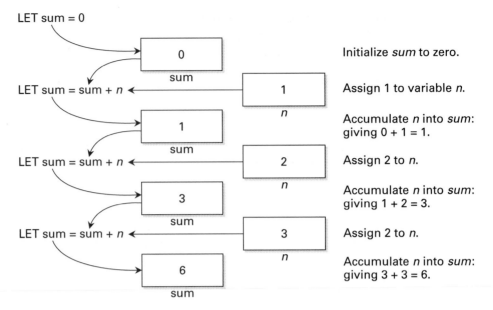

LET sum = 0	0 sum		Initialize *sum* to zero.
LET sum = sum + *n*		1 *n*	Assign 1 to variable *n*.
	1 sum		Accumulate *n* into *sum*: giving 0 + 1 = 1.
LET sum = sum + *n*		2 *n*	Assign 2 to *n*.
	3 sum		Accumulate *n* into *sum*: giving 1 + 2 = 3.
LET sum = sum + *n*		3 *n*	Assign 2 to *n*.
	6 sum		Accumulate *n* into *sum*: giving 3 + 3 = 6.

Figure 7.2

▶ *Example*

Suppose you want to add the first five counting numbers: 1, 2, 3, 4, 5. A simple variation on the incrementing statement above will allow this.

LET n = 1	Initialize *n* to the first value to be accumulated.
LET sum = 0	Initialize *sum* so accumulation will be zero before *n* is added.
DO WHILE n < = 5	The loop condition begins as true since $1 < = 5$.
LET sum = sum + n	**Accumulate *n* into *sum* each time through the loop.**
LET n = n + 1	Increment *n* by 1 each time through the loop.
LOOP	This statement marks the bottom of the loop.
OUTPUT sum	The output at the end will be 15.

Notice that the variable *sum* in the statement "LET sum = sum + n" is incremented by a variable amount according to the value assigned to the variable *n*. Notice also that *sum* begins at zero because of the initial value set in the second line of the program.

 Accumulation is like incrementing by a variable amount rather than a constant amount.

Practice Problems

7-4.18 In what way is accumulating similar to incrementing? In what way are they different?

7-4.19 Modify the example program immediately preceding these practice problems to sum the integers from 1 to 10.

7-4.20 Modify the example program immediately preceding these practice problems to sum the integers from 6 to 10.

7-4.21 Modify the example program immediately preceding these practice problems to sum the integers from 1 to 100.

7-4.22 Modify the example program immediately preceding these practice problems to sum the squares of the integers from 1 to 5.

7-4.23 Modify the example program immediately preceding these practice problems to sum the squares of the integers from 5 to 10.

Order of Statements Is Important

It should be clear by now that the order of statements in a program can significantly change the output of a program. Notice in the previous example that because the OUTPUT statement is below the loop, only the final value of the accumulation will be shown. If the OUTPUT statement had been placed inside the loop, just below the accumulation statement, the output would become: 1, 3, 6, 10, 15. In this case no harm was done; in fact, you may appreciate seeing the intermediate values of the accumulation.

Perhaps two OUTPUT statements would be a better choice: one inside the loop to show the step-by-step accumulation and another after the loop to show the final result. Perhaps adding a message would make things clear.

OUTPUT "The sum of the first five integers is:", sum

The situation above is a change for the better. But in some cases, moving statements within a program will cause the program to give erroneous results. For example, placing the statement "LET sum = sum + n" below the statement "LET n = n + 1" will cause the values 2, 3, 4, 5, and 6 to be summed. This will change the final output from 15 to 20, which is not the sum of the first five counting numbers.

 The order of statements is very important.

Desk Checking

Before a program is run on a computer, you can check its operation at your desk. This is done with a table that shows statements as they are executed (down the rows) and individual values (across the columns).

Statement	n	x	s	n < 3	av	Comments
LET n = 0	0					Initial value of *n* makes *n < 3* true.
LET s = 0	0		0			Initial value of variable *s*
DO WHILE n < 3	0		0	True		Condition is true. Do statements in loop.
INPUT x	0	10	0	True		Input first data item from list into *x*.
LET s = s + x	0	10	10	True		Accumulate *x* into *s*.
LET n = n + 1	1	10	10	True		Increment *n* to count first data item.
LOOP	1	10	10	True		Return to top of loop.
DO WHILE n < 3	1	10	10	True		Condition evaluates as true; repeat loop.
INPUT x	1	15	10	True		Input second data item from list into *x*.
LET s = s + x	1	15	25	True		Accumulate *x* into *s*.
LET n = n + 1	2	15	25	True		Increment *n* to count first data item.
LOOP	2	15	25	True		Return to top of loop.
DO WHILE n < 3	2	15	25	True		Condition evaluates as true; repeat loop.
INPUT x	2	20	25	True		Input third data item.
LET s = s + x	2	20	45	True		Accumulate sum of third data item.
LET n = n + 1	3	20	45	True		Count the third data item.
LOOP	3	20	45	True		Return to top of loop.
DO WHILE n < 3	3	20	45	False		Condition evaluates as false; exit loop.
LET av = s / n	3	20	45	False	15	Compute average.
OUTPUT av	3	20	45	False	15	Output average and end program.

Consider a program that outputs the average of a list of input data. In this example we use only three values in the data list to keep things simple. Notice that the variable n is used to count the looping.

```
Data list: 10, 15, 20
        LET n = 0
        LET s = 0
        DO WHILE n < 3
                INPUT x
                LET s = s + x
                LET n = n + 1
        LOOP
        LET av = s / n
        OUTPUT av
```

Several important things about the program can be seen from the preceding table:

- The initial value of 0 of the variable n ensures that the count will start at 1.

- The initial value of 0 of the variable n ensures that the condition is true the first time through the loop.

- The initial value 0 of the variable s avoids a program error when "LET $s = s + x$" is executed the first time.

- Statements inside the loop appear in the desk check table as many times as they are executed.

- The statement "LET $n = n + 1$" inside the loop provides a means to terminate the loop after the third data item is processed.

- After the third time around the loop the value of n is 3. Then, the condition at the top of the loop is tested again. The condition now tests false, terminating the loop.

- The loop does not terminate when the value of n becomes 3. It terminates when the condition tests false in the DO WHILE statement.

- If the program checks out for three data items, it will work for 1000 data items, as long as the condition in the DO WHILE is changed to $n < 1000$.

Quick Check

A faster way to check a program is the quick check. Just draw a box to hold the value for each variable and conditional expression. Fill in the values as you move through the program. As each new value is assigned, place it to the right of the old value. That way you see all the old values, and the current value is always the one on the right. While not as detailed as the full desk check, it is much faster.

Here is the quick check for the previous example. Again, the data list is: 10, 15, 20.

Code	Box	Variable
LET n = 0	0 1 2 3	n
LET s = 0	0 10 25 45	s
DO WHILE n < 3	T F	n < 3
INPUT x	10 15 20	x
LET s = s + x	45 / 3 = 15	av
LET n = n + 1		
LOOP		
LET av = s / n		
OUTPUT av		

Practice Problems

7-4.24 Desk check the example program for this section using the data list 25, 17, 3.

Unless directed otherwise by your instructor, quick check each of the following problems.

7-4.25 Write a simple program that uses a loop to sum the even integers from 2 to 10 and outputs the final result. Assign the even integers by incrementing by two. Check the program to ensure that it works correctly. Ensure that initial values are assigned to all variables that require them.

7-4.26 Write a simple program that uses a loop to sum the odd integers from 1 to 9 and outputs the final result. Assign the odd integers by incrementing by two. Check the program to ensure that it works correctly.

7-4.27 Write a simple program that uses a loop to sum every third integer from 1 to 10 and outputs the final result. Assign every third integer by incrementing by three so that the values 1, 4, 7, 10 can be summed. Check the program to ensure that it works correctly.

7-4.28 How can the program in problem 7-4.26 be modified to output the subtotal after each value is added instead of just the final result? Include messages in the OUTPUT statements to distinguish between the subtotal and the grand total. Check the program to ensure that it works correctly.

Evolution of a Simple Program

Stepwise Refinement

Stepwise refinement is a program development approach that starts with a simple and limited solution to a problem. Then, through a series of refinement steps, the program evolves into a fully functioning version. This step-by-step approach usually arrives at a much better result than could have been achieved in a single step.

The following six examples show how a program that computes the average of a set of numbers can evolve through some simple steps to make it less restrictive and allow it to more easily process large amounts of data. Notice how the changes from one example to the next are almost trivial, yet the difference in functionality between the first example and the last is quite profound.

Later, in Chapter 9, a program-diagramming tool called a *flowchart* is introduced. Flowcharts for these six programs are given as examples. You may find it helpful to look ahead now, as you examine these programs, to see the processing displayed in a diagram.

Six Examples

1. Average three numbers assigned inside the program.

Let's begin with a program that averages just a few numbers, for example, three numbers. The simplest program to average three numbers uses LET statements inside the program to assign each number into a separate variable.

First write out the algorithm in structured English:

> Average three numbers
> > Assign three numbers
> > Compute the average
> > Output the average

Now convert the structured English into a computer program using the simple programming language. Here is an example averaging the numbers 10, 15, and 20:

> LET num1 = 10
> LET num2 = 15
> LET num3 = 20
> LET sum = num1 + num2 + num3
> LET average = sum / 3
> OUTPUT average

2. Average three numbers provided in a data list.

While program 1 works perfectly when your task is to average the three numbers 10, 15, and 20, its design has two major limitations. First, should you want to average three different numbers, you will need to modify the LET statements in your program. Second, should you want to average more than three numbers, you will have to add one new variable for each new data item. Let's examine the first problem here and put off the second problem until program 3.

It is not a big deal to have to modify the three LET statements when you are using the program yourself. As the programmer, you understand how the program is designed and how to use the programming language. But it is more often the case that someone else, a nonprogrammer, will be running the program. It's not practical to require a programmer to have to modify a program each time someone wants to do a calculation with different data.

Therefore, it is common practice to design programs in which the data are kept separate from the program instructions. Then the program will then work with any data that are supplied, and nonprogrammers can supply the data with ease.

An improved design is accomplished using an INPUT statement inside a LOOP. The INPUT statement allows data to be read from a data list that is kept separate from the programming instructions. Each time the INPUT statement is encountered, the next available data item in the data list is assigned into a variable. The data in that variable is then processed and, once done, the variable is available to be used for the next data item. The processing includes accumulating the data item and counting the data item. The LOOP repeats these steps (input, accumulate, count) until all of the data have been processed. Then the program can compute and output the average.

Here is the structured English for this approach to average any three data items:

Average three numbers
    Initialize accumulation and counting variables
    Repeat while count is less than 3
        Input one data item
        Accumulate this data item
        Count this data item
    Loop back
    Compute the average
    Output the average

Here is the program code for this design. Note that the data list has the same three numbers as in program 1. However, any three-number data list can be averaged without having to modify the program.

Data List: 10, 15, 20

LET sum = 0	Initialize *sum* before it is used to accumulate.
LET count = 0	Initialize *count* before it is used below.
DO WHILE count < 3	Test loop condition; repeat while count < 3.
INPUT num	Input the next value into *num* from the data list.
LET sum = sum + num	Accumulate *num* into the variable *sum*.
LET count = count + 1	Count this data item.
LOOP	Return to top of loop.
LET average = sum / count	After all data is processed, compute the average.
OUTPUT average	Output the result.

3. Average numbers from a list in which the first value is the count of values to be averaged.

The next improvement will be to address the second major limitation of program 1—that is, to allow it to average more than three data items. While the program 1 design could easily be made to work for four or five numbers, making it work for 40 or 50 data items would be a major hassle, and making it work for 400 or 500 data items would be completely impractical.

The design approach used in program 2, however, is ideally suited for handling large data lists. All that is necessary is to change the logical condition in the DO WHILE statement so the loop will repeat one time for each data item in the list.

This might be accomplished by changing the DO WHILE condition each time the program is run with a different number of data items. However, this violates the approach used in program 2 that made the program instructions independent of the data. A better approach would be to include the precount at the beginning of the data list as a number to be read as input before the loop begins. Thus, the design now incorporates two INPUT statements. The new one, above the loop, reads the precount value. The second one, inside the loop, reads the data values just like program 2.

Here is the structured English for this design. Notice that, except for two statements, it is identical to program 2.

> Average a list of numbers
>> Initialize accumulation and counting variables
>> Input the precount   *This is a new statement.*
>> Repeat while count is less than the precount   *Note the change here.*
>>> Input one data item
>>> Accumulate this data item
>>> Count this data item
>> Loop back
>> Compute the average
>> Output the average

The following program code can be used to average *any* number of data items that have been precounted. The example data list contains 10 values, a precount (9) followed by the nine data items to be averaged. The precount tells the program how many times to loop as the remaining values are read from the list and processed. Notice that the precount value is not included in the average.

> Data List 9, 10, 15, 20, 33, 52, 17, 62, 22, 76

```
LET sum = 0
LET count = 0
INPUT precount
DO WHILE count < precount
    INPUT num
    LET sum = sum + num
    LET count = count + 1
LOOP
LET average = sum / count
OUTPUT average
```

4. Average numbers from a list using a *sentinel value* to terminate the list.

While program 3 is a major improvement over the previous versions, it is still a hassle to have to precount the data. With a simple modification, the program can be made to automatically stop the processing when the data list is at the end. One way to accomplish this is by removing the precount from the beginning of the data list and adding a *sentinel value* at the end.

A sentinel value is a special number that is different from the rest of the data. For example, if the data consists of persons' ages, then a sentinel value of 9999 will be appropriate since no one will have an age that is anywhere near that number. If the data consist of monthly sales amounts for a large company that range into the millions, then a negative sentinel value will work nicely. Just pick a sentinel value that will never occur in the data you are averaging.

You do not want the sentinel value to be included in your average. This can be avoided by doing the accumulation and count conditionally (in an IF statement). (In program 5 an even better way to handle this will be shown.)

Here is the structured English for an averaging program that uses 9999 as the sentinel value. Except for three statements (annotated), it is the same as program 3.

> Average a list of numbers
>> Initialize accumulation and counting variables
>> Initialize the data item variable   *Since the precount input is gone.*
>>
>> Repeat while data item is less than 9999   *Note the new condition.*
>>> Input one data item

If data value is less than 9999 then: *Avoid processing the 9999.*
      Accumulate this data item
      Count this data item
Loop back
Compute the average
Output the average

Here is the program code for this version. Notice that each data item is stored in variable *num* and it must have an initial value so the condition "num < 9999" will test true at the start of the loop.

Data List: 10, 15, 20, 33, 52, 17, 62, 22, 76, 38, 9999

```
LET sum = 0
LET count = 0
LET num = 0
DO WHILE num < 9999
        INPUT num
        IF num < 9999 THEN
                LET sum = sum + num
                LET count = count + 1
        END IF
LOOP
LET average = sum / count
OUTPUT average
```

5. Use a *priming read* to avoid processing the sentinel value.

Program 4 was a major improvement over the previous versions, but the need to do the processing (accumulate and count) conditionally to avoid processing the sentinel value was a bit awkward. There is a simple way around this. Put the INPUT statement as the last thing to be done inside the loop before the condition is tested. This will work for every data item except the first. Since the input is done after the processing, there must be some way to "prime" the variable *num* with the first data item, before the loop begins.

The technique to do this, named the *priming read,* is a standard part of every programmer's tool box. It simply places an initial INPUT statement just before the loop begins. This statement is used to initialize the variable *num* to the value of the first data item. Once in the loop, the processing of that value is done first and the input on the next value is done last. When the loop repeats, the processing of the data read on the previous loop is accomplished. When the sentinel value is finally read, the loop condition is tested in the next statement. This test terminates the loop, thus avoiding the processing of the sentinel value.

Here is the structured English for this design. Changes from program 4 are noted.

Average a list of numbers
    Initialize accumulation and counting variables
    Input first data item   *Priming read initializes data item.*

    Repeat while count is less than sentinel value
        Accumulate this data item
        Count this data item
        Input next data item   *Input is moved to bottom of loop.*

    Loop back
    Compute the average
    Output the average

Here is the program using a sentinel value of 9999:

Data List: 10, 15, 20, 33, 52, 17, 62, 22, 76, 38, 9999

```
LET sum = 0
LET count = 0
INPUT num
DO WHILE num < 9999
     LET sum = sum + num
     LET count = count + 1
     INPUT num
LOOP
LET average = sum / count
OUTPUT average
```

Note: The INPUT statement above the loop acts as an initialization statement, assigning the first data item to the variable *num*. Thus, there is no need for a LET statement to do this. Inside the loop the processing precedes the input of the next data item. Testing the loop condition happens immediately after each INPUT. Thus, when the sentinel value is finally encountered, the loop terminates immediately and the sentinel value is not processed.

6. Allow the program to handle the special case of an empty data list.

Program versions 3–5 were touted as being able to average *any* number of data items. Strictly speaking, this is not exactly true since there is some maximum number of data items that will fit in the data list. But there is also a minimum number of data items that will cause problems—a data list with a sentinel value and no data.

Consider how program 5 will process an empty data list. Since the sentinel value is encountered by the priming read statement, the loop is never entered. Thus, no data is processed and the program skips to the statement below the loop that computes the average. This is where the problem arises. The variable count is still zero since no data was processed. Thus, when the average statement tries to divide by zero, an error occurs and the program stops.

This may not seem like a big deal, but programmers do not like their programs to end with an error. They would rather write programs that avoid all error situations, even ones that occur infrequently. To fix the problem, just do the average computation on the condition that the count is greater than zero. It would also be nice to issue a message if there is no data to average.

Here is the structured English. Except as noted, it is the same as program 5.

```
Average a list of numbers
     Initialize accumulation and counting variables
     Input first data item                          Priming read initializes data
                                                     item

     Repeat while count is less than sentinel value
          Accumulate this data item
          Count this data item
          Input next data item
     Loop back
     If count is greater than zero then:             This condition is new.
          Compute the average
          Output the average
     Otherwise output an "empty data list" message   This message is new.
```

Here is the program code with an empty data list (containing only a sentinel value):

Data List: 9999

LET sum = 0

```
LET count = 0
INPUT num
DO WHILE num < 9999
        LET sum = sum + num
        LET count = count + 1
        INPUT num
LOOP
IF count > 0 THEN
      LET average = sum / count
      OUTPUT "The average is ", average
ELSE
      OUTPUT "There is no data to average"
END IF
```

The simple programming language introduced in this chapter is quite similar to the BASIC computer language, especially to older versions that ran under DOS. If you have access to a computer with any DOS versions of BASIC, for example QBASIC that Microsoft used to package with DOS Version 5.0, you can run any of the example programs with only minor changes.

The most important change is that each OUTPUT statement needs to become a PRINT statement, since in QBASIC the key word PRINT displays information on the screen.

The INPUT statement will work if you type in the values from the keyboard. The program will pause and display a question mark on the screen when it is ready to accept an input value.

If you wish to use a list of data values, QBASIC allows you to use the READ and DATA statements. Place your comma-delimited data in a DATA statement. The DATA statement can be placed anywhere in the program, but is usually placed at the bottom. Then, use a READ statement in place of each INPUT to be able to "read" data from the list.

Also note that underscores in variable names are not allowed in QBASIC. However, dots (periods) may be used instead. Stick to using only numeric data, as special variables with a $ sign at the end of their names are needed to store text, for example, Name$ = "Wanda." Here is an example using only numeric data:

```
READ num
DO WHILE num <> 9999
        PRINT num
        READ num
LOOP
DATA 3, 7, 32, 64, 9999
```

Each time a READ statement is encountered in your program, the next value from your DATA statement will be assigned to the variable *num*. In the previous example, the values 3, 7, 32, and 64 will print to the screen before the sentinel value is read to 9999 and ends the loop.

Practice Problems

In the following problems you need only write each original statement that needs modification, followed by its modified version.

7-5.1 Modify example program 1 to average 33, 47, and 29, instead of the given data.

7-5.2 Modify example program 2 to average four numbers read from a list.

7-5.3 Modify example program 3 to average the squares of the data given.

7-5.4 Modify example program 4 so it can accommodate data values up to one million.

7-5.5 Modify example program 5 so it can accommodate any positive data value up to the maximum that the computer can accept. Your solution should not depend on knowing the value of that maximum number.

Chapter Summary

A computer variable is a named area of memory where the results of one instruction can be stored until needed by another instruction.

Arithmetic expressions combine constants, variables, and arithmetic operators and evaluate to some numeric value.

Conditional expressions evaluate logically to either a true or false value.

Order of precedence rules for evaluating an arithmetic expression ensure that there is only one possible result. Components within an expression are evaluated in the following order:

Parentheses:	Innermost first, then left to right
Exponentiation:	Left to right
Multiplication or division:	Left to right
Addition or subtraction:	Left to right

A simple programming language needs only five statements:

INPUT

OUTPUT

LET (Assignment)

IF-THEN-ELSE (Condition)

LOOP (Repetition)

Computer programs combine these statements in different ways to do a variety of computational tasks.

Within a computer program, the **assignment statement** can be used to:

Initialize	Example:	LET count = 0
Count	Example:	LET count = count + 1
Increment	Example:	LET number = number + 2
Accumulate	Example:	LET sum = sum + number

Desk checking a program is an important tool that allows you to trace the step-by-step progress of a program before it is run on a computer and to confirm that the program is written correctly.

Quick checking is faster than desk checking. Although not as detailed and clear, it still provides a fairly good way to check a program.

Review Questions

1. Define the following terms: *computer, computer program, programming language.*
2. Define the following terms: *variable, arithmetic expression, conditional expression.*
3. Discuss naming variables.
4. What are the rules for order of precedence in evaluating arithmetic expressions?
5. Name the five instructions introduced for the simple programming language.
6. Which two programming language instructions allow data to be assigned to memory?
7. Which two programming language instructions include a conditional expression?
8. Which programming language instruction displays data stored in memory?

9. Which two programming language instructions may involve blocks of statements?
10. Discuss the similarities and differences among the four variations of the LOOP statement.
11. Define the purpose, and give an example, of a statement that will accomplish each of the following: assignment, initialization, counting, incrementing, accumulation.
12. Discuss the details of how an incrementing statement is evaluated.
13. What are the similarities and differences among counting, incrementing, and accumulating?
14. Where in a DO WHILE loop does the conditional expression get evaluated?
15. Where in a LOOP WHILE loop does the conditional expression get evaluated?
16. What is an infinite loop?
17. How are infinite loops avoided?
18. What is the significance of the last statement of a block of statements inside a loop?
19. What is a sentinel value and why is it useful?
20. What is priming read and why is it useful?
21. Use examples in Section 7-5 to discuss how the order of program statements affects output.
22. Why is stepwise refinement helpful in attacking a challenging programming problem?
23. Use stepwise refinement to design a program that finds the maximum value from among a list of data. Refine your design along the lines of the examples in Section 7-5 until you arrive at a version similar to example 6. Be sure your final design can handle a mixture of positive and negative data between the values +1000 and −1000.
24. How would you modify the final design of question 23 to find the minimum value?

Problem Solutions

Section 7-3

7-3.2   INPUT x, y
7-3.4   INPUT car_manufacturer
7-3.6   INPUT item_number, number_on_hand, number_ordered
7-3.8   Data List: "Impala", "Camaro", "Corvette"
7-3.10 Data List: "Bill", 32, 70, "Sally", 36, 68, "Jill", 24, 64
7-3.12 OUTPUT "The average is "
7-3.14 OUTPUT num1
        OUTPUT num2
        OUTPUT num3
7-3.16 Bill
7-3.18 The age of Bill is 31
7-3.20 Bill has reached the age 31
7-3.22 LET name = "Bill"
7-3.24 LET total = num1 + num2 + num3
7-3.26 LET age = previous_age
7-3.28 LET age = age + 1
7-3.30 The answer to life, the universe, and everything is 42
7-3.32 IF count > 99 THEN
                LET count = 0
        END IF
7-3.34 IF count > 99 THEN
                OUTPUT "Count exceeds 99"
        END IF
7-3.36 IF x = 1 THEN
                OUTPUT "Yes"
        ELSE
                OUTPUT "No"
        END IF

7-3.38   IF age > = 65 THEN
                         LET answer = "No"
              ELSE
                         LET answer = "Yes"
              END IF
              Equivalent Version:
                         IF age < 65 THEN
                                      LET answer = "Yes"
                         ELSE
                                      LET answer = "No"
                         END IF
7-3.40   Single-line version:
                         IF x < 0 THEN LET answer = "The value of x is less than zero."
                         IF x = 0 THEN LET answer = "The value of x is equal to zero."
                         IF x > 0 THEN LET answer = "The value of x is greater than zero."
              Combined version:
                         IF x < 0 THEN
                                      LET answer = "The value of x is less than zero."
                         ELSE
                                      IF x > 0 THEN
                                                   LET answer = "The value of x is greater than zero."
                                      ELSE
                                                   LET answer = "The value of x is equal to zero."
                                      END IF
                         END IF
7-3.42   n > 0
7-3.44   DO WHILE n = 0
                         OUTPUT "Once more around the loop"
              LOOP
7-3.46   DO WHILE n < > 99
                         INPUT n
                         OUTPUT n
              LOOP
7-3.48   One of several possible solutions:
              DO WHILE count < = 5
                         OUTPUT count
                         LET previous_count = count
                         LET count = previous_count + 1
              LOOP
7-3.50   DO WHILE diff < > 0
                         INPUT num1, num2
                         LET diff = num1 − num2
              LOOP
7-3.52   DO
                         INPUT num1, num2
                         LET diff = num1 − num2
              LOOP WHILE diff < > 0

Section 7-4

7-4.2   LET count = 1
7-4.4   LET n = 99   (or any value above 99)
7-4.6   Step 1: Add 1 to the current value of the counting variable.
              Step 2: Assign the result as a new value of the counting variable.
7-4.8   LET number = number + 1
7-4.10   Replace LET n = 6   with LET n= 10
              Replace DO WHILE n < = 10   with DO WHILE n < = 21

7-4.12 Incrementing increases or decreases the value of a variable by some constant amount. Counting is a special case in which the incrementing constant amount is + 1.

7-4.14 LET n = n − 1

7-4.16 Replace LET n = 200   with LET n = 15
     Replace DO WHILE n > = 0   with DO WHILE n > = −15
     Replace LET n = n − 25   with LET n = n − 5

7-4.18 Similarities: Both are assignment statements. Both have the variable appearing on the left side of the equal sign also appearing on the right. Each time one of these statements is executed, an amount is added to the old value of the variable on the left and the result is saved as the new value of that variable.

     Differences: Incrementing adds a constant value. Accumulating adds a variable amount. The amount to be accumulated is stored in a variable and is changed before each accumulation step.

7-4.20 Replace LET n = 1   with LET n = 6
     Replace DO WHILE n < = 5   with DO WHILE n < = 10

7-4.22 Replace LET sum = sum + n   with LET sum = sum + n * n

7-4.24 See Student Solution Manual.

7-4.26 LET sum = 0 | 0 1 4 9 16 25 | sum

     LET n = 1 | 1 3 5 7 9 11 | n

     DO WHILE n < = 9 | T F | n < = 9
          LET sum = sum + n
          LET n = n + 2
     LOOP
     OUTPUT sum   The output is 25

7-4.28 LET sum = 0 | 0 1 4 9 16 25 | sum

     LET n = 1 | 1 3 5 7 9 11 | n

     DO WHILE n < = 9 | T F | n < = 9
          LET sum = sum + n
          OUTPUT "The subtotal is", sum
          LET n = n + 2
     LOOP
     OUTPUT "The grand total is ", sum

     The output would look like this:
         The subtotal is 1
         The subtotal is 4
         The subtotal is 9
         The subtotal is 16
         The subtotal is 25
         The grand total is 25

Section 7-5

7-5.2 Add a fourth value to the data list
     Replace DO WHILE count < 3   with   DO WHILE count < 4

7-5.4 Replace sentinel data value 9999   with   1000000
     Replace DO WHILE num < 9999   with   DO WHILE num < 1000000
     Replace IF num < 9999   with   IF num < 1000000

CHAPTER

8

Computer Logic

Everything should be made as simple as possible, but not one bit simpler.

—Albert Einstein

8-1 ■ Introduction

How to Use This Chapter

The most powerful and enabling part of any computer language are the statements that involve logic. Without logic, computers would be little more than expensive adding machines. With logic, programs can do certain processing conditionally (that is, do it or not do it), depending on the outcome of some previous processing. Logic is what allows the same program to work in different ways according to the data supplied.

For the beginner, computer logic presents a conceptual challenge that is a bit more difficult than the other parts of the language. Although statements that involve logic, such as the LOOP and the IF statement, have already been introduced in the previous chapter, the details of the conditional part of these statements that involve logic were put aside until the other aspects of simple computer programs were covered.

Understanding computer logic is the key to understanding computer programming. Until computer logic is mastered, you will be limited in the types and complexity of the programs you can write. This chapter will give you insight into how logical statements are used to make a program do things that otherwise would not be possible. The basic concepts are introduced in the first few sections. In Section 8-5, Problem Solving with Computer Logic, these basic concepts are applied.

Background

The central processor unit (CPU) of a computer contains special circuitry that allows it to do calculations. This circuitry is called the arithmetic logic unit (ALU). The arithmetic part is used to evaluate the familiar arithmetic expressions involving addition, subtraction, multiplication, division, and exponentiation. The logic part is used to evaluate conditional expressions that may be less familiar.

Unlike the arithmetic expressions, which can evaluate to many different numeric values, conditional expressions have only two outcomes: true or false. No other outcome is possible. Conditional expressions are found in two of the simple programming statements introduced in Chapter 7, the conditional statement IF-THEN-ELSE and the looping statement DO WHILE.

Chapter Objectives

In this chapter you will learn:

- About relational operators such as > (greater than), < (less than), and = (equal to)
- About conditional expressions that evaluate to true or false
- About logical operators such as AND, OR, and NOT
- How to remove the NOT operator from a conditional expression
- How to test to see if a value is within a range of values
- About flags and Boolean variables that indicate whether some processing has been done

Introductory Problem

The conditional expression that evaluates as *true* for all values of the variable x that are either 1 or greater and evaluates as *false* for all other values assigned to the variable x is

$$x > = 1$$

Another expression that evaluates as *true* for all values of the variable x that are either 5 or less and evaluates as *false* for all other values of the variable x is

$$x < = 5$$

The range of values in which *both* expressions are true is between 1 and 5 (including the end points). The way to say this using the language of logic is to use the word AND to imply that *both* the expression ($x > = 1$) *and* the expression ($x < = 5$) must be true for the combined expression to be true:

$$(x > = 1) \text{ AND } (x < = 5)$$

Given this background information, your assignment is to do the following:

- Write another conditional expression that is *false* for all values in the range 1 to 5 (including end points) and *true* for all other values of x
- Test your expression for several values to confirm that it works correctly.

Relational Operators

Recall that the simplest arithmetic expression may be just a single constant or a variable that has been assigned a numeric value. More complex expressions combine constants and variables with arithmetic operators such as $+$, $-$, $*$, $/$, and $\wedge$. An arithmetic expression will evaluate to a single value.

 Relational operators are used to compare two arithmetic expressions and determine whether the value of the first is less than, equal to, or greater than the value of the second.

Less than, Equal to, Greater than
Consider the following statements:

$3 < 5$	Reads: "Three is less than five."
$5 = 5$	Reads: "Five is equal to five."
$5 > 3$	Reads: "Five is greater than three."

The symbols that appear between the numbers are called relational operators. The three basic operators are:

$=$	Equal to
$>$	Greater than
$<$	Less than

Number Line Comparisons:

- For two numbers to be **equal to** one another, they must be **at exactly the same point** on the number line.

- For a number to be **greater than** another, it must be to the **right** of the other number on the number line.

- For a number to be **less than** another, it must be to the **left** of the other number on the number line.

Opposite Operators
In addition there are three other relational operators that are the logical opposites of those already mentioned:

- If a quantity is not equal to another quantity, it must be *either* less than the other quantity *or* greater than the other quantity. The new symbol $<>$ is used to denote this.

  $<>$ Not equal to is the opposite of the equal operator.

- If a quantity is not greater than another quantity, it must be *either* less than the other quantity *or* equal to the other quantity.

  $<=$ Less than or equal to is the opposite of the greater than operator.

- If a quantity is not less than another quantity, it must be *either* greater than *or* equal to the other quantity.

  $>=$ Greater than or equal to is opposite of the less than operator.

Note: Expressions that use these opposite operators must satisfy *two* conditions. To be true, *either* condition may be true. However, to be false, *both* conditions must be false.

Opposite Operators Using the Number Line

Pick some number, such as the number 5. Divide the number line into two sections at 5 so that

- The left section includes numbers *less than or equal to 5*.

- The right section includes numbers *greater than 5*.

less than or equal to 5   ←|→   greater than 5

Note: The parenthesis after the 5 on top of the number line indicates that 5 is included in the left section.

Consider the following conditional statement using some variable a and the greater than operator:

$$a > 5$$

This expression will evaluate as true for any value of variable a that is to the right of 5 on the number line. (All of these numbers are in the *right section*).

The opposite relational operator would be the less than or equal to operator:

$$a <\, = 5$$

This expression will evaluate as true for any value of variable a that is either exactly 5 or to the left of 5 on the number line. (All of these numbers are in the *left section*.)

 Opposite operators correspond to opposite sections of the number line.

Summary of Relational Operators

The six operators presented in this section are often encountered in computer programs. An alternative computer notation, less often encountered, makes use of letters to state the relationship. In addition, another notation is used in mathematics. All of these are summarized in the following table.

Relational Operators	Computer Symbol	Computer Example	Math Symbol	Math Example
Equal to	= EQ	5 = 5 5 EQ 5	=	5 = 5 a = b
Greater than	> GT	5 > 4 5 GT 4	>	5 > 4
Less than	< LT	3 < 4 3 LT 4	<	3 < 4
Not equal to	< > NE	3 < > 5 3 NE 5	≠	5 ≠ 4
Less than or equal to	< = LE	3 < = 4 3 LE 5 3 LE 3	≤	3 ≤ 4 or 3 ≤ 3
Greater than or equal to	> = GE	3 > = 1 3 GE 1 3 GE 3	≥	3 ≥ 1 or 3 ≥ 3

Practice Problems

Use the *computer symbols* for relational operators given in the previous table to write a conditional expression for each of the following conditions. The expression should be true for all values of the variable *x* that meet the condition and false for all other values.

8-2.1 Above 4

8-2.2 Below 4

8-2.3 4 and above

8-2.4 4 and below

8-2.5 Equal to 4

8-2.6 Not equal to 4

Conditional Expressions

A conditional expression can be evaluated as either true or false. No other outcome is possible. This type of expression is also referred to as either a Boolean expression (George Boole developed the algebra of logic in the 19th century) or a logical expression.

▶ Examples Using Numbers

5 = 6	evaluates to false	(Because 5 and 6 are not of equal value)
5 < 6	evaluates to true	(Because 5 is to the *left* of 6 on the number line)
−5 < 2	evaluates to true	(Because −5 is to the *left* of 2 on the number line)
5 < = 6	evaluates to true	(Although 5 is not equal to 6, 5 *is* less than 6.)
5 < = 5	evaluates to true	(Although 5 is not less than 5, the two values *are* equal.)
6 < = 5	evaluates to false	(Both the < condition and the = condition are false.)

▶ Examples Using Arithmetic Expressions

(5 + 3) < 10	evaluates to true	(Because 8 is less than 10)
(2 / 10) < = (3 / 18)	evaluates to false	(Because 1 / 5 is greater than 1 / 6)
(3 + 4 + 1) / (1 + 2 + 1) = 2	evaluates to true	(Because 8 / 4 is equal to 2)

▶ Examples Using Variables

When variables are included in a conditional expression, they must have values assigned before the expression can be evaluated. For example, when

$$\boxed{4} \quad\quad \boxed{2}$$
$$\text{a} \quad\quad\quad\quad \text{b}$$

a > b	evaluates to true	(Because 4 is greater than 2)
a > = b	evaluates to true	(Because 4 is greater than 2)
a = b	evaluates to false	(Because 4 is not equal to 2)

Practice Problems

Evaluate the following conditional expressions as either *true* or *false* for the data values given.

Introductory Problems

8-2.7 (x < 4)	when x = 5
8-2.8 (x < 4)	when x = 4
8-2.9 (x < 4)	when x = 3
8-2.10 (x < 4)	when x = −3
8-2.11 (x − 5) > = 2	when x = 10

8-2.12 $(x - 5) > = 2$ when x = 8

8-2.13 $(x - 5) > = 2$ when x = 7

8-2.14 $(x - 5) > = 2$ when x = 6

8-2.15 $(x - 5) > = 2$ when x = -6

Intermediate Problems

8-2.16 $(x * 3) > = (x * x)$ when x = 2

8-2.17 $(x * 3) > = (x * x)$ when x = 3

8-2.18 $(x * 3) > = (x * x)$ when x = 4

8-2.19 $(x * 3) > = (x * x)$ when x = -2

8-2.20 $(x * 3) < = (x * x)$ when x = 2

8-2.21 $(x * 3) < = (x * x)$ when x = 3

8-2.22 $(x * 3) < = (x * x)$ when x = -1

Logical Operators

Logical operators translate the logical input of one or more conditional expressions, according to specific rules, to produce a logical output. Each operator has an associated truth table that specifies the various logical outputs for all combinations of the logical inputs. The most frequently encountered logical operators are NOT, AND, and OR.

The NOT Operator

The NOT operator negates the logical outcome of the expression that follows it. If the expression that follows is true, NOT makes it false. If the expression that follows is false, NOT makes it true. Here is a truth table for the NOT operator:

Condition A	NOT (A)
True	False
False	True

The NOT operator provides a way to reverse the logical outcome of any conditional expression. Consider a variable named *more_data* that is set to true when there is more data to process and set to false when all the data has been processed. The expression NOT (more_data) will be true when all the data has been processed and false otherwise.

However, it should be pointed out that many people in the information technology industry find the NOT operator confusing and try whenever possible to avoid it. Luckily it is always possible to remove a NOT from a conditional expression and replace it with an expression that is logically equivalent. This will be covered a little later in this chapter in the section called Removing NOTs.

▶ *Examples Using NOT.* Each of the following combined expressions contains a NOT operator. The expressions are evaluated in a step-by-step manner (line by line) until a T (true) or F (false) result is obtained.

1. Evaluate NOT $(5 > 6)$ Evaluate $(5 > 6)$ as F.
   NOT (F)   Use the truth table to evaluate NOT (F) as T.
   T   Thus, the combined expression is **true.**

2. Evaluate NOT $(x < 1)$ Evaluate the combined expression when x = 0.
   NOT (T)   Evaluate $(0 < 1)$ as T.
   F   Use the truth table to evaluate NOT (F) as T.
                              Thus, the combined expression is **false.**

3. Evaluate NOT (x + y = z)

  NOT (T)

  F

Evaluate the expression when $x = 1$, $y = 2$, and $z = 3$.
Evaluate $(1 + 2 = 3)$ as T.
Use the truth table to evaluate NOT (T) as F.
Thus, the combined expression is **false.**

4. Evaluate NOT (x > z)

  NOT (F)

  T

Evaluate the expression when $x = 1$ and $z = 3$.
Evaluate $(1 > 3)$ as F.
Use the truth table to evaluate NOT (F) as T.
Thus, the combined expression is **true.**

The AND Operator

The AND operator always appears between *two* conditional expressions. Let the capital letters A and B represent conditional expressions that can be evaluated as either true or false.

  A AND B

The combined expression A AND B will evaluate as true only when A and B are both true. In all other cases, the combined expression A AND B will evaluate as false.

Since the two variables each have two possible outcomes, there are four possible outcomes for the combined expression. An effective way to display all of the possible logical outcomes of a logical operator is by using a truth table. Here is a truth table for the AND operator:

Condition A	Condition B	A AND B
True	True	True
True	False	False
False	True	False
False	False	False

▶ *Examples Using AND.* Each of the following combined expressions contains an AND operator. The expressions are evaluated in a step-by-step manner (line by line) until a T (true) or F (false) result is obtained.

1. Evaluate (5 < 6) AND (6 = 5)

     T   AND   F

          F

Evaluate the given combined expression.
Evaluate $(5 < 6)$ as T; evaluate $(6 = 5)$ as F.
Use the truth table to evaluate (T AND F) as F.
Thus, the combined expression is **false.**

2. Evaluate (x < 1) AND (y > 1)

     T   AND   T

          T

Evaluate the expression when $x = 0$ and $y = 2$.
Evaluate $(0 < 1)$ as T; evaluate $(2 > 1)$ as T.
Use the truth table to evaluate (T AND T) as T.
Thus, the combined expression is **true.**

3. Evaluate (x + y = z) AND (x < y)

     T   AND   T

          T

Evaluate the expression when $x = 1$, $y = 2$, and $z = 3$.
Evaluate $(1 + 2 = 3)$ as T; evaluate $(1 < 2)$ as T.
Use the truth table to evaluate (T AND T) as T.
Thus, the combined expression is **true.**

4. Evaluate (x + y = z) AND (x < y) | Evaluate the expression when $x = 2$, $y = 1$, and $z = 3$.

     T   AND   F

              F

Evaluate $(2 + 1 = 3)$ as T; evaluate $(2 < 1)$ as F.
Use the truth table to evaluate (T AND F) as F.
Thus, the combined expression is **false.**

5. Evaluate (x < z) AND (x = y)

     T   AND   F

         F

Evaluate the expression when $x = 1$, $y = 2$, and $z = 3$.
Evaluate $(1 < 3)$ as T; evaluate $(1 = 2)$ as F.
Use the truth table to evaluate (T AND F) as F.
Thus, the combined expression is **false.**

The OR Operator

The OR operator always appears between *two* conditional expressions. Let the letters A and B represent conditional expressions that can be evaluated as either true or false. The combined expression A OR B will evaluate as true when either A or B evaluate as true. Only when both conditions A and B evaluate as false will the combined expression evaluate as false. Here is a truth table for the OR operator:

Condition A	Condition B	A OR B
True	True	True
True	False	True
False	True	True
False	False	False

▶ *Examples Using OR.* Each of the following combined expressions contains an OR operator. The expressions are evaluated in a step-by-step manner (line by line) until a T (true) or F (false) result is obtained.

1. Evaluate (5 < 6) OR (6 = 5)

     T   OR   F

        T

Evaluate the given combined expression.
Evaluate $(5 < 6)$ as T; evaluate $(6 = 5)$ as F.
Use the truth table to evaluate (T OR F) as T.
Thus, the combined expression is **true.**

2. Evaluate (x < 1) OR (y > 1)

     T   OR   T

        T

Evaluate the expression when $x = 0$ and $y = 2$.
Evaluate $(0 < 1)$ as T; evaluate $(2 > 1)$ as T.
Use the truth table to evaluate (T OR T) as T.
Thus, the combined expression is **true.**

3. Evaluate (x + y = z) OR (x < y)

     T   OR   T

        T

Evaluate the expression when $x = 1$, $y = 2$, and $z = 3$.
Evaluate $(1 + 2 = 3)$ as T; evaluate $(1 < 2)$ as T.
Use the truth table to evaluate (T OR T) as T.
Thus, the combined expression is **true.**

4. Evaluate (x > z) OR (x = y)

     F   OR   F

        F

Evaluate the expression when $x = 1$, $y = 2$, and $z = 3$.
Evaluate $(1 > 3)$ as F; evaluate $(1 = 2)$ as F.
Use the truth table to evaluate (F OR F) as F.
Thus, the combined expression is **false.**

Practice Problems

Use the step-by-step method shown in the previous examples to evaluate the following conditional expressions as either true (T) or false (F), when $x = 1$ and $y = 2$.

8-2.23 x = 1 OR y = 2

8-2.24 x = 1 OR y = 3

8-2.25 x = 3 OR y = 2

8-2.26 x = 3 OR y = 3

8-2.27 x = 1 AND y = 1

8-2.28 x = 1 AND y = 2

8-2.29 x = 3 AND y = 2

8-2.30 NOT (x = 2)

8-2.31 NOT (x = 1)

8-2.32 NOT (x = 1 AND y = 2)

8-2.33 NOT (x = 3 OR y = 4)

Combined Expressions

Now that the various components that may be part of a conditional expression have been introduced, expressions that combine these components will be discussed. Conditional expressions may include.

- Arithmetic expressions (variables, constants, arithmetic operators $(+, -, *, /, ^\wedge)$
- Relational operators $(<, =, >,$ etc.)
- Logical operators (NOT, AND, OR)

Order of Precedence

When both the AND operator and the OR operator are included in an expression, the order in which the parts of the expression are evaluated *may*, in certain cases, change the logical outcome. For example, using T for true and F for false, the expression F AND F OR T will yield a different logical outcome when the AND is evaluated first than it will when the OR is evaluated first.

This can be demonstrated using parentheses to indicate which operator is evaluated first:

Evaluate the AND first:	(F AND F) OR T	= F OR T	= T
Evaluate the OR first:	F AND (F OR T)	= F AND T	= F

Just as it does for evaluating arithmetic, the computer uses an order of precedence for evaluating logic. This way every conditional expression gives a single unambiguous result.

Order of Precedence Rules for Logic

Conditional expressions are evaluated **left to right,** in the following order:
- Expressions within parentheses ( ), innermost first
- Expressions with NOT operators
- Expressions with AND operators
- Expressions with OR operators

Although conditional expressions may be unambiguous to the computer, they are often confusing to humans who do not always apply the rules of precedence consistently.

 To avoid confusion, use parentheses to make your conditional expressions perfectly clear. In certain cases parentheses are *required* to avoid errors. (See "Removing NOTs" at the end of this section.)

Evaluating Combined Expressions

A combined conditional expression, no matter how complex, evaluates to only one of two possible logical outcomes: true or false. The logical evaluation of a conditional expression proceeds as a series of steps:

1. Assign values.

   Determine the value assigned to each variable.

2. Add parentheses.

   Break the combined expression into subexpressions using parentheses.

   • Identify arithmetic expressions on either side of relational operators.

   • Identify relational expressions or either side of logical operators.

   • Leave the grouping of logical operations until step 5.

   • Use nested parentheses for expression within expressions.

3. Do arithmetic.

   Replace each arithmetic expressions with its arithmetic outcome.

4. Evaluate relational expressions.

   Evaluate the subexpressions that include *relational* operators.

   Replace each relational expression with its logical outcome (T or F).

5. Evaluate logical expressions.

   Work through the remaining logical subexpressions one by one according to the following rules for order of precedence. Remember, work left to right when an operator occurs more than once.

   > Expressions in parentheses
   >
   > Then, NOT
   >
   > Then, AND
   >
   > Finally, OR

   As you do each one:

   • Indicate which subexpression is being evaluated.

   • Replace that expression with its logical outcome (T or F).

6. Get Result.

   After all levels of subexpressions have been evaluated, the logical outcome of the combined expression will be either T or F.

Note: Like the IPO worksheet, this step-by step procedure should be followed carefully as you are learning. Once learned, you can use it as a mental checklist and take shortcuts by doing some steps in your head.

▶ Simple Example

Using the previous step-by-step rules, evaluate each component of the conditional expression:

a < 3 AND (a = 2 OR b = 2)

Note: The logical result for the same expression depends on the values assigned. Gray shading is used to indicate the next logical subexpression being evaluated.

1. a < 3 AND (a = 2 OR b = 2)
    Assign values.
    Add parentheses.
    Evaluate arithmetic.
    Evaluate relational expressions.
    Evaluate logical expressions.
        Parentheses First
        AND next
    Get result.

    For a = 2 and b = 1
    2 < 3 AND (2 = 2 OR 1 = 2)
    (2 < 3) AND ((2 = 2) OR (1 = 2))
    None needed
    T AND (T OR F)

    T AND (T OR F) = T AND T
    T AND T = T
    TRUE

2. a < 3 AND (a = 2 OR b = 2)
    Assign values.
    Add parentheses.
    Do arithmetic.
    Evaluate relational expressions.
    Evaluate logical expressions.
        Parentheses first
        AND next
    Get result.

    For a = 0 and b = 0
    0 < 3 AND (0 = 2 OR 0 = 2)
    (0 < 3) AND ((0 = 2) OR (0 = 2))
    None needed
    T AND (F OR F)

    T AND (F OR F) = T AND F
    T AND F = F
    FALSE

3. a < 3 AND (a = 2 OR b = 2)
    Assign values.
    Add parentheses.
    Do arithmetic.
    Evaluate relational expressions.
    Evaluate logical expressions.
        Parentheses first
        AND next
    Get result.

    For a = 1 and b = 1
    1 < 3 AND (1 = 2 OR 1 = 2)
    (1 < 3) AND ((1 = 2) OR (1 = 2))
    None needed
    T AND (F OR F)

    T AND (F OR F) = T AND F
    T AND F = F
    FALSE

4. a < 3 AND (a = 2 OR b = 2)
    Assign values.
    Add parentheses.
    Do arithmetic.
    Evaluate relational operators.
    Evaluate logical operators.
        Parentheses first
        AND next
    Get result.

    For a = 3 and b = 2
    3 < 3 AND (3 = 2 OR 2 = 2)
    (3 < 3) AND ((3 = 2) OR (2 = 2))
    None needed
    F AND (F OR T)

    T AND (F OR F) = T AND F
    T AND F = F
    FALSE

▶ Combined Example

Use the step-by-step method outlined earlier to evaluate the following combined expression:

NOT x * y / z < z AND z > = 2 OR z = 0

for the values x = 1, y = 2, and z = 3.

Step-by-step solution:

1. Assign values.

   NOT 1 * 2 / 3 < 3 AND 3 > = 2 OR 3 = 0

2. Add parentheses.

   NOT (1 * 2 / 3 < 3) AND (3 > = 2) OR (3 = 0)

3. Do arithmetic.

   NOT (2 / 3 < 3) AND (3 > = 2) OR (3 = 0)

4. Evaluate relational expressions.

   NOT (T) AND T OR F

5. Evalute logical expressions.

   Evaluate the NOT operator first:

   NOT (T) AND T OR F
   F AND T OR F

   Evaluate the AND operator next:

   F AND T OR F
   F OR F

   Evaluate the OR operator last:

   F OR F
   F

6. Get result.

   For the given data, the entire expression evaluates to **FALSE.**

Simple Shortcuts

All ANDs. When a combined expression involves several AND operators (and no OR or NOT operators), each part of the expression must be true for the combined expression to be true. Should any part of the expression be false, then the entire combined expression will be false. Therefore, once you discover part of the expression to be false, there is no need to look any further; you know the combined expression will be false.

 It takes only one false part to make the all AND's combined expression false.

All ORs. When a combined expression involves several OR operators (and no AND or NOT operators), only one part of the expression needs to be true for the combined expression to be true. Therefore, once you discover part of the expression to be true, there is no need to look any further; you know the combined expression will be true.

 It takes only one true part to make the all OR's combined expression true.

Practice Problems
Evaluate the following conditional expressions in a step-by-step manner as shown in the previous examples when $x = 1$ and $y = 2$.

8-2.34 x $<$ 3 AND x $>$ 0 AND y $>$ = 1

8-2.35 x $<$ 3 AND x $>$ 0 AND y $>$ 2

8-2.36 x = 1 OR x = 2 OR x = 3

8-2.37 x = 2 OR y = 2 OR y = 3

8-2.38 x = 1 AND y = 2 OR y = 3

8-2.39 x = 2 OR x = 3 AND y = 2

Intermediate Problems

8-2.40 (x + 1) / 2 $>$ 0 AND y $-$ 2 = 0 OR y = 0

8-2.41 y $-$ 3 = 0 OR x $-$ 3 = 0 AND y = 2

8-2.42 (x $-$ 1) * y $<$ 1 AND y $>$ 0 OR x $>$ 1

8-2.43 NOT (x = 1 AND y $-$ 2 = 0)

8-2.44 NOT (x = 1) AND y = 2

8-2.45 NOT (x = 3) AND NOT (x = 4) AND NOT (x = 5)

8-2.46 x $<$ $>$ 3 OR x $<$ $>$ 4 OR x $<$ $>$ 5

8-2.47 NOT ((x $-$ 2) $>$ = 1 OR (x $-$ 1) $<$ $>$ 0 AND NOT (x = 1))

Removing NOTs

The NOT operator makes conditional expressions confusing and difficult to understand. Therefore, it is best to avoid writing expressions that include them. Sometimes, however, it is convenient to flip the logical outcome of an entire expression by placing a NOT in front of it. Then, as a second step, the expression can be rewritten in an equivalent form without the NOT. Regardless of the complexity of the original expression, or the number of NOT operators it contains, it is always possible to translate it into an equivalent expression that has the same logical outcome but is free of NOT operators. This section shows you how to do this translation.

Flipping Relational Operators

It is easy to remove NOT operators from in front of any conditional expression when the only operators are *relational operators* ($<$, $=$, $>$, etc.). Each of the six relational operators has a corresponding operator with the opposite logical outcome, as shown in the following table.

NOT (Expression)	Equivalent Expression
NOT (a $<$ b)	a $>$ = b
NOT (a = b)	a $<$ $>$ b
NOT (a $>$ b)	a $<$ = b
NOT (a $<$ = b)	a $>$ b
NOT (a $>$ = b)	a $<$ b
NOT (a $<$ $>$ b)	a = b

Note: In the table, *a* and *b* represent any *arithmetic* expression.

The following procedure will remove the NOT operator from in front of any conditional expression containing only *relational* operators. (Should the expression also contain *logical* operators, use the procedure given later in this section.)

1. Remove the NOT operator.

2. Flip each relational operator the NOT acts on, using the previous table.

Examples are presented in the following section.

▶ Examples of Flipping Relational Operators

Using the rules for flipping relational operators given, rewrite each of the following conditional expressions in an equivalent form that excludes the NOT operator yet evaluates to the same logical outcome as the original expression.

 The examples evaluate the expressions for $a = 4$ and $b = 2$. However, both expressions will have the same logical outcome for *any* other values of variables a and b as well. This is left for the student to show.

T represents *true* and F represents *false*.

1. Original expression: NOT (a < b)   For a = 4, b = 2; evaluates to NOT (F) = T

   Equivalent expression: a > = b   For a = 4, b = 2; evaluates to T

   Note: The equivalent expression removes the NOT and at the same time replaces < with its negative operator > =, thus keeping the logical outcome the same as the original.

2. Orginal expression: NOT (a = b)   For a = 4, b = 2; evaluates to NOT (F) = T

   Equivalent expression: a < > b   For a = 4, b = 2; evaluates to T

3. Original expression: NOT (a > b)   For a = 4, b = 2; evaluates to NOT (T) = F

   Equivalent expression: a < = b   For a = 4, b = 2; evaluates to F

4. Orginal expression: NOT (a < = b)   For a = 4, b = 2; evaluates to NOT (F) = T

   Equivalent expression: a > b   For a = 4, b = 2; evaluates to T

5. Original expression: NOT (a > = b)   For a = 4, b = 2; evaluates to NOT (T) = F

   Equivalent expression: a < b   For a = 4, b = 2; evaluates to F

6. Original expression: NOT (a < > b)   For a = 4, b = 2; evaluates to NOT (T) = F

   Equivalent expression: a = b   For a = 4, b = 2; evaluates to F

Flipping Logical Operators

Removing NOT operators from conditional expressions containing both *logical* operators (AND, OR) and *relational* operators (<, =, >, etc.) can be accomplished using the relationships in the following table.

The table shows equivalent expressions with the logical operator flipped. It is based on a famous relationship of logical mathematics called De Morgan's theorem (see the sidebar on Boolean algebra near the end of Section 8-2). Notice that the equivalent expressions on the right of the table still contain NOT operators. Additional flips are needed to remove these. This is handled by the procedure that follows.

NOT (Expression)	Equivalent Expression
NOT (A AND B)	(NOT A) OR (NOT B)
NOT (A OR B)	(NOT A) AND (NOT B)

Note: A and B represent any *conditional* expression containing relational operators. Flipping AND to OR (and vice versa) does *not* imply that AND is the logical opposite of OR.

The following procedure can be used to implement the relationships in the table. However, the procedure goes one step further. By flipping both logical and relational operators, an expression logically equivalent to the original is obtained that is free of NOT operators.

1. Remove the NOT.

2. Flip each *logical* operator acted on by that NOT.

      Each AND is replaced by OR.

      Each OR is replaced by AND.

3. Flip each *relational* operator acted on by that NOT.

Notes:

- All parentheses in the original expression must be maintained in the equivalent expression. See example 8 on page 276 for more details.

- Should the original expression contain more than one NOT operator, apply the procedure once for each, starting with the innermost.

▶ A simple example:

Start with an original expression containing a NOT operator:

$$NOT(x \geq 2 \text{ AND } x \leq 3)$$

Apply three steps: Remove the NOT, flip *logical* operators, flip *relational* operators.

$$(x < 2) \text{ OR } (x > 3)$$

The final expression is logically equivalent to the original, but omits the NOT operator.

▶ **Examples of Flipping Logical Operators**

Using the rules for flipping logical and relational operators, rewrite each of the following conditional expressions in an equivalent form that excludes NOT operators and yet evaluates to the same logical outcome as the original. As shown in the previous example, all the flips will be done in a single step.

 The examples evaluate the expressions for $a = 4$ and $b = 2$. However, both expressions should have the same logical outcome for *any* other values of the a and b as well. This is left to the student to show.

T represents *true* and F represents *false*.

1. Original expression: NOT (a > b AND b = 2)

    For a = 4, b = 2; evaluates to NOT (T AND T) = NOT (T) F

    Equivalent expression: (a $\leq$ b OR b $<>$ 2)

    For a = 4, b = 2; evaluates to F OR F = F

2. Original expression:
NOT (a $\leq$ 4 AND b $\geq$ 2)

    For a = 4, b = 2; evaluates to NOT (T AND T) = NOT (T) = F

    Equivalent expression: (a > 4 OR b < 2)

    For a = 4, b = 2; evaluates to F OR F = F

3. Original expression: NOT (a > 4 AND b < 2)

    For a = 4, b = 2; evaluates to NOT (F AND F) = NOT (F) = T

    Equivalent expression: (a $\leq$ 4 OR b $\geq$ 2)

    For a = 4, b = 2; evaluates to T OR T = T

4. Original expression: NOT (a < 1 AND a > 4) For a = 4, b = 2; evaluates to
NOT (F AND F)
= NOT (F) = T

  Equivalent expression: (a > = 1 OR a < = 4) For a = 4, b = 2; evaluates to
T OR T = T

5. Original expression:
NOT (a < = 4 AND NOT b > = 2) For a = 4, b = 2; evaluates to
NOT (T AND NOT(T))
= NOT (T AND F)
= NOT (F) = T

  Equivalent expression:

    First remove innermost NOT: NOT (a < = 4 AND b < 2)

    Then remove remaining NOT: (a > 4 OR b > = 2)

For a = 4, b = 2; evaluates to
F OR T = T

In the following examples the NOT operator applies to only *part* of the expression.

6. Original expression: NOT (a < = 4) AND NOT (b > = 2)

Note: The first NOT applies only to (a < = 4) and the second NOT applies only to (b > = 2). Neither of the NOT operators apply to the AND operator.

For a = 4, b = 2; evaluates to
NOT (T) AND NOT (T)
= F AND F
= F

  Equivalent expression: (a > 4) AND (b > = 2) For a = 4, b = 2; evaluates to
F AND T = F

7. Original expression: NOT (a < = 4) OR (b > = 2)

Note: The NOT applies only to (a < = 4). The NOT operator does not apply to the OR operator or to (b > = 2).

For a = 4, b = 2; evaluates to
NOT(T) OR (T)
= F OR T
= T

  Equivalent expression: (a > 4) OR (b > = 2) For a = 4, b = 2; evaluates to
F OR T = T

The following example requires *both* AND and OR operators to be flipped.

8. Original expression: NOT ((a < 4 AND b > = 2) OR a > 3)

For a = 4, b = 2; evaluates to
NOT ((F AND T) OR T))
= NOT (F OR T)
= NOT (T)
= F

The equivalent expression will be found by removing the NOT and flipping all the relational and logical operators at the same time. *Notice that parentheses are used to force the order of evaluation.*

Equivalent expression: $((a >= 4) \text{ OR } (b < 2)) \text{ AND } (a <= 3)$

For a = 4, b = 2; evaluates to
(T OR F) AND F = (T AND F)
= F

Beware: When both AND and OR appear in the original expression, it is essential to use parentheses to dictate which logical operator will be evaluated first. The equivalent expression must keep the parentheses in place to force the flipped operators to be evaluated in the same order as before. Relying on the default order of precedence may result in errors in some cases. For example, the values *a* = 4, *b* = 2 will produce errors when the default is used.

Aside: Boolean Algebra

In the 19th century, a mathematician named George Boole created an algebra of logic. Variables *a, b,* and *c* below may only be true or false:

a = 1 is true   a = 0 is false

a * b	The logical product of two variables is equivalent to the AND operator.
a + b	The logical sum of two variables is equivalent to the OR operator.
$\overline{a}$	Overscore indicates negation (the NOT operator).
$\overline{\overline{a}}$	Double overscore indicates double negation.

Postulates

a = 1 (if a ≠ 0)	a = 0 (if a ≠ 1)
0 * 0 = 0	0 + 0 = 0
1 * 1 = 1	1 + 1 = 1
1 * 0 = 0	1 + 0 = 1
$\overline{1} = 0$	$\overline{0} = 1$

Algebraic Properties

Commutative	a * b = b * a	a + b = b + a
Distributive	a * (b * c) = (a * b) * c	a + (b + c) = a + (b + c)
Associative	a * (b + c) = (a * b) + (a * c)	a + b * c = (a + b) * (a + c)

Theorems

a * 0 = 0	a + 0 = a
a * 1 = a	a + 1 = 1
a * a = a	a + a = a
a * $\overline{a}$ = 0	a + $\overline{a}$ = 1
$\overline{\overline{a}}$ = a	a = $\overline{\overline{a}}$

De Morgan's Theorem

$\overline{a} * \overline{b} = \overline{a + b}$	$\overline{a} + \overline{b} = \overline{a * b}$
$\overline{a} * \overline{b} * \overline{c} = \overline{a + b + c}$	$\overline{a} + \overline{b} + \overline{c} = \overline{a * b * c}$

Note: De Morgan's theorem is the basis for removing NOTs from a conditional expression.

Practice Problems

Rewrite the following conditional expressions without any NOT operators so that the logical outcome is unchanged for all values of the variable x. Then evaluate the original expression and the equivalent expression using the values $x = 1, y = 1, z = 1$. Compare the results.

Notes:

- The NOT operator applies to the part of the expression enclosed in parentheses that follows immediately.

- Evaluating the original and the equivalent expressions for the values given is only a partial check since both expressions must be equivalent for all values of the variables.

Introductory Problems

8-2.48 NOT $(x = 1)$ AND NOT $(x = 2)$

8-2.49 NOT $(x < 1)$ OR NOT $(x > 2)$

8-2.50 $(x > = 3)$ OR NOT $(x \neq 1)$

8-2.51 NOT $(x < = 4)$ AND $(x = 2)$

8-2.52 NOT $(x < 1)$ OR NOT $(x > 5)$

Intermediate Problems

8-2.53 NOT $(x < = 1$ OR $x > = 5)$

8-2.54 NOT $(x = 1$ OR $x = 2)$

8-2.55 NOT $(x > = 1$ AND $x < = 4)$

8-2.56 NOT $((x - 1) < 2$ AND $(x - 2) = 3)$

8-2.57 NOT $(x < > 2$ AND $x < > 3)$

More Difficult Problems. Be sure to apply the correct order of precedence.

8-2.58 NOT $(x = 1$ OR $x = 2$ OR $x = 3)$

8-2.59 NOT $(x < > 1$ AND $x < > 2$ AND $x < > 3)$

8-2.60 NOT $(x = 0$ OR $(y > = 1$ AND $z = 5))$

8-2.61 NOT $((x = 1$ OR $y = 1)$ AND $z = 1)$

Testing Ranges

One of the most important uses of conditional expressions is to determine if the value assigned to a variable is within a given range of values. For example, you may need to test if

Age is 21 or over

Page is between 15 and 32

Cost is $10 or more, but less than $100

Conditional expressions can be constructed for each of the above tests that will be

- True when the value of the variable is *within the range*

- False when the value is *outside the range*

For the condition "age is 21 or over", the conditional expression is

age > = 21

Any age equal to or greater than 21 will make this expression true. However, be aware that even an impossible age such as 9999 will give a true outcome.

For the condition "page is between 15 and 32", the conditional expression is

page > = 15 AND page < = 32

Notes:

- The variable *page* is repeated on both sides of the AND operator. This is because both sides must be valid conditional expressions.

- The left side is true for all pages 15 and over.
  The right side is true for all pages 32 and under.
  To be in range, both sides need to be *true*.
  This will make the entire expression *true*.

- The word *between* is interpreted to include the end points.

For the condition "cost of $10 or more, but less than $100", the conditional expression is

cost > = 10 AND cost < 100

Again the variable *cost* is repeated on both sides. Notice that 100 is not in the *true* range.

If the preceding condition is modified to say "cost is $10 or more, up to $100", the expression becomes

cost > = 10 AND cost < = 100

Now 100 is in the true range.

Confirming Range Expressions

Unlike conditional expressions in which the variable is assigned a specific value, range expressions must be true for all values in a given range and false for all values outside the range. To confirm that the expression is designed correctly, the statement must be evaluated several times, with a different value assigned to the variable each time.

Evaluate the range expression at the following values:

- A point below the lowest end point

- The lowest end point

- Some point between the end points
- The highest end point
- A point above the highest end point

End Points

Write a conditional expression that is true for all numbers between 10 and 20, *including* the end points, and false for all other values. This expression will do it:

number > = 10 AND number < = 20

Now, write a conditional expression that is true for all numbers between 10 and 20, *excluding* the end points, and false for all other values. Consider the following expression:

number > = 11 AND number < = 19

This will work if the variable *number* is assigned only integers. But the range specified *all* numbers. Will the decimal number 10.3 make the statement true? No, the statement is false for 10.3 because 10.3 is less than 11. The correct expression is

number > 10 AND number < 20

This statement will be true for numbers slightly larger than 10 or slightly smaller than 20, within the accuracy of the computer.

Testing for Numbers outside a Range

Suppose you are asked to design a conditional expression that is false for all values between 10 and 20 (including the end points) and true for all other values. It has already been shown that the expression

number > = 10 AND number < = 20

is true for values in the range 10 to 20 (including end points) and false for values outside this range. Therefore, the NOT operator can be used to flip the logical outcome so the expression is false for all values in the range and true for values outside the range:

NOT (number > = 10 AND number < = 20)

But, remembering the rule to avoid NOTs, it should now be removed. The equivalent expression, without the NOT operator, is

number < 10 OR number > 20

Notice that the NOT operator acts on all the operators within the parentheses. In order to remove the NOT, you must flip three operators to their corresponding operators.

number > = 10	flips to	number < 10
AND	flips to	OR
number < = 20	flips to	number > 20

All three flips are needed to produce an equivalent expression without a NOT. (Review the portion of Section 8-2 entitled "Removing NOTs" for additional details.)

Check the equivalent expression to make sure it will be true for all values outside 10 to 20 and false for values between (and including) 10 to 20.

Summary: It is often helpful to use a NOT operator to flip the logical outcome of an entire expression. However, once done, the NOT should be removed.

Practice Problems

Avoid using the NOT operator in your answers.

Introductory Problems. Write a conditional expression that will be *true* for all values of the variable *x* in the given range and *false* for all values outside the given range. Note that the variable *x* can take on any value, not just integer values.

8-3.1   Between 1 and 5 including end points $(x >= 1)$ and $(x <= 5)$

8-3.2   Between 1 and 5 excluding end points $(x > 1)$ and $(x < 5)$

8-3.3   Between 5 and 10 including end points

8-3.4   Between -1 and $+1$ including end points

8-3.5   Between 20 and 30 excluding end points

Intermediate Problems. Write a conditional expression that will be *false* for all values of the variable *x* in the given range and *true* for all values outside the given range. Note that the variable *x* can take on any value, not just integer values. (This is the reverse of the previous problems.) You may use a NOT operator to help arrive at your answer. However, the NOT operator must be removed from your final answer.

> **Note:** Although the following ranges are identical to those in problems 8-3.1 to 8-3.5, the problem statement is reversed.

8-3.6   Between 1 and 5 including end points

8-3.7   Between 1 and 5 excluding end points

8-3.8   Between 5 and 10 including end points

8-3.9   Between -1 and $+1$ including end points

8-3.10 Between 20 and 30 excluding end points

More Difficult Problems. The C programming language, as well and C++ and Java, use the following symbols to represent logical and relational operators:

&&	is used for AND	==	is used for equal to
‖	is used for OR	>	is used for greater than
!	is used for NOT	>=	is used for greater than or equal to
		<	is used for less than
		<=	is used for less than or equal to
		!=	is used for not equal to

8-3.11 The following conditional expression is written in the C language. It is true when the value of the variable *a* is outside the range of 20 to 30 (including end points) and false otherwise.

!(a < = 20 && a > = 30)

Which of the following conditional expressions is equivalent?

a) a > 20 && a < 30
b) a < 20 && a > 30
c) a > = 20 && a < = 30
d) a > 20 ‖ a < 30
e) a < = 20 ‖ a > = 30
f) a < 20 ‖ a > 30

8-3.12 The following conditional expression is written in the C language. It is false when variable *a* is assigned either 10 or 20 and true otherwise.

!(a = = 10 ‖ a = = 20)

Which of the following conditional expressions is equivalent?

a) a ! = 10 || b != 20
b) a < = 10 || b > = 20
c) a < 10 || a > 20
d) a = = 10 && a = = 20
e) a ! = 10 && a ! = 20
f) !(a = = 10) && !(a = = 20)

Boolean Variables and Flags

So far you have seen variables that may be used to store either numbers or text. But, variables can also be used to store logic. Such variables are called Boolean variables after George Boole, who developed the algebra of logic. Many computer languages include Boolean variables as a special variable type, reserved solely to store true/false logical outcomes.

Simulating Boolean Variables

Our simple programming language does not distinguish among different types of variables. However, a regular variable can be made to serve as a Boolean variable by assigning numeric values to represent logical outcomes. The numeric value zero will be used to represent *false* and the numeric value negative one will be used to represent *true*.[1]

To facilitate this, these two numeric values can be assigned to the variables *true* and *false* at the beginning of any program that requires Boolean variables:

```
LET true = −1
LET false = 0
```

The variables *true* and *false* provide a convenient way to assign and compare logic of Boolean variables throughout the program. Here is an example using the Boolean variable *more_data:*

```
LET more_data = true
IF more_data = true THEN . . .
```

Flags

One of the best uses of a Boolean variable is the flag. A flag is a programming technique that communicates an event recorded in one part of a program to another part of the program. It functions like the red flag on many rural postboxes, hence the name. The postbox flag is raised to indicate that there is mail to be picked up. Later the postman lowers the flag to indicate the mail has been picked up.

Within a program an event such as reading a sentinel data value (to indicate the end of a data list) needs to be communicated to the conditional statement in a loop so that the loop will stop reading data. Flags provide a convenient way to do this.

Two approaches to using flags will be shown. The first approach makes use of the variables *true* and *false* to assign and later compare the logical outcome of Boolean variables. The second approach shows how to do this more directly, without using the variables *true* and *false*.

---

[1] Computer languages such as BASIC and Visual Basic allow this, but only evaluate logic consistently when −1 is used to represent *true* rather than some other nonzero value. This is because numeric values are stored in 2's complement binary notation. Negative one (−1) has all bits set to 1. NOT (−1) flips all bits and evaluates as zero so that NOT (true) becomes false. Negative one is the only nonzero value that will do this.

Using Boolean Variables as Flags

Here are the steps involved in using a simulated Boolean variable as a flag.

1. The values −1 and zero are assigned to the variables *true* and *false*.

2. A Boolean variable serving as the flag is initialized to one of these values.

3. The flag is set to the opposite logical value after some event occurs.

4. The flag is tested in a conditional statement to determine if the event has occurred.

Let's see how this works.

1. At the beginning of the program, initialize variables *true* and *false* as follows:

   LET true = −1
   LET false = 0

   These variables are not flags, but contain the logical values for *true* and *false* that will be assigned later to the flag variable.

2. Next, assign an initial value to your flag variable. In this case we will create a flag variable named *more_data* that will be used to indicate whether more data remains to be processed. At the beginning of the program all of the data remains to be processed, so assign the variable *true* to the variable *more_data* (remember that the variable *true* contains −1).

   LET more_data = true

3. When data is to be processed in a loop, there needs to be a statement at the bottom of the loop to set the flag to false at the appropriate time. For, example, if your data list contains a sentinel value of 9999, the following statement will set the flag to false when 9999 has been supplied as input to the variable *num*:

   IF num = 9999 THEN LET more_data = false

4. Test the flag in the conditional statement at the top of a loop:

   DO WHILE more_data = true

▶ **Example of Using a Simulated Boolean Variable as a Flag**

```
'Initial values
      LET sum = 0
      LET count = 0
      LET true = −1
      LET false = 0
      LET more_data = true

'Main processing
      INPUT num
      IF num = 9999 THEN LET more_data = false      'Assign false on 9999
      DO WHILE more_data = true                      'Loop while flag is true
            LET sum = sum + num
            LET count = count + 1
            INPUT num
            IF num = 9999  THEN LET more_data = false   'Assign false on 9999
      LOOP

'Final processing
      LET average = sum / count
      OUTPUT average
```

Note: The statements highlighted in gray will be changed in the next example, using direct assignment and comparison of logic. Also note that comment statements (preceded by a single quote) explain what is happening.

Direct Assignment of Logic

Assignment statements can assign the logical outcome of a conditional statement *directly* into a Boolean variable, without using the variables *true* and *false*.

Rather than

IF num = 9999 THEN LET more_data = false,

the logic can be assigned directly:

LET more_data = num < > 9999

This more streamlined statement will assign *false* to the flag *more_data* when *num* becomes 9999; otherwise, it assigns *true*. There are two reasons why this is simpler. First, there is no need to use the variables *true* and *false*. Second, there is no need to initialize *more_data* since the assignment happens in either logical case. (Recall that the IF statement assigned false only when the condition was true.)

However, notice that the conditional statement was reversed from that used with the IF statement:

- With the IF statement: num = 9999

- With direct assignment: num < > 9999

Direct Comparison of Logic

There is an alternate approach for logical comparisons, which does not use *true* and *false*. In the previous example, statements that compared flags to *true* and *false* were used:

IF more_data = true THEN . . .
DO WHILE more_data = true

Using direct comparison, these statements are simplified to

IF more_data THEN . . .
DO WHILE more_data

Both the true/false form and the direct form evaluate logically in an identical way. However, the statements that use direct comparison are easier to read.

▶ Example with Direct Assignment and Comparison

Modify example program 5 from Section 7-5 to include the *more_data* flag. Notice the comment statements (preceded by a single quote) that explain what is happening. Comment statements are not executable and are added only for readability.

```
'Initial values
    LET sum = 0
    LET count = 0                          'True and false variables gone

'Main processing
    INPUT num
    LET more_data = num < > 999            'Direct assignment of logic
    DO WHILE more_data                     'Direct comparison of logic
        LET sum = sum + num
        LET count = count + 1
        INPUT num
        LET more_data = num < > 999        'Direct assignment of logic
    LOOP

'Final processing
    LET average = sum / count
    OUTPUT average
```

Naming Boolean Variables

Boolean variables must be named carefully. Names must reflect the true state of the logic they carry. For example, the variable *more_data* is intended to be true while there is more data and false when the last data item has been read.

Suppose you want to use a Boolean variable that will be true after the last data item has been read. The name *more_data* is not appropriate in this case. You might consider the name *no_more_data*. However, stating this in a negative fashion is best avoided since it often leads to confusion. The name *end_of_data* is a more appropriate choice.

Here are two naming rules:

- Always use a name that reflects the *true* state of the Boolean variable.

- State the name positively and avoid negatives such as *no* or *not* in the name.

▶ Example with Multiple Boolean Variables

Modify the previous example program to validate the data before it is summed for the average. Valid data include values in the range 100 and 999, except for the values 500 and 700.

Rather than using a single lengthy conditional expression to test for all the conditions of valid data at once, separate tests are done. The logical outcome of each test is evaluated and the logical outcome is assigned to its own flag variable. For this example, two Boolean variables with positive names are needed:

    in_range   good_value

These variables are then combined logically into one flag variable named *valid_data*. In this case both in_range and good_value must be true for the data to be valid, so the AND operator is used:

    valid_data = in_range AND good_value

Then *valid_data* is used to direct the processing as follows:

- When *valid_data* is true, the data item *num* is counted and summed.

- When *valid_data* is false, the data item *num* is skipped.

The new statements to accomplish this are shaded in gray in the following example. Notice the comment statements (preceded by a single quote) that explain what is happening. Comment statements are not executable and are added only for readability.

```
'Initial values
      LET true = −1                    'Initial variable true to −1
      LET false = 0                    'Initial variable false to zero
      LET sum = 0
      LET count = 0
      LET valid_data = true            'Initial valid_data flag to −1
'Main processing
      INPUT num
      LET more_data = num < > 9999     'Assign zero on sentinel value
      DO WHILE more_data               'Test the flag for true or false

          LET in_range = num > = 100 AND num < = 999
          LET good_data = num < >500 AND num < > 600
          LET valid_data = in_range AND good_data
          IF valid_data THEN
                LET sum = sum + num
                LET count = count + 1
          END IF
```

```
        INPUT num
        LET more_data = num < > 9999    'Assign zero on sentinel value
LOOP
```

'Final processing
```
IF count > 0
    LET average = sum / count
    OUTPUT count
ELSE
    OUTPUT "There is no valid data to average"
END IF
```

Notes:

- Assigning a conditional expression to a variable causes the condition to be evaluated and its logical outcome to be stored in the variable.

- The final processing needs to be done conditionally to avoid division by zero when there is no valid data.

Practice Problems

Introductory Problems. Create suitable names for Boolean variables to be used as flags. The variables will be *true* for each of the following situations and *false* otherwise. State names in a positive way and avoid including the terms *not* or *no* as part of the name.

8-3.13 When there are more data items available for input

8-3.14 When there are no more data items available for processing

8-3.15 When the input is in range

8-3.16 When the input is out of range

8-3.17 When the variable *value* is not equal to 100

8-3.18 When the variable *value* is equal to 100

8-3.19 When an error has occurred in the processing

8-3.20 When no error has occurred in the processing

Intermediate Problems. For the following problems, write a LET statement that will assign the logical outcome of a conditional expression directly to a Boolean variable to be used as a flag. Notice that variable names are italicized. Assume that the variables *true* and *false* have been initialized appropriately.

8-3.21 Write a statement that sets the flag *valid_id* to *true* when *both* of the following conditions are true. Assume *valid_id* has been initialized as *false.*

- *id* is above 999

- *id* is below 10000

8-3.22 Write a statement that sets the flag *stop_processing* to *true* when *either* of the following conditions are true. Assume that *stop_processing* has been initialized as *false.*

- *end_of_data* is *true*

- *processing_error* is *true*

More Difficult Problems. For the following problems, write a LET statement that will assign the logical outcome of a conditional expression directly to a Boolean variable to be used as flag. Notice that variable names are italicized. Assume that the variables *true* and *false* have been initialized appropriately. The NOT operator is forbidden in the final answer.

8-3.23 Write a statement that sets the flag *valid_data* to *false* when *both* of the following conditions are true. Assume *valid_data* has been initialized as *true*.

- *large_data_set* is *true*

- *count* is greater than 100

Hint: Build an expression that evaluates to *true* for the conditions given. Temporally place a NOT operator in front of it to make the expression false for the given conditions. Finally, remove the NOT using the method given in Section 8-2.

8-3.24 Write a statement that sets the flag *ok_to_continue* to *false* when *either* of the following conditions are true. Assume *ok_to_continue* has been initialized as *true*.

- *error_count* is above zero

- *data_count* is above 1000

Hint: The hint given in the problem also applies here.

If large_data_set = true and Count >100
    Then LET valid_data = false

End IF

Chapter Summary

Arithmetic expressions can be evaluated as true or false depending on numeric evaluation:

- Zero evaluates as *false.*
- All other values evaluate as *true.*

Conditional expressions combine any of the following to evaluate to either *true or false:*

- Constants
- Variables
- Arithmetic expressions
- Relational operators
- Logical operators

Relational Operators

Relational Operator		*Logically Opposite Relational Operator*	
Greater than	>	Less than or equal to	< =
Less than	<	Greater than or equal to	> =
Equal to	=	Not equal to	< >

Logical Operators (where A and B are conditional expressions)

NOT A is true only when A is false.

For A AND B to be true, both A and B must be true.

For A OR B to be true, either A or B must be true.

Logically Equivalent Expressions (where *a* and *b* are variables. A and B are conditional expressions.)

NOT a < b is equivalent to a > = b.

NOT a < = b is equivalent to a > b.

NOT a = b is equivalent to a < > b.

NOT a > = b is equivalent to a < b.

NOT a > b is equivalent to a < = b.

NOT (A OR B) is equivalent to NOT(A) AND NOT(B).

NOT (A AND B) is equivalent to NOT(A) OR NOT (B).

Removing NOT Operators

- Remove the NOT and flip each relational and logical operator to which it applied.
- The resulting expression will be logically equivalent to the original.

Boolean or Flag Variables

- May be assigned logical values (either -1 for *true* or 0 for *false*)
- Are used to communicate processing events from one part of a program to another

Review Questions

1. Define all of the relational operators and give an example of each.
2. Which relational operators are the logical opposites of one another?
3. How does the math notation for relational operators differ from the computer notation?
4. What is a conditional expression? Give two alternative names for conditional expressions.
5. Define the logical operators AND, OR, and NOT. Make a truth table for each.
6. Describe how to evaluate conditional expressions that combine arithmetic expressions, relational operators, and logical operators.
7. When is the NOT operator useful and when is it not useful?
8. What are the rules for removing NOT operators from a conditional expression?
9. How do you write a conditional expression that will be true for values within a given numeric range and false for values outside that range?
10. How do you write a conditional expression that will be false for values within a given numeric range and true for values outside that range?
11. How do you test range expressions to confirm they are written correctly?
12. Define the terms *Boolean variable* and *flag*. How are they used? Give examples.
13. How can the logical outcome of a conditional expression be assigned directly to a flag?
14. How can the logic stored in a Boolean variable be compared directly?

Problem Solutions

Section 8-2
8-2.2 $x < 4$	8-2.4 $x < = 4$	8-2.6 $x < > 4$
8-2.8 False	8-2.10 True	8-2.12 True
8-2.14 False	8-2.16 True	8-2.18 False
8-2.20 False	8-2.22 True	8-2.24 True
8-2.26 False	8-2.28 True	8-2.30 True
8-2.32 False	8-2.34 True	8-2.36 True
8-2.38 True	8-2.40 True	8-2.42 True
8-2.44 False	8-2.46 True	

8-2.48 $(x < > 1)$ AND $(x < > 2)$ 8-2.50 $(x > = 3)$ OR $(x < > 1)$
8-2.52 $(x > = 1)$ OR $(x < = 5)$ 8-2.54 $(x < > 1)$ AND $(x < > 2)$
8-2.56 $((x - 1) > = 2)$ OR $((x - 2) < > 3)$
8-2.58 $(x < > 1)$ AND $(x < > 2)$
           AND $(x < > 3)$
8-2.60 $(x < > 0)$ AND $((y < 1)$ Parentheses for OR are required.
           OR $(z < > 5))$

Section 8-3
8-3.2 $(x > 1)$ AND $(x < 5)$ 8-3.4 $(x > = -1)$ AND $(x < = 1)$
8-3.6 $(x < 1)$ OR $(x > 5)$ 8-3.8 $(x < 5)$ OR $(x > 10)$
8-3.10 $(x < = 20)$ OR $(x > = 30)$ 8-3.12 e)
8-3.14 end_of_data   8-3.16 out_of_range
8-3.18 value_100   8-3.20 valid_processing
8-3.22 LET stop_processing = end_of_processing OR processing_error
8-3.24 LET ok_to_continue = (error_count < = 0)
           AND (data_count < = 1000)

CHAPTER

9

Structured Program Design

Don't lose your confidence if you slip, be grateful for a pleasant trip, pick yourself up, dust yourself off and start all over again.
—Dorothy Fields lyrics from the 1936 film *Swing Time*

9-1 ■ Introduction

How to Use This Chapter

Now that you have an understanding of the basic concepts of computer programming and are able to write some very simple programs, it is time to consider more ambitious programs. However, the nature of algorithms is that as they grow in length; they can quickly become complicated and confusing. To overcome this confusion, you will need some programming design tools to help you see where you are going as the design of your program progresses.

Background

The process of designing and coding computer programs is fraught with peril. Unless great care is taken, the resulting programs will usually have errors that prevent them from working at all. Or, if they do work, they will often have "bugs" that prevent them from carrying out the intended tasks correctly. Even if they do work correctly, they often are difficult to understand, making maintenance and modification extremely hard and sometimes impossible. To avoid these common pitfalls, programmers must employ systematic techniques and design tools to help them create error-free and maintainable programs.

Program Development Tools

You have already seen how structured English can be used to outline and organize the components of an algorithm, such as making tea, into different levels. In this chapter other design tools will be introduced.

IPO charts:	A way to display the input, processing, and output that make up a program.
Pseudocode:	A technique similar to structured English. However, pseudocode is more closely aligned to the computer language that will eventually be used to "code" the program. Pseudocode avoids much of the rigor and precise grammar (syntax) of a real programming

<table>
<tr><td></td><td>language. In this regard, pseudocode and the simple programming language introduced in Chapter 7 are much the same.</td></tr>
<tr><td>Flowcharts:</td><td>A graphic presentation of the statements in a program that clearly shows how one part of the process flows into another.</td></tr>
<tr><td>Structure charts:</td><td>A graphic presentation that shows how the different components (or modules) of a program relate to one another.</td></tr>
</table>

Program Design Concepts

In addition to tools to help design and plan your programs, a method called *structured programming* will be introduced to help you organize your programs. The structured programming approach employs three interrelated components:

Top-Down development:	Program development starts with a general (or top-level) solution to the problem. This is then broken down into more detailed (midlevel) steps through a process called *functional decomposition.* The midlevel steps are also broken down into still simpler (lower-level) steps. As the decomposition process continues, things become defined at lower and lower levels. Eventually, the detailed understanding of the problem is at a point that can be "understood" by a computer, and it is time to begin coding with a programming language.
Modular design:	Functional decomposition organizes the processing steps that make up a complex task into several groups of instructions called modules. Each module contains the instructions for a specific subtask that represents a small piece of the overall processing. Modules are related to one another in a hierarchical fashion. The lowest-level tasks combine to make midlevel tasks. These, in turn, combine to make higher-level tasks.
Structure theorem:	Programming statements are assembled into three basic control structures:

- Sequence structure

- Repetition structure

- Selection structure

The structure theorem shows that these three basic structures are all that are needed to write any program.

Programs that employ this *structured* approach are:

- Easier to code

- Easier to understand

- Easier to read

- Easier to maintain

Object-oriented design:	Programs that employ a graphical user interface, such as those that run under Microsoft Windows, have added yet another layer of complexity to the design process. To help deal with this complexity, a technique that employs program *objects* has become important. One characteris-

tic of objects is that they can be developed and tested independently of one another and are reusable in many different programs.

Chapter Objectives

In this chapter you will learn:

- How to write algorithms
- How to employ algorithm design tools such as IPO charts, pseudocode, flowcharts, and structure charts
- Some basic concepts of top-down design, modular design, structure theorem, and object-oriented design

Introductory Problem

Write an algorithm to sort a deck of any number of cards, given the following constraints:

- Each card in the deck has a number between 1 to 10. Duplicates are OK.
- Identify three spots on your desk where cards will be placed. These spots, whether having cards or being empty, will be called:

Unsorted deck	U (for short)
Intermediate deck	I (for short)
Sorted deck	S (for short)

Your algorithm may use only the following *three* permitted actions and *two* permitted conditions:

Permitted Actions:

MOVE CARD moves the top card from one deck to the top of another deck.

Example: MOVE CARD from I to S

Error: Attempting to move a card *from* an empty deck stops the processing.

SWAP CARDS exchanges top cards between one deck and another.

Example: SWAP CARDS between I and S

Swapping cards when one deck is empty will cause a MOVE CARD statement to the empty deck to be carried out instead.

Error: Swapping cards *between* two empty decks stops processing.

MOVE DECK moves all the cards from one deck to the top of another deck.

Example: MOVE DECK from I to U

It is permitted to move a deck *to* an empty deck. After the move, the cards of the first deck belong to the second deck and the first deck is empty.

Error: A MOVE DECK *from* an empty deck stops the processing.

Permitted Conditions:
Any of the previous actions may be performed conditionally using an IF statement or in a loop using a DO WHILE statement.

The *has cards* condition tests whether a specified deck has cards or is empty.

IF example: IF I has cards THEN
         MOVE DECK from I to U
   END IF

The condition will be true if the intermediate deck has one or more cards and will be false if the intermediate deck is empty.

LOOP example: DO WHILE U has cards
       some statements inside the loop
   LOOP

The *card* condition compares the values of the *top cards* of any two decks.

The top cards from two specified decks are compared using a relational operator ($=, <, >, <=, >=, <>$) as being true or false.

The value of a card in an empty deck will compare as *less than* a card in a deck that is not empty. The values of cards in two empty decks will compare as equal.

Example: IF I card $<$ U card THEN
         SWAP CARDS between I and U
   END IF

At the Start:
The deck is shuffled and placed on the spot reserved for the unsorted deck. At this time, the intermediate and sorted decks are empty.

Sort Objectives:
Using only the actions and conditions given previously and *avoiding any errors,* move all the cards originally in the unsorted deck to the sorted deck. After the sort, the cards in the sorted deck should be in *ascending order,* from the bottom up. Thus, the sorted deck will have the lowest value on the bottom and the highest value on the top.

Desk Check:
Verify your algorithm using a sample deck of five cards with duplicate numbers. Make sure that the algorithm stops the processing when the sorted deck contains all the cards in correct sorted order.

Hint:
First, write a procedure to move the lowest card in the unsorted deck to the sorted deck. Then repeat this "lowest card" procedure over and over until all the cards are in the sorted deck. Conditional IF statements may be needed to avoid errors.

Algorithms

An algorithm is a set of instructions that tells how to accomplish some process. If the instructions are telling a cook how to bake a cake, the algorithm is called a recipe. If the instructions are telling a computer how to produce an annual sales report, the algorithm is called a computer program. Notice, in the planning stage, that the output should be defined first and the process last.

The algorithms for baking a cake and producing a sales report have many things in common.

Output: Both must describe what the algorithm will accomplish.

> The recipe may produce a German chocolate cake.
>
> The program may produce an annual report that summarizes regional sales for the fiscal year just concluded.

Input: Both must describe the input needed to accomplish the process.

> The recipe tells the cook what raw materials to assemble before starting: 2 cups sugar, 2 cups flour, 2 eggs . . .
>
> The computer program must specify what raw data must be assembled before the report can be processed: January sales, February sales . . .

Process: Both must specify the step-by-step process needed to convert input into output.

> The recipe tells how to assemble the ingredients, how long to bake, and how to make and apply the frosting.
>
> The program tells how to read the input data, what calculations are needed to average the data, and how to present the processed information on the printed report.

In both cases, processing instructions must be ordered, detailed, and unambiguous.

> The recipe must make it clear that it is necessary to bake the cake and let it cool before the frosting is applied.
>
> The program must make it clear that the input data is to be arranged in records, with each record giving 12 months' worth of sales data for one region. The process must proceed as follows: input a record, process a record, output a record. The process must be repeated for seven regions. At the end a national summary of all regions is produced.

Audience of Algorithms

The complexity of algorithms varies considerably depending on the process being described and the audience being addressed. In the case of an algorithm for making a German chocolate cake, the entire recipe, including making the frosting, might require half a page in a cookbook. In the case of the algorithm for producing an annual sales report, the entire program might require 10 pages of instructions written in a computer language.

Is making a sales report 20 times as complex as making a cake? An argument could be made that making a cake is more complex than making a sales report. It is the audience that the algorithm addresses that makes the instructions for the report more complex.

The audience for the recipe is a person who has had some prior cooking experience. Thus, when the recipe states: "beat at high speed for 2 minutes," it is not necessary to explain what an electric beater is, how to plug it into an electrical outlet, and how to operate it. Nor is it necessary to explain how to accomplish the innate physiological actions of lifting the beater, placing it in the bowl, and turning it on.

The audience for the report algorithm is a computer, a machine with no intelligence at all except that imparted by its software. If that software is based on a language similar to the one presented in the previous two chapters, it should be clear that many instructions are needed to deal with an annual regional sales report.

Application to Information Technology

New Programming Languages

New "intelligent" languages are becoming available that simplify the programming process. Many of the instructions needed to make an annual regional sales report, for example, have been anticipated and programmed ahead of time. The *intelligence* of this software enables it to match the requests of a person wanting to produce the report with the preprogrammed routines that will actually instruct the computer how to produce the report.

In this case the "programmer" needs only to tell the computer what is wanted. For example, instructions might be reduced to the following:

Report Regional Sales with Heading "ABC Regional Sales for 2001",
using data for 7 regions, each for 12 months, end with summary totals.

The intelligent software determines *how* to accomplish the task and sends the appropriate instructions to the computer.

It is interesting to note that these new languages require many more instructions than does the simple programming language you are now learning. The extra complexity has been hidden from the person producing the report in much the same way that the complexity of physiological actions such as lifting the beater are hidden from the cook who is making the cake.

Algorithm Design Tools

Algorithms that will become computer programs require a good deal of planning outside the computer before they are coded using a computer language. Several tools are available to make the design stage easier. Three of these will be presented here:

- IPO charts
- Pseudocode
- Flowcharts

IPO Charts

You were first introduced to IPO charts in Chapter 1 where they were used as the basis for the IPO method of solving problems. We will now see IPO charts in their original context as a tool for planning the design of computer programs.

All computer programs can be viewed as being composed of three separate components:

Input:	External data that comes into the program at the time it is run
Process:	Steps needed to convert the input into the output
Output:	Processed information that comes out of the program

The IPO chart is the first step in the design process. It presents an overview in a three-column chart of the input, process, and output that will be needed by the program. The information in this chart will be expanded later into a complete algorithm via either pseudocode or flowcharts.

 Because the IPO chart is the initial definition of the program, it is sometimes referred to as a *defining diagram*.

Less Is More

The key thing about the IPO chart is that it is simple. Its purpose is to describe *what* the major input-process-output components of the program will be. It is not intended to describe *how* to incorporate these components in a specific program design. That is done later using pseudocode.

Input: The input section lists the carefully chosen variable names (descriptive nouns) that help us to understand the input data that will be assigned to them.

Process: The process section lists, in general terms, the major steps that will be needed to convert the input into the output. For the IPO chart to be successful, you must keep these steps simple. They are by no means a complete algorithm. Thus, you will not indicate here the details that will be needed later to describe how the program will operate. Do not give specific formulas. Do not indicate looping. Do not provide specific logical expressions. *At this stage less is better.* You may, however, indicate what is done first, middle, and last.

Output: The output section lists the variable names that will be displayed as output.

 The usefulness of the IPO chart relies on its simplicity. It only sketches the major components of *what* is needed. The details of *how* to implement this processing remains to be described using pseudocode or flowcharts.

A Few Examples

▶ *Example 1*

Compute the average of three numbers provided as input.

IPO Chart

Input	Process	Output
num1 num2 num3	Accept the three numbers Compute the sum Compute the average Display the average	average

The algorithm would look like this:

```
INPUT num1, num2, num3
LET sum = num1 + num2 + num3
LET average = sum / 3
OUTPUT average
```

Note: The input and output use the variable names (descriptive nouns) that will be used to store the data. These names should provide a meaningful description of the data they will hold. The process section indicates that the input is read first, the calculations are done next (although no formulas are given), and the output is done last.

► **Example 2**

Compute the area of a room given its length and width.

IPO Chart

Input	Process	Output
length width	Accept length and width Compute area Display area	area

The algorithm for this program would look like this:

```
INPUT length, width
LET area = length * width
OUTPUT area
```

► **Example 3**

Given the dimensions of the rooms, compute the area of a four-room house and output the result.

IPO Chart

Input	Process	Output
length width	Accept length and width Compute the area Accumulate total area Display total area	total_area

Note: The IPO chart is not the place to show details of how looping is accomplished. This is done in the pseudocode.

The algorithm for this program might look like this:

```
LET count = 1
LET total_area = 0
DO WHILE count < = 4
        INPUT length, width
        LET area = length * width
        LET total_area = total_area + area
        LET count = count + 1
LOOP
OUTPUT total_area
```

Note: This is one of several possible ways to design the program. This example uses a loop. Another way might be to input the length and width of each room into separate variables, compute the area of each room in separate variables, and sum the areas of the four rooms in one step. The advantage of the loop is that that the same algorithm, with only minor modifications, will work for any number of rooms.

Practice Problems

Prepare an IPO chart for each of the following tasks.

9-2.1 Compute the area of a triangle given the length of its altitude and base.

9-2.2 Compute the average grade on a quiz for a class of four students.

9-2.3 Compute the average grade on a quiz for a class of any number of students.

9-2.4 Given a temperature in the Fahrenheit scale, display its equivalent in the Celsius scale.

9-2.5 Prompt the user to input a temperature in the Fahrenheit scale. Then display its equivalent in the Celsius scale. Repeat until a temperature of 9999 is entered.

Pseudocode

The next step in the development of a computer program is to turn the information in the IPO chart into a complete algorithm. Ultimately, when the design of the algorithm is fully understood, it will be written out (or "coded") in a computer programming language. However, a real programming language is very awkward to use in the planning stage. Unlike the simplified one we have been learning, real programming languages must adhere to strict rules of grammar called syntax. Only valid key words are allowed. Certain additional information (called parameters) must be provided. Every comma, semicolon, and period must be precisely in place. Therefore, a real programming language is not an appropriate algorithm design tool.

You have already seen an example of an algorithm design tool in Chapter 7. The instructions for the "Enjoy a Cup of Tea" algorithm were written in English language phrases arranged in a hierarchical structure much like an outline. This *structured English* is appropriate for general-purpose algorithms like making a cup of tea and can even be used for an initial "sketch" for a computer program. However, before an algorithm is transcribed into a real programming language, a special form of structured English called *pseudocode* should be employed.

Pseudocode should have the "flavor" of the real programming language that it will become. However, because pseudocode is for human designers rather than computers, it can avoid much of the complexity of the real language. For programs that will be transcribed into Visual Basic, the simple programming language introduced in Chapter 7 is the perfect pseudocode. For other languages such as JAVA or C, the pseudocode may require variations to the simple programming language.

> ❗ The simple programming language in the previous two chapters is an example of pseudocode.

► Example of Pseudocode

Although we didn't call it pseudocode at the time, you have already seen many examples in Chapters 7 and 8. Here is one example for a program that will compute the average of three numbers provided as input:

```
INPUT num1, num2, num3
LET sum = num1 + num2 + num3
LET average = sum / 3
OUTPUT average
```

Practice Problems

9-2.6   Explain how processing described in pseudocode differs from that in an IPO chart.

9-2.7   Write the pseudocode for problem 9-2.1.

9-2.8   Write the pseudocode for problem 9-2.2.

9-2.9   Write the pseudocode for problem 9-2.3.

9-2.10 Write the pseudocode for problem 9-2.4.

9-2.11 Write the pseudocode for problem 9-2.5.

Flowcharts

An alternative to using pseudocode as a program design tool, flowcharts provide a means to diagram an algorithm. Using just a few standard symbols, a step-by-step process can be displayed in a graphic format that is easy to understand.

Flowchart Advantages and Disadvantages

Advantages: Graphic format is easy to understand. An entire process can be grasped in one glance.

Disadvantages: Flowcharts are tedious and time consuming to draw. Software tools, such as VISIO, help. However, flowcharts still take longer to create than pseudocode.

Overall: Pseudocode is the tool of choice in many cases. However, flow-charts do have their place. It is important to master the basis of this diagramming tool.

Basic Flowchart Symbols

Consider the algorithm that inputs two numbers into variables a and b, adds them together, stores the result in variable s, and outputs the value s. The flowchart for such an algorithm is shown in Figure 9.1.

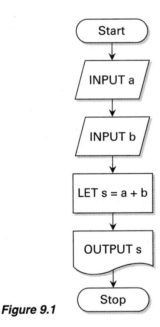

Figure 9.1

Seven Flowchart Symbols

You can purchase a plastic template to assist you in drawing flowcharts. The template will have dozens of symbols, each for a specific task. However, at this stage of learning to program, a simplified set of symbols will work better.

> This book will use a simplified set of symbols to match our simple programming language.

1. Input Symbol

   Used for input statements.

2. Output Symbol

   Used for output statements.

3. Process Symbol

   Used for assignment statements and to reflect other processing.

4. Decision Symbol

   Used for both conditional and looping statements.

5. Terminator Symbol

   Used to mark the beginning and end of the program.

6. Connector Symbol

   Used to connect diagrams that require more than one page.

7. Flow Symbol

   Used to connect the other symbols. The arrow indicates the direction in which the processing proceeds.

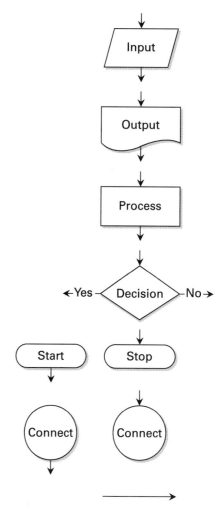

Figure 9.2

 You can diagram any program using these seven symbols.

IF-THEN-ELSE Statements

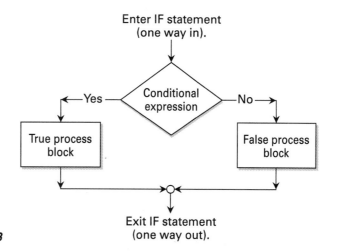

Figure 9.3

LOOP Statements

There is no single symbol for the LOOP statement. Instead, the flowchart for a loop statement is built from several symbols. Here are two examples:

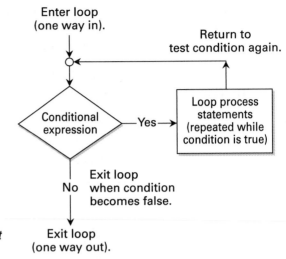

Figure 9.4 *DO WHILE-LOOP Statement (in which the condition is tested at the top of the loop)*

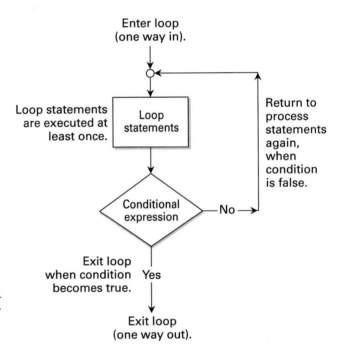

Figure 9.5 *DO-LOOP UNTIL Statement (in which the condition is tested at the bottom of the loop)*

Flowchart Standards

While there is general agreement about basic flowchart conventions, throughout the literature you will encounter many variations in the details. For example, this book uses a restricted set of symbols to keep things simple. The symbol used here for general output is one intended for printed reports, but others are also available. The Yes/No labels on the decision symbol can easily be replaced with True/False (or T/F) with equally good effect. It is best, therefore, to adopt a set of conventions that makes sense to you and then be consistent.

The following conventions are used in this book to diagram decisions:

- IF statements branch horizontally, Yes to the left and No to the right.

- Loops branch to the right to loop back and to the bottom to exit. The orientation of the Yes/No labels depends on whether WHILE or UNTIL is used.

Whatever conventions you adopt, the overriding guideline when drawing a flowchart is to ensure that your diagrams add clarity and not confusion.

Example Flowcharts

Following are flow charts for the six examples from Section 7-5, Evolution of a Simple Program.

1. Average three numbers assigned inside the program.

    LET num1 = 10
    LET num2 = 15
    LET num3 = 20
    LET sum = num1 + num2 + num3
    LET average = sum / 3
    OUTPUT average

Input	Process	Output
num1 num2 num3	Assign data values Compute sum Compute average Display average	average

Note: Although num1, num2, and num3 are assigned inside the program, they can be changed to average other numbers. Thus, we will list them in the input column.

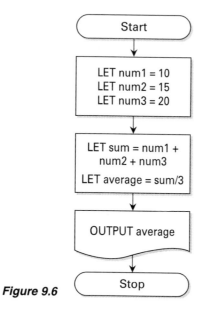

Figure 9.6

2. Average three numbers inputted from a list.

    LET sum = 0
    LET count = 0
    DO WHILE count < 3
            INPUT num
            LET sum = sum + num
            LET count = count + 1
    LOOP
    LET average = sum / count
    OUTPUT average

Input	Process	Output
num	Accept data items Accumulate sum Count data items Compute average Display average	average

Note: Even though the number of data items is fixed at three, each item is counted inside the loop.

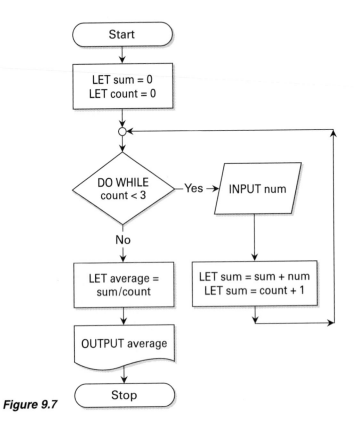

Figure 9.7

3. Average numbers from a list in which the first item is the count of the other data items.

```
LET sum = 0
LET count = 0
INPUT precount
DO WHILE count < precount
        INPUT num
        LET sum = sum + num
        LET count = count + 1
LOOP
LET average = sum / count
OUTPUT average
```

Input	Process	Output
precount num	Accept precount Accept data items Accumulate sum Count data items Compute average Display average	average

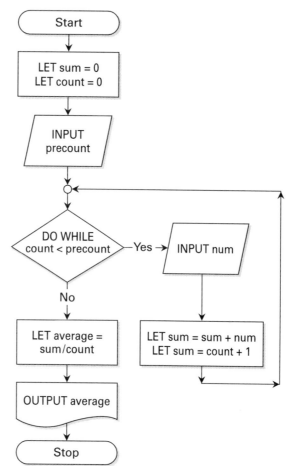

Figure 9.8

4. Average numbers from a list using a sentinel value to terminate the list.

```
LET sum = 0
LET count = 0
LET num = 0
DO WHILE num < 9999
        INPUT num
        IF num < 9999
                LET sum = sum + num
                LET count = count + 1
        END IF
LOOP
LET average = sum / count
OUTPUT average
```

Input	Process	Output
num	Accept data items Accumulate sum Count data items Compute average Display average	average

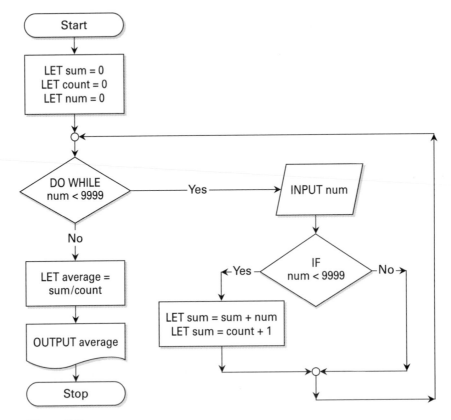

Figure 9.9

5. Use a priming read to avoid processing the sentinel value.

```
LET sum = 0
LET count = 0
INPUT num
DO WHILE num < 9999
        LET sum = sum + num
        LET count = count + 1
        INPUT num
LOOP
LET average = sum / count
OUTPUT average
```

Input	Process	Output
num	Accept data items Accumulate sum Count data items Compute average Display average	average

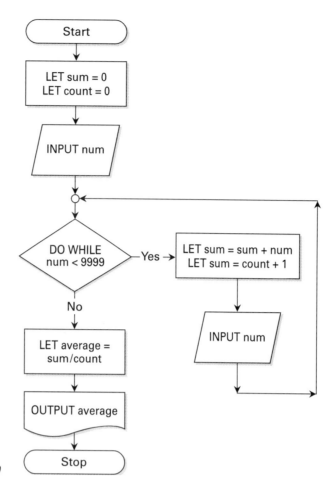

Figure 9.10

6. Allow the program to handle the special case of an empty data list.

```
LET sum = 0
LET count = 0
INPUT num
DO WHILE num < 9999
        LET sum = sum + num
        LET count = count + 1
        INPUT num
LOOP
IF count > 0 THEN
        LET average = sum / count
        OUTPUT "The average is", average
ELSE
        OUTPUT "There is no data to average"
END IF
```

Input	Process	Output
num	Accept data items Accumulate the sum Count data items Compute average Display average or message	average or message

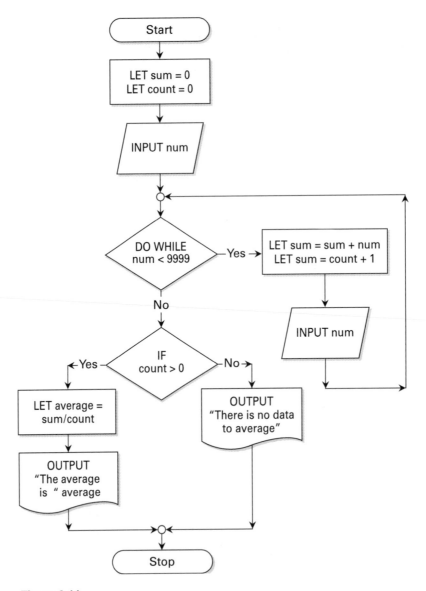

Figure 9.11

Practice Problems

Draw a flowchart for the following program fragments (there is no need to show start and stop).

9-2.12 LET n = 0

9-2.13 LET sum = 0
        LET n = 0

9-2.14 INPUT number

9-2.15 INPUT value

9-2.16 OUTPUT number

9-2.17 OUTPUT value

9-2.18 IF value = 0 THEN OUTPUT "Value equals zero"

9-2.19 IF value = 0 THEN
                OUTPUT "Value equals zero"
        END IF

9-2.20 IF value = 0
                OUTPUT "Value equals zero"
        ELSE
                OUTPUT "Value not zero"
        END IF

9-2.21 IF count > 0 THEN
                LET av = sum / count
                OUTPUT "The average is", av
        ELSE
                OUTPUT "There is no data to average"
        END IF

9-2.22 DO WHILE n < 3
                LET sum = sum + n
                LET n = n + 1
        LOOP

9-2.23 DO UNTIL n = 3
                LET sum = sum + n
                LET n = n + 1
        LOOP

9-2.24 DO
                LET sum = sum + n
                LET n = n + 1
        LOOP WHILE n < 3

9-2.25 DO
                LET sum = sum + n
                LET n = n + 1
        LOOP UNTIL n = 3

9-2.26 DO UNTIL value = 9999
                INPUT value
        IF value < 9999 THEN
                        LET sum = sum + value
                        LET n = n + 1
                END IF
        LOOP

9-2.27 LET sum = 0
        LET count = 0
        LET item = 0
        DO UNTIL item = 9999
                INPUT item
                IF item < 9999 THEN
                        LET sum = sum + item
                        LET count = count + 1
                END IF
        LOOP
        LET average = sum / count
        OUTPUT average

Background

This section will introduce some of the design concepts that have become known as structured programming. These concepts will help you write well-designed programs that are easier to code, easier to understand, and easier to maintain.

In the early days of computing, the programming tasks that were attempted were relatively simple by today's standards. This was because the capability of the computer hardware at that time had limited memory, disk storage, and processing speed. During the 1960s and 1970s, as hardware capability advanced, programmers attempted to tackle more difficult problems. The programs that resulted often had major problems. The added complexity proved difficult to read and understand. Programs had numerous errors (or bugs) that were difficult to eliminate. In some cases the programs never became operational and had to be abandoned and begun anew, at great expense.

To attack these problems, new programming concepts had to be developed. One of the earliest was the structure theorem. In 1964 two Italian mathematicians (Bohm and Jacopine) published a paper that presented a way to write any algorithm using only three basic ways to assemble instructions into what were called *control structures*. Others picked up and extended this idea. New concepts emerged from the work of Edsgar Dijkstra, Niklaus Wirth, Ed Yourdon, and Michael Jackson. Techniques such as *top-down development and modular design* were added to the basic structure theorem. Collectively, these programming techniques have become known as *structured programming*.

More recently, the concept of *object-oriented* programming has been added to the list of concepts to be considered in the design of a program. Each of these components of a "well-designed program" will be briefly introduced. If you wish to pursue computer programming beyond the introductory level, you will need to delve deeper into these important concepts.

Top-Down Development

Top-down development is a program design technique that takes a big task (the overall problem) and breaks it down into a number of smaller tasks. These smaller tasks are called modules. If necessary, any of these smaller tasks can be broken down into still smaller tasks. The process of breaking down a problem into modular components is called *functional decomposition*.

The objective of top-down development is to arrive at a set of related modules in which each module does some specific subprocessing and collectively they accomplish the entire task. While the name implies that the design process proceeds from the whole to the parts, the approach is not that rigid. If it makes more sense to proceed from the bottom up by assembling several modules to accomplish a larger task, this is permitted.

However, whether you start at the top (the big picture), at the bottom (the smallest components), or in the middle (the intermediate components), the end objective is to define a structure of modules that make up the entire process.

This structure of modules is often shown graphically in a special diagram called a *structure chart*. In the end, no matter how you arrive at the final design, the structure chart shows the hierarchical relationship that flows from the top down, hence the name *top-down development*.

Modular Design

During top-down development, the decomposition of tasks into subtasks is not haphazard. The division of the entire task into smaller processing modules is done along functional lines, with like functions grouped together into the same module. Different approaches are used to accomplish this.

The simplest approach, yet one that is very effective, is to divide the entire task along temporal lines into three modules. The first module includes the steps that are done in the beginning, the initial processing. The second module contains steps that are done in the middle, the detailed processing. The third module contains the steps that are done at the end, the final processing.

For example, consider the production of a report. At the beginning, variables need to be initialized and page headings printed. In the middle, the detailed data are processed. This involves reading one data record, processing the data, and printing the results as a line on the report. The detailed processing is repeated until all the data have been printed. Then the final processing can be accomplished. This might involve printing totals.

An example using this temporal approach is shown in Section 9-4, Problem Solving with Structured Design.

The Structure Theorem

The structure theorem states that any algorithm, and hence any computer program, can be written using only three basic ways to assemble instructions into what are known as control structures:

- Sequence (one step after another)
- Selection (also called IF-THEN-ELSE)
- Repetition (looping)

A key point is that each of these structures may contain substructures composed of any of the other structures. Thus:

A sequence structure may be composed of sequence structures, selection structures, and repetition structures.

A selection structure may be composed of sequence structures, selection structures, and repetition structures.

A repetition structure may be composed of sequence structures, selection structures, and repetition structures.

All of these structures can be built from one or more of the five program language statements that we have been using.

Sequence Structure

A sequence structure is composed of several statements that are processed one after the other. There may be only one statement or there may be many statements. Any of the three control structures may be included in the sequence. Therefore, the sequence structure may include within it any of the five programming statements: INPUT, OUTPUT, LET, IF-THEN-ELSE, and LOOP. Here is an example of a sequence structure with three assignment statements:

```
LET a = 1
LET b = 2
LET c = 3
```

Here is an example of a sequence structure with input, assignment, and output statements:

```
INPUT num
LET square = num * num
OUTPUT square
```

Selection Structure

A selection structure is composed of an IF-THEN-ELSE statement. The THEN and the ELSE blocks are usually composed of sequenced statements. However, it is also possible to include other selection structures or a repetition structure inside the selection structure. Here is an example of a simple selection structure:

```
IF purchase_price > 1000 THEN
        LET discount_rate = 0.10
ELSE
        LET discount_rate = 0.05
END IF
```

Repetition Structure

A repetition structure is composed of a LOOP statement. The block of statements inside the loop is usually a sequence structure. However, a selection structure may also be included. It is also possible to include repetition structures inside other repetition structures. This special case is called a nested loop. Here is an example of a simple repetition structure:

```
DO WHILE count < 5
        INPUT n
        LET sum = sum + n
        LET count = count + 1
LOOP
```

Beyond Structured Programming

Structured programming came about in response to a need for better-designed software in the era of text processing. It was an excellent tool for designing programs of that era. However, with the advent of the graphic user interface, the type of interface used by Microsoft Windows or the Macintosh operating system, programming has taken a giant step into the realm of ever more complexity.

Traditional Design

Traditional design views the program as a system of modules with interacting functions. Traditional modules contain instructions for processing data that are structured in a specific way. The instructions define exactly how the data will be used. The data are stored separately from the module, and the two are combined only when the program is run. Thus, if the structure of the data is changed, the instructions must also be changed. Traditional modules are not reusable with data having even a slightly different structure.

Structured programming techniques of the 1970s and 1980s are no longer sufficient to cope with this new complexity. In response to this need, a new design methodology has emerged called *object-oriented design*.

Object-Oriented Design

A radical departure from traditional design, object-oriented design views the program as a set of interacting objects having properties quite different from modules. Objects combine both data and programming instructions (called *methods*) in a way that allows them to be reused in many programs. One object can be built from several other objects. The methods of the related objects are shared in a process called *inheritance,* so that the parent object assumes the combined characteristics of its component objects.

 Programming objects are reusable, saving time and money in the overall software development process.

Learn Structured Programming First

The object-oriented design technique is very powerful and allows important economic benefits. However, it does not entirely replace structured programming design techniques. Instead, the new techniques are built on top of the old. Therefore, learn structured design first. Then, if you want to continue on to more advanced programming techniques, you will probably want to learn object-oriented design.

Program Development

The solution to many problems often involves computer programs, so it seems natural to have a problem-solving methodology for designing and building computer programs. The IPO method introduced in Chapter 1 was purposely made very general so that it could be used to solve a wide variety of problems. It will now be modified to be more appropriate for program development. Notice, however, the similarities between the following method and the IPO method that you have already learned.

Step-by-Step Program Development

1. Define the problem.
   - Understand the problem (usually provided in a design specification).
   - Identify the inputs and outputs.
   - Identify the processes needed to convert inputs into outputs.

2. Outline the solution.
   - Use an IPO chart to present the inputs, outputs, and processes.
   - Use a top-down approach to break program tasks into modules.
   - Draw a structure chart to show how the various modules are related.
   - Outline the modules using structured English; redo structure chart if needed.

3. Design an algorithm.
   - Convert the outline into pseudocode (or draw flowcharts).
   - Make sure you employ only the three basic control structures.

4. Desk check the algorithm.
   - Ensure that the algorithm works according to the design specifications.
   - Identify and fix any logic errors.

5. Code and document the program.
   - Convert the pseudocode into a specific programming language.
   - Ensure that proper syntax is employed.
   - Add documentation as you go.

6. Run the program on the computer.
   - Ensure that the program works as designed.
   - Fix any problems.

7. Maintain the program.
   - This is a natural part of program development.
   - Features not included in the initial design are added at this stage.

Modular Design

So far the algorithms used for examples have been fairly short. Most have been less than a dozen lines of code. Many problems require hundreds of instructions. Really tough problems may require many thousands or even millions of instructions.

As programs grow to more than about 20 lines of instructions, they become confusing. To offset this, larger programs need to be broken down into modules. Modules are subprograms (small programs within a larger program) that accomplish just a part of the overall processing.

You have already seen how modular design works in the "Enjoy a Cup of Tea" algorithm. Recall the different levels in the structured English outline. The various level 1 statements represented all of the major tasks needed for the entire process. The various level 2 statements described the steps needed for each major task. The hierarchy continued with several other levels of detail.

This same hierarchy is used for modular program design. All the major tasks are contained in a main module. The details for each major task are contained in submodules. They, in turn, can refer to lower-level submodules as needed to get the job done.

Structure Charts

Structure charts provide a way to show the relationships among the modules of a program. Every computer program has a beginning, middle, and end. The simplest structure chart might look something like Figure 9.12.

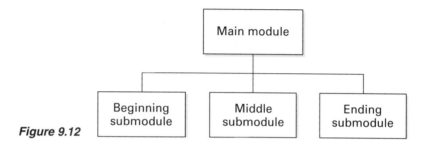

Figure 9.12

Naming Modules

Modules have meaningful names, just like variables. When used in pseudocode or actual language statements, naming conventions follow the same rules as those for variables.

Calling Modules

We need to add a new statement to the simplified programming language that will transfer processing to a module. The CALL statement will be used for this purpose. For example, when

CALL initial_processing

is placed in the main module, processing is transferred to the initial processing submodule.

 Once processing in the submodule is completed, processing continues in the main module at the point immediately following the CALL statement.

Modules may be called from any point in the program. Main modules can call submodules, and submodules, in turn, can call lower-level submodules.

 Whatever the modular complexity, a well-designed program will eventually have the processing return to the main module where the program can terminate normally.

Inventory Report Example

This example shows the development of an inventory report program that reads data from a data list. The data list contains four pieces of data for each inventory item in the report. The following four variables will be used to store the data for each inventory item:

item_number   item_description   quantity_on_hand   unit_price

Another variable will store a calculated value for each inventory item:

item_value = unit_price * quantity_on_hand

The report should have a title and column headings at the top, detailed information (one line of five columns for each inventory item) in the middle, and total inventory information at the end.

IPO Chart for Example

Input	Process	Output

item_number
item_description
quantity_on_hand
unit_price
	Initial values
	Print headings
	Accept item records
	Calculate item value
	Accumulate total inventory value
	Output item details
	Output total inventory value
		Report title
		Column headings
		Detail for each item
		    item_number
		    item_description
		    quantity_on_hand
		    unit_price
		    item_value
		Total heading
		    total_inventory_value

Structure Chart for Example

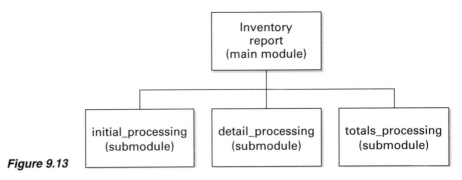

Figure 9.13

Structured English for Example

Main module: Inventory report
                  Call initial processing
                  Read data for first item
                  Repeat until item number = −9999 (sentinel value)
                        Call detail processing
                  End of loop
                  Call totals processing

Submodules: Initial processing
                  Set initial values
                  Print report title
                  Print column headings

               Detail processing
                  Compute item value
                  Accumulate total inventory value
                  Output detail data for one item
                  Read data for next item

               Totals processing
                  Print total heading
                  Print total inventory value

Pseudocode for Example (without Modules)

```
LET item_value = 0
OUTPUT "INVENTORY REPORT"
LET total_inventory_value = 0
LET total_inventory_value = 0
OUTPUT "Item Number Item Description Quantity on Hand Unit Price Item Value
INPUT item_number, item_description, quantity_on_hand, unit_price
DO UNTIL item_number = −9999
        LET item_value = quantity_on_hand * unit_price
        LET total_inventory_value = total_inventory value + item_value
        OUTPUT item_number, item_description, quantity_on_hand, unit_price,
                item_value

        OUTPUT "Item Number Item Description Quantity on Hand Unit Price Item
                Value"
        INPUT item_number, item_description, quantity_on_hand, unit_price
LOOP
OUTPUT "Total Value for Entire Inventory"
OUTPUT "$", total_inventory_value
```

Pseudocode for Example (with Modules)

Main Module

```
inventory_report
        CALL initial_processing
        DO UNTIL item_number = −9999
                CALL detail_processing
        LOOP
        CALL totals_processing
```

Submodules

```
initial_processing
        LET total_inventory_value = 0
        OUTPUT "INVENTORY REPORT"
        OUTPUT "Item Number Item Description Quantity on Hand Unit Price Item
                Value"
        INPUT item_number, item_description, quantity_on_hand, unit_price

detail_processing
        LET item value = quantity_on_hand * unit_price
        LET = total_inventory_value = total_inventory_value + item_value
        OUTPUT item_number, item_description, quantity_on_hand, unit_price,
                item_value
        INPUT item_number, item_description, quantity_on_hand, unit_price

totals_processing
        OUTPUT "Total Value for Entire Inventory"
        OUTPUT "$", total_inventory_value
```

Notes for Modular Pseudocode

- Modules organize the program. They are especially useful for large programs. However, even for a simple program like the example, they help make the program easier to understand. This will be especially true as new features are added.

- Notice that the module names must now conform to the naming convention for variables (spaces are not allowed).

- Notice that a sentinel value for the item number is used to stop the detail processing and begin the final processing. Because all of the item numbers are positive numbers, a negative sentinel value avoids stopping the processing prematurely. The DO UNTIL statement will continue to loop until the sentinel value is encountered.

New Flowchart Symbols for Modules. Two new flowchart symbols are needed for modules: a predefined process symbol to indicate that a submodule is being called, and a return terminator symbol to indicate that a submodule flowchart has terminated and control is being returned to the calling flowchart.

The *predefined process* symbol is used to indicate the points in the program at which a submodule is called. In the example program this will happen in the main module. At this point, processing will be transferred to the submodule named inside the box.

Figure 9.14 *Predefined Procedure Symbol*

The *return terminator* symbol is used in each submodule flowchart. Upon encountering the return terminator, program control is transferred back to the flowchart that called the submodule, in this case the main module. Processing resumes there at the exit point of the predefined process symbol that called the submodule.

Figure 9.15 *Return Terminator*

Flowchart for Main Module

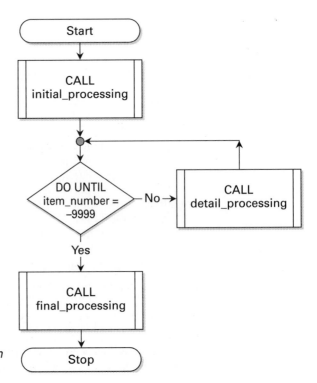

Figure 9.16 *Inventory Report Main Module*

Upon encountering a predefined process symbol, program control is transferred to the appropriate submodule flowchart. When processing in the submodule is complete, control returns to the calling module at a point immediately below the predefined procedure that called it.

In this example, the initial and final processing modules are called only once, while the detail processing module is called once for each inventory item in the data list.

Flowchart for One Submodule

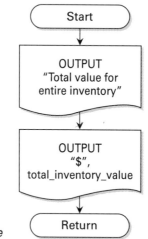

Figure 9.17 *Totals Processing Submodule*

Upon encountering the return terminator, program control is transferred back to the flowchart that called the submodule. In this example, processing returns to the main module just following the predefined procedure for totals processing. The processing ends at this point.

Flowcharts should also be drawn for the other submodules, but they will not be included in this example.

▶ Example Input and Output

Sample Data List: 1155, "Wrench", 1, 25, 2233, "Hammer", 2, 40, −9999, "NA", 0, 0

Inventory Report

Item Number	Item Description	Quality on Hand	Unit Price	Item Value
1155	Wrench	1	25	25
2233	Hammer	2	40	80

Total Value for Entire Inventory
$105

Adding an Additional Module. The example program can be refined to include additional features such as multiple pages. The number of lines per page would need to be set as a variable and a line count accumulated as each detailed line is printed. When the line count exceeds the lines per page, a page break would need to be initiated.

The headings need to be printed at the beginning of the program and again whenever a page becomes full. Instead of repeating the instructions for the headings several times

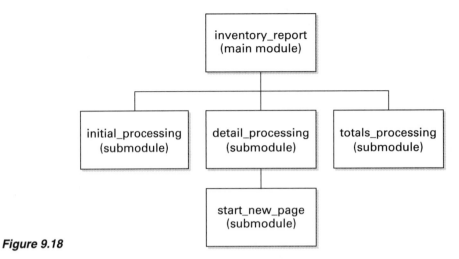

Figure 9.18

within the program, a new submodule called *start_new_page* will be created. This submodule can be invoked whenever it is needed during the detail processing.

Pseudocode for Revised Example. The new instructions to handle the multiple-page feature are shown in bold.

```
inventory_report
        CALL initial_processing
        DO UNTIL item_number = −9999
             CALL detail_processing
        LOOP
        CALL final_processing

initial_processing
        LET total_inventory_value = 0
        LET lines_per_page = 50
        LET page_break = Chr(12)
        LET line_count = 99
        INPUT item_number, item_description, quantity_on_hand, unit_price

detail_processing
        LET item value = quantity_on_hand * unit_price
        LET total_inventory_value = total_inventory_value + item_value
        LET line_count = line_count + 1
        IF line_count > lines_per_page THEN CALL start_new_page
        OUTPUT item_number, item_description, quantity_on_hand, unit_price,
                   item_value
        INPUT item_number, item_description, quantity_on_hand, unit_price

start_new_page
        LET line_count = 0
        OUTPUT page_break
        OUTPUT "INVENTORY REPORT"
        OUTPUT "Item Number Item Description Quantity on Hand Unit Price Item
                   Value"

totals_processing
        OUTPUT "Total Value for Entire Inventory"
        OUTPUT "$", total_inventory_value
```

Notes: The title and heading statements have been moved to the new module.

- By setting *line_count* to 99 in *initial_processing,* the *start_new_page* module will be called at the beginning. Then *line_count* is set to zero. As each inventory line is printed, *line_count* is incremented by 1. Each time *line_count* exceeds *lines_per_page,* the *start_new_page* module is called and *line_count* is reset again.

- The special code Chr(12), assigned to the variable *page_break,* will cause a new page to be started when sent to the printer via an OUTPUT statement. For now, don't worry about how to make the output go to the printer.

As more features are added to the program, other modules should be created. The idea of modular programming is to keep the complexity of any one module fairly simple. If a module becomes too complex, it should be broken down into smaller modules.

Practice Problems

For each of the following problems, design a program incorporating modules similar to those shown in the previous example. You are free to use any of the design tools presented in this chapter to assist you. However, you need only turn in the final version of the pseudocode.

9-4.1 Redesign the program that computes an average from a list of data, shown in Section 7-5, example 6. The new program should have a main module called *compute_average* and three subordinate modules: *initial_processing, detail_processing,* and *final_processing.*

9-4.2 Design a program that produces a "Parts to Reorder" report. Data for the report should be read in pairs from a data list. Each pair of data items includes a part ID followed by the number of that part in stock. Assume that the data list gives the part IDs in ascending order in the range 1000 to 5999. The data include parts that need to be reordered as well as parts that do not. The final part ID, 9999, is a sentinel value. A sample data list looks like this: 1001, 50, 1002, 120, 1003, 5, . . . 9999, 0.

  The report should include (1) a report title on the first line; (2) reorder information, showing ID and number on hand, one ID per line, for those parts that have fewer than 20 in stock; and (3) a message on the last line telling the number of IDs in the report.

  There is no need to handle page breaks. Let the report run over as many pages as needed without printing additional headings.

9-4.3 Design a program that will read a list of student data and produce a report. The data for each student include three items: a name, a code indicating the sex of the student (M for male, F for female), and a code indicating the status of the student (F for full-time, P for part-time). A sample data list looks like this: "Jones, John", "M", "F", "Smith, Sally", "F", "P", . . . "End of Data", "X", "X".

  a) The report should include only the names of full-time female students, one per line. Do not include the codes. Use the name "End of Data" as a sentinel value to end the processing. At this point include no title or summary lines.

  Work the following options as directed by your instructor:

  b) Add a *final_processing* module that outputs a summary line just below the final detail line of the report. The summary line should indicate the total number of students printed.

  c) Add a *start_new_page* module that allows the report to continue over several pages using a page break code as shown on page 320. Each page should begin with a title line followed by a blank spacing line. To allow for ample top and bottom margins, limit the printed information on each page to a maximum of 50 lines. This maximum should include the title and spacing lines. Let the summary line fall where it will.

  d) Now include the summary line (preceded by a blank line) on the last page, if both lines can fit within the maximum. Otherwise, put the summary on a separate page.

  e) Add a feature to print a page number as part of each title line.

  f) Suppress the page number on the first page.

Chapter Summary

Algorithm Design Tools

Structured English: A technique using the English language in a hierarchical outline to show the relationship among the components of an algorithm.

IPO charts: A way to display the input, process, and output that make up a program.

Pseudocode: A technique similar to structured English, except that it is more closely aligned to the computer language that will eventually be used to "code" the program. Pseudocode avoids much of the rigor and precise grammar (syntax) of a real programming language.

Flowcharts: A graphic presentation of the statements in a program that clearly shows how one part of the process flows into another. In modular design, like functions are grouped together.

Structure charts: A graphic presentation that shows how the different components (or modules) of a program relate to one another.

Program Design Concepts

Top-down development: The top-level solution to a problem is broken down into more and more detailed steps through a process called *functional decomposition.* Top-level steps are broken down into several midlevel steps. Midlevel steps are broken down into lower-level steps. The decomposition process continues until everything is completely defined. Only then does coding using a programming language begin.

Modular design: Functional decomposition produces groups of related instructions called modules, thus making programs easier to understand.

Structure theorem: Programming statements are assembled into three basic control structures:

- Sequence
- Repetition
- Selection

The structure theorem shows that these three basic structures are all that are needed to write any program.

Review Questions

1. What is an algorithm?
2. Name three algorithm design tools.
3. Discuss the purpose, advantages, and restrictions of each of the algorithm design tools.
4. Draw the seven flowchart symbols presented in this chapter. What is the purpose of each?
5. Define *top-down development.*
6. Define *modular design.*
7. Define the structure theorem.
8. What are the three basic control structures?
9. Discuss how one control structure may have other control structures embedded within it.
10. Define structured design. How does it differ from object-oriented design?
11. What is a structure chart?

12. Discuss the idea that processing in programs can be divided into a beginning, middle, and end.
13. What additional symbols are used in flowcharts for modules?
14. What additional programming statement allows modules to be called within a program?
15. What happens when a module is called?
16. What happens when the processing in a called module is complete?

Problem Solutions

9-2.2

Input	Process	Output
grade1 grade2 grade3 grade4	Accept values for the four grades Compute sum of four grades Compute average Display average	average

9-2.4

Input	Process	Output
f_degree	Compute c_degree Display c_degree	c_average

9-2.6 The IPO chart briefly sketches only the major components of the processing. At this stage less is more. For example, looping is not shown. The process shown in pseudocode must be in explicit detail so it can be converted directly into a program language.

9-2.8 INPUT grade1
INPUT grade2
INPUT grade3
INPUT grade4
LET sum = grade1 + grade2 + grade3 + grade4
LET average = sum / 4
OUTPUT average

9-2.10 INPUT f_degree
LET c_degree = (5 / 9) * (f_degree − 32)
OUTPUT c_degree

9-2.12 LET n = 0

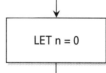

9-2.14 INPUT number

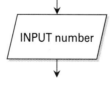

9-2.16 OUTPUT number

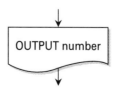

9-2.18 IF value = 0 THEN OUTPUT "Value equals zero"

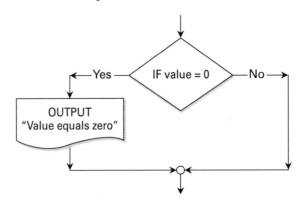

9-2.20 IF value = 0
        OUTPUT "Value equals zero"
ELSE
        OUTPUT "Value not zero"
END IF

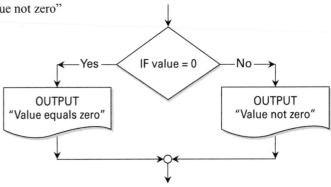

9-2.22 DO WHILE n < 3
        LET sum = sum + n
        LET n = n + 1
LOOP

9-2.24 DO
        LET sum = sum + n
        LET n = n + 1
LOOP WHILE n < 3

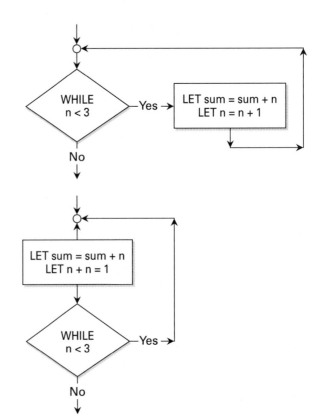

9-2.26 DO UNTIL value = 9999
        INPUT value
        IF value < 9999 THEN
                LET sum = sum + value
                LET n = n + 1
        END IF
LOOP

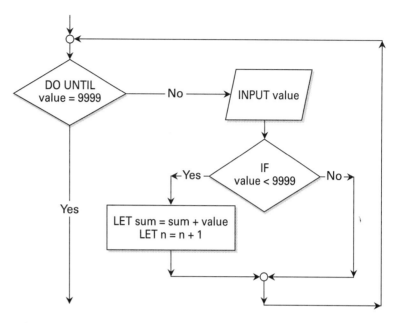

9-4.2   produce_reorder_report
        CALL initial_processing
        DO WHILE part_id < 9999
                CALL detail_processing
        LOOP
        CALL final_processing
   initial_processing
        LET count = 0
        OUTPUT "Parts to Reorder"
        INPUT part_id, num
   detail_processing
        IF num < 20 THEN
                OUTPUT part_id, num
                LET count = count + 1
        END IF
        INPUT part_id, num
   final_processing
        OUTPUT "Total IDs in report are", count

APPENDIX A

Arithmetic Review

A-1 ■ Introduction

How to Use This Appendix

Foundation

The material in this appendix is the foundation on which the topics in Chapters 1 through 6 are based. It is important that you have a clear understanding of the material in Sections A-1 through A-5. The material in Section A-6 is optional.

Self-Test

For most students, much of the material in this appendix will be a review since it is material that is prerequisite to this course. A self-test is presented at the beginning of the appendix so you can assess your skills in applying these prerequisite tools. To help focus your study, the self-test indicates the sections in which concepts underlying each problem are covered. Use the self-test to determine what sections of the appendix you need to work on. Answers are provided.

New Material

Since each student's background is slightly different, it is possible that a few topics in this appendix will be new to you. Go over the appendix carefully to determine if you are unfamiliar with any of the topics. These new topics will require careful work until you understand them. Practice problems are presented in each section to help you master the material.

Sidebars

Many sections have sidebars (boxes with gray backgrounds) that discuss how that topic is applied to computers. These sidebars present background material that will be useful in later courses.

Some sidebars are used to extend the material in a section beyond the scope of the section. This material may be of interest to some students but is not required to be able to work the practice problems.

Background

Arithmetic involves the study of numbers. The numbers used in this book are what mathematicians refer to as *real* numbers. These are the numbers with which you are already familiar and use every day.

Arithmetic also involves operations on numbers. Basic operations such as addition, subtraction, multiplication, and division are covered here. Topics developed from the basic operations are then introduced. These include signed numbers, fractions, ratios, reciprocals, rounding, and averages, all of which are important problem-solving tools. Additional operations such as raising a number to a power (squaring, cubing, etc.) or taking a root (square root, cube root, etc.) are not included here, but are covered in detail in Chapter 2.

For the student beginning this book, a thorough understanding of fractions is especially important. Therefore, an entire section is devoted to manipulating fractions. As you begin this course, you should be comfortable with adding, subtracting, multiplying, and dividing fractions.

Self-Test

This self-test will help you assess your skills and determine topics of this review material that remain to be mastered. Solutions are given on page 361.

Evaluate the following:

1. a) $((2 + 3) / 5) * 2$ b) $(3 \wedge 2) * 4$ c) $2 + 24 / 2 \wedge 3$

2. a) Reciprocal of 2   b) $|-2|$ c) $|+3|$ d) Ratio of 4 and 8

3. a) $-2 + (-3)$ b) $-2 * (-3)$ c) $-2 * (0)$

4. a) Denominator of $\frac{3}{4}$ b) Numerator of $\frac{1}{4}$ c) $\frac{1}{2}$ of $\frac{3}{4}$

5. a) True or False: $6 > 5$ b) True or False: $-6 < +6$ c) True or False: $5 \geq 5$

6. a) $(2 / 3) * (1 / 5) = ?/?$ b) $(2/3) \div (1/5) = ?/?$ c) $(2/3) + (1/5) = ?/?$

7. a) $1/(2a - b)$ when $a = 2$ and $b = 1$ b) $1/(2a - b)$ when $a = 2$ and $b = 4$

8. a) 8/24 in lowest terms   b) 18/48 in lowest terms

9. a) Prime factors of 60   b) Prime factors of 96

10. a) $\frac{3}{4}$ as a percent   b) $\frac{1}{4}$ as a decimal   c) 20% as a fraction

11. a) 12,745 rounded to nearest 1000   b) 0.12756 rounded to nearest 1/1000

Notation

The table below shows the notation used in this text. It includes notation for computers and notation for mathematics. The chapters on computers (Chapters 7–9) use computer notation exclusively. The chapters on mathematics use mathematics notation for the most part, but you will encounter computer notation as well. Since you will use all of these notations on the job in IT, begin now becoming comfortable with the different ways the same expression can be written.

For example, the area (a) of a rectangle with a height (h) and width (w) may be written:

$$a = hw \qquad \text{or} \qquad a = h * w \qquad \text{or} \qquad area = height * width$$

The area (a) of a square with four equal sides (s) may be written:

$$a = s^2 \qquad \text{or} \qquad a = s * s \qquad \text{or} \qquad a = s \wedge 2$$

Arithmetic Operators	Computer Notation	Examples	Math Notation	Examples
Addition	+	5 + 7	Same	Same
Positive numbers	None or +	5 or +5	Same	Same
Subtraction	−	7 − 5	Same	Same
Negative numbers	−	−5	Same	Same
Multiplication	*	5 * 7	× or dot adjacent letters	5 × 7 or 5·7 ab or $5a$
Division and fractions	/	5/7	Horizontal bar or / or ÷	$\frac{3}{4}$ or 3/4 or 3 ÷ 4
Exponentiation (powers)	∧	5 ∧ 2	Superscript	5^2
Parentheses	( ) or (( ))	(3 + 1) / 2	Same	Same
Nested Parentheses	(( ))	5 * (2 + (3 + 1) / 2)	[( )]	5[2 + (3 + 1) / 2]

Relational Operators	Computer Notation	Examples	Math Notation	Examples
Equal	=	a = b	Same	Same
Greater than	>	5 > 4	Same	Same
Less than	<	3 < 4	Same	Same
Greater than or equal	> =	5 > = 4 5 > = 5	≥	5 ≥ 4 5 ≥ 5
Less than or equal	< =	3 < = 4 3 < = 3	≤	3 ≤ 4 3 ≤ 3
Not Equal	< >	5 < > 4	≠	5 ≠ 4

Order of Precedence

Expressions using arithmetic operations give different results depending on the order that operations are carried out. Consider the expression

$$6 + 4/2$$

Does this evaluate as 8 or 5? If the division is done first, the expression evaluates to 8. However, if the addition is done first, the expression evaluates to 5. To make calculation unambiguous, computers follow specific rules when evaluating expressions using arithmetic operations. The various parts of an expression are evaluated in the following order:

1. Parentheses   Innermost first, then left to right

2. Exponentiation   Left to right

3. Multiplication or division   Left to right

4. Addition or subtraction   Left to right

Using this order of precedence, even complex calculations done on a computer will always result in a single result. Although complex calculations do not confuse the computer, humans reading these expressions often are confused. To avoid confusion, you should make extensive use of parentheses. For example, depending on your intentions, the previous expression is easier to read if written

$$(6 + 4) / 2 \quad \text{(evaluates to 5)} \quad \text{or} \quad 6 + (4 / 2) \quad \text{(evaluates to 8)}$$

 Learn the rules for order of precedence, then use parentheses to avoid confusion.

▶ *Example Problems without Parentheses*

1. $6 + 4 / 2$ $= 6 + \underline{2}$ (Underscore indicates most recent operation.)
   $= 8$

2. $6 + 3 \wedge 2 / 3$ $= 6 + \underline{9} / 3$
   $= 6 + \underline{3}$
   $= 9$

3. $6 + 2 - 3 * 2 / 3$ $= 6 + 2 - \underline{6} / 3$
   $= 6 + 2 - \underline{2}$
   $= \underline{8} - 2$
   $= 6$

4. $6 / 2 + 3 \wedge 2 * 3 / 3$ $= 6 / 2 + \underline{9} * 3 / 3$
   $= \underline{3} + 9 * 3 / 3$
   $= 3 + \underline{27} / 3$
   $= 3 + \underline{9}$
   $= 12$

Nested Parentheses

For added clarity, it is sometimes necessary to use parentheses inside parentheses. There are two conventions for such use. In mathematics, square brackets are often used to differentiate an inner set of parentheses from an outer set. For example:

$$3 + [6 * (3 + 2)]$$

The convention used with computers is to nest several levels of curved parentheses ( ), one set within another. Before an expression containing nested parentheses can be evaluated, the parentheses within it must be in balance. That means:

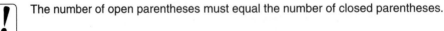 The number of open parentheses must equal the number of closed parentheses.

▶ Example:

$$(((6 + 4)/(5 - 2)) \wedge (2 + 1)) / 5 \quad \text{There are 5 open parentheses symbols (.}$$
There are 5 closed parentheses symbols).

Note: Because this book is intended for students who will work in the field of Information Technology, students should become familiar with balancing sets of parentheses.

▶ **Example Problems with Parentheses.** Remember, according to the rules of precedence, expressions within the innermost level of parentheses are evaluated first. When two sets of parentheses are at the same level, they are evaluated left to right. Here are some examples that use the computer notation:

1. $(3 + 6) / 3$ $= \underline{9} / 3$ (Underscore indicates most recent operation.)
                                          $= 3$

2. $(3 + 6) / (2 + 1)$ $= \underline{9} / (2 + 1)$
                                           $= 9 / \underline{3}$
                                           $= 3$

3. $((3 + 6) * 2) / ((2 + 1) * 2)$ $= (\underline{9} * 2) / ((2 + 1) * 2)$
                                       $= (9 * 2) / (\underline{3} * 2)$
                                       $= \underline{18} / (3 * 2)$
                                       $= 18 / \underline{6}$
                                       $= 3$

4. $((3 + 9) / (2 + 1)) \wedge 2$ $= (\underline{12} / (2 + 1)) \wedge 2$
                                       $= (12 / \underline{3}) \wedge 2$
                                       $= \underline{4} \wedge 2$
                                       $= 16$

Removing Parentheses

Although parentheses are often used to clarify the order in which operations are to be performed, too many parentheses add their own style of confusion. Therefore, it is sometimes desirable to remove parentheses from an expression. As you have seen above, this can be done by evaluating a part of the expression within an inner set of parentheses:

$$3 + [6 * (3 + 2)] \quad \text{simplifies to} \quad 3 + (6 * 5)$$

Another way to simplify an expression is to remove *unnecessary* parentheses. For example:

$(3 + 2)$	simplifies to $3 + 2$	Parentheses do no good.
$4 + (3 + 2)$	simplifies to $4 + 3 + 2$	You can add in any order.
$(4 * 3) * 2$	simplifies to $4 * 3 * 2$	You can multiply in any order.

 The associative law of arthmetic allows a string of addition (or multiplication) to be evaluated in any order. (See Properities of Numbers at the end of Section A-3).

Practice Problems
Evaluate the following expressions using the rules for order of precedence.

A-2.1 $4 + 2 / 2$

A-2.2 $4 + 2 \wedge 2$

A-2.3 $4 * 5 - 4$

A-2.4 $4 * 5 \wedge 2$

A-2.5 $4 + 5 * 3 - 2$

A-2.6 $4 + (2 * 2)$

A-2.7 $(4 + 2) * 2 + 1$

A-2.8 $4 + 3 - 1 / 2$

A-2.9 $(4 + 3 - 1) / 2$

A-2.10 $(((((4 - 2) * (3 + 1)) / 2) \wedge 2) / 4)$

This appendix discusses the properties of the numbers that we use every day, numbers that mathematicians refer to as *real numbers*.

The Number Line

The relative values of real numbers can be represented graphically by using a *number line*. Take a horizontal straight line extending to infinity toward the right and the left. Identify a point on the line to represent zero. The positive integers are positioned to the right of zero, at points equidistant from one another. Each positive integer has a corresponding negative integer at the same distance from zero, but to the left of zero.

$$\dots -5 \quad -4 \quad -3 \quad -2 \quad -1 \quad 0 \quad 1 \quad 2 \quad 3 \quad 4 \quad 5 \dots$$

Rational and Irrational Numbers

Between zero and 1 are an infinite number of fractional numbers (1/2, 1/3, 1/4, 1/5 . . .), each at the appropriate distance from zero. Similarly, other fractions populate the space between other consecutive integers. For example, between 1 and 2 there are the fractional counterparts to each of the fractions between zero and 1 (3/2, 4/3, 5/4, 6/5 . . .).

The integers and the fractions are called *rational numbers* because they can be expressed as the ratio of two integers. (Every integer can be expressed as a ratio of itself divided by 1.) Other numbers fall among the rational numbers that cannot be expressed as a ratio of integers. These are called *irrational numbers*. Examples of such numbers are the square root of 2 ($\sqrt{2}$) and pi (π), the ratio of the circumference to the diameter of a circle. Irrational numbers have endless decimals with no regular repeating groups, while rational numbers have no decimal part or have decimal parts that either terminate or repeat.

Of course, numbers that fall at exactly the same point on the number line are neither greater than nor less than one another. They are said to be equivalent, for example,

$$\frac{5}{1} = \frac{10}{2} = \frac{15}{3} = 5$$

These three fractions are equivalent to the integer 5 and occupy the same point on the number line.

One Point for Each Number

It is important to note that because the real numbers are ordered, each number corresponds to one and only one point on the number line. This allows us to say that one number is greater than or less than another number. For example, 1 is greater than $\frac{1}{2}$, which is greater than zero, which is greater than -1. This *relationship* among numbers is covered in the next two sections.

Less Than

If one number falls to the *left* of another on the number line, it is said to be less than the other number. The $<$ symbol is used to denote this. Thus, "5 is less than 7" can be written as $5 < 7$. Note that one number being less than another does not depend on the absolute values of the numbers, but rather on the relative position on the number line. Thus, -1000 is less than $+3$.

| $a < b$ if a is to the *left* of b on the number line.

For example, 3 is less than 5 because 3 is to the *left* of $+5$ on the number line.

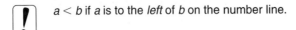

Similarly, -2 is less than 3 because -2 is to the *left* of $+3$ on the number line.

$$-2 \leftarrow \leftarrow \leftarrow \leftarrow \leftarrow 3$$

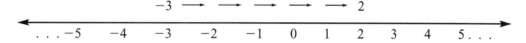

$$\dots -5 \quad -4 \quad -3 \quad -2 \quad -1 \quad 0 \quad 1 \quad 2 \quad 3 \quad 4 \quad 5 \dots$$

Other examples:

5 is less than (to the left of) 10	$5 < 10$
-3 is less than (to the left of) 3	$-3 < 3$
-10 is less than (to the left of) 0	$-10 < 0$
-10 is less than (to the left of) -2	$-10 < -2$
-10000 is less than (to the left of) 1	$-10000 < 1$

Note: Very large negative numbers are less than very small positive numbers.

Greater Than

Similarly, if one number falls to the *right* of another on the number line, it is said to be greater than the other number. The $>$ symbol is used to denote this. Thus, "7 is greater than 5" can be written as $7 > 5$. As before, one number being greater than another does not depend on the absolute values of the numbers, but rather on the relative position on the number line. Thus -3 is greater than -1000.

 $a > b$ if a is to the *right* of b on the number line

For example, 5 is greater than 3 because $+5$ is to the *right* of $+3$ on the number line.

$$3 \longrightarrow \longrightarrow 5$$

$$\dots -5 \quad -4 \quad -3 \quad -2 \quad -1 \quad 0 \quad 1 \quad 2 \quad 3 \quad 4 \quad 5 \dots$$

Similarly, 2 is greater than -3 because $+2$ is to the *right* of -3 on the number line.

$$-3 \longrightarrow \longrightarrow \longrightarrow \longrightarrow \longrightarrow 2$$

$$\dots -5 \quad -4 \quad -3 \quad -2 \quad -1 \quad 0 \quad 1 \quad 2 \quad 3 \quad 4 \quad 5 \dots$$

Other examples:

100 is greater than (to the right of) 10	$100 > 10$
10 is greater than (to the right of) 1	$10 > 1$
1 is greater than (to the right of) -1	$1 > -1$
1 is greater than (to the right of) -10000	$1 > -10000$
-1 is greater than (to the right of) -10000	$-1 > -10000$

Note: Very small positive numbers are greater than very large negative numbers.

Adding Numbers

The addition of one number to another can be defined by locating the starting number on the number line and then moving from that point a distance equal to the number added. The direction you move from the starting number is indicated by the sign of the number to be added. If the sign of the number to be added is

- **negative,** move to the LEFT
- **positive,** move to the RIGHT

This operation gives the same result if the two numbers you are adding are reversed.

To add 3 to 2, start with 2 and move right 3 whole numbers to arrive at 5:

$$2 \longrightarrow \longrightarrow \longrightarrow 5$$

$$\ldots -5 \quad -4 \quad -3 \quad -2 \quad -1 \quad 0 \quad 1 \quad 2 \quad 3 \quad 4 \quad 5 \ldots$$

▶ *Examples*

To add 5 to 10, start at 10 and move right 5 whole numbers to give 15.

$$10 + 5 = 15$$

To add 10 to 5, start with 5 and move right 10 whole numbers to give 15.

$$5 + 10 = 15$$

To add 2 to -5, start with -5 and move right 2 whole numbers to give -3.

$$-5 + 2 = -3$$

To add -5 to 2, start with 2 and move left 5 whole numbers to give -3.

$$2 + (-5) = -3$$

To add -2 to -5, start with -5 and move left 2 whole numbers to give -7.

$$-5 + (-2) = -7$$

To add -5 to -2, start with -2 and move left 5 whole numbers to give -7.

$$-2 + (-5) = -7$$

Note: Addition works in reverse. 2 + 3 gives the same result as 2 + 3. Both are 5.

Subtracting Numbers

Before you subtract, you must first decide which is the starting number and which is the number to "take away" from the starting number. For addition, the order of these numbers makes no difference. For subtraction, the order matters. Locate the starting number on the number line. Then move a distance equal to the number you are taking away. The direction you move is indicated by the sign of the number you are taking away. If the sign of the number to be taken away is

- **negative,** move to the RIGHT
- **positive,** move to the LEFT

Note: The direction is reversed from that employed in addition. Subtraction will give *a different result* if the two numbers you are subtracting are reversed.

To subtract 2 from 5, start at 5 and move left 2 whole numbers from 5 to arrive at 3:

$$3 \longleftarrow \longleftarrow 5$$

$$\ldots -5 \quad -4 \quad -3 \quad -2 \quad -1 \quad 0 \quad 1 \quad 2 \quad 3 \quad 4 \quad 5 \ldots$$

To subtract 10 from 100, start at 100 and move left 10 whole numbers to give 90.

$$100 - (+10) = 90$$

To subtract 100 from 10, start at 10 and move left 100 whole numbers to give -90.

$$10 - (+100) = -90$$

To subtract 5 from -10, start at -10 and move left 5 whole numbers to give -15.

$$-10 - (+5) = -15$$

To subtract -10 from 5, start at 5 and move right 10 whole numbers to give 15.

$$5 - (-10) = 15$$

To subtract -4 from -6, start at -6 and move right 4 whole numbers to give -2.

$$-6 - (-4) = -2$$

To subtract -6 from -4, start at -4 and move right 6 whole numbers to give 2.

$$-4 - (-6) = 2$$

Note: Reversing the numbers in a subtraction reverses the sign of the result.

Other Definitions

Zero

Zero is the number that represents the absence of anything. It has the following properties:

- It is the number at the center of the number line.
- Any number subtracted from itself is zero.
- When zero is added to another number, that number is unchanged.
- The product of zero and any number is zero.
- Any number (except zero) raised to the power of zero is 1.

Division of any number by zero is undefined. Beware, division by zero is a mistake commonly made by students. The violation of the rule gives unusual and misleading results.

Unity

Unity represents the number 1. It has the following properties:

- Any number multiplied by unity remains unchanged.
- Any number divided by unity remains unchanged.

Positive Numbers

Numbers that are to the right of zero on the number line are positive and are said to be greater than zero. This can be denoted by the $+$ sign, as in $+5$. However, the plus sign is usually omitted.

Thus, any number without a sign is understood to be positive. In certain cases, zero is also considered a positive number. In general, though, positive numbers are greater than zero. Here are some examples:

$$0.0001 \qquad 1 \qquad 2 \qquad 2.567 \qquad 1000$$

Negative Numbers

Negative numbers are to the left of zero on the number line and are said to be less than zero. To indicate this, they are preceded by a minus sign. The numbers -5, -2.6, and -1001 are all negative. Negative numbers are less than zero. Here are some examples:

$$-0.00001 \qquad -1 \qquad -2 \qquad -2.567 \qquad -1000$$

Absolute Value

Every positive number has a corresponding negative number. Both numbers are said to have the same absolute value. On the number line, two numbers with the same absolute value are the same distance from zero. Absolute value represents the value of a number without regard for its sign.

The absolute value of a number is expressed by including the number between vertical bars. For example:

$$
\begin{aligned}
|-1| &= 1 & \text{and} && |+1| &= 1 \\
|-10| &= 10 & \text{and} && |+10| &= 10 \\
|-1000| &= 1000 & \text{and} && |+1000| &= 1000
\end{aligned}
$$

Additive Inverses

Numbers that have the same absolute value but opposite signs are called additive inverses. Adding a number to its additive inverse results in zero. This can easily be understood by looking at the number line and noting that both the number and its additive inverse are the same distance from zero. For example:

10 and -10 are additive inverses.   Thus, $10 + (-10) = 0$.
-10 and $+10$ are additive inverses.   Thus, $-10 + 10 = 0$.
50 and -50 are additive inverses.   Thus, $50 + (-50) = 0$.
-50 and $+50$ are additive inverses.   Thus $-50 + 50 = 0$.

Integers

Integers are whole numbers such as -3, -2, -1, 0, 1, 2, 3 . . . 100, 101, 102, etc. Integers can be positive or negative. Zero is grouped with the positive integers. The positive integers (0, 1, 2, 3 . . .) are called natural numbers.

Fractions

 Dividing one integer into another integer forms a fraction.

Fractions are usually written with the dividend (the number divided by) and the divisor (the number divided into) separated by a fraction bar or a slash.

$$\frac{2}{3} \quad \text{or} \quad 2/3 \qquad\qquad \frac{a}{b} \quad \text{or} \quad a/b$$

$$\frac{2}{3000} \quad \text{or} \quad 2/3000$$

The number on the top of the bar (or left of the slash) is called the **numerator.** The number on the bottom of the fraction bar (or right of the slash) is called the **denominator.** In the fraction $\frac{3}{4}$ the number 3 is the numerator and the number 4 is the denominator.

$$\text{fraction} = \text{numerator/denominator}$$

$$\text{fraction} = \frac{\text{numerator}}{\text{denominator}}$$

Fractions can also be thought of representing parts of a whole. The denominator represents the whole and the numerator represents the number of parts of that whole. The fraction $\frac{3}{4}$ represents 3 parts of a whole that has been divided into 4 equal parts.

▶ Examples:

1 part in 2	can be represented by the fraction	$\frac{1}{2}$
3 parts in 10	can be represented by the fraction	$\frac{3}{10}$
7 parts in 14	can be represented by the fraction	$\frac{7}{14}$
	and also by	$\frac{1}{2}$

The denominator of the fraction $\frac{1}{2}$ is 2
The denominator of the fraction $\frac{2}{3}$ is 3
The numerator of the fraction $\frac{2}{5}$ is 2
The numerator of the fraction $\frac{3}{7}$ is 3

 The denominator of a fraction can never be assigned the value zero.

Ratios

Ratios can be used to measure the relative size of two quantities (in similar units) expressed as a proportion. Ratios, like fractions, involve dividing one number by another. But ratios are not restricted to dividing integers.

Ratios can be expressed like fractions, using a slash or a fraction bar:

$$5/7 \quad \text{or} \quad \frac{5}{7}$$

$$3.1/5.3 \quad \text{or} \quad \frac{3.1}{5.3}$$

Ratios can also be expressed like proportions, using a colon:

$$5:7 \quad \text{and} \quad 3.1:5.3$$

For a rectangle with a length of 5 feet and a width of 7 feet, the ratio of length to width can be written as 5:7 or $\frac{5}{7}$.

Reciprocals

The **reciprocal** of a number is 1 divided by that number. Another term for the reciprocal is the multiplicative inverse. All numbers have reciprocals except zero, since division by zero is undefined. Thus, the reciprocal of n is $1/n$ for all $n \neq 0$.

A quantity multiplied by its reciprocal is unity:

$$n * \left(\frac{1}{n}\right) = 1$$

▶ *Examples*

The reciprocal of 2 is $\frac{1}{2}$.

The reciprocal of 3 is $\frac{1}{3}$.

The reciprocal of 10 is $\frac{1}{10}$.

The reciprocal of $\frac{1}{2}$ is 2.

The reciprocal of $\frac{1}{10}$ is 10.

Prime Numbers

When integers are multiplied together to make larger integers, the numbers multiplied are called *factors*. For example, the factors of 24 are 8 and 3:

$$8 * 3 = 24$$

In turn, some of the factors may themselves have factors. The factor 8 can be broken into 4 * 2:

$$4 * 2 * 3 = 24$$

Finally, the factor 4 can be broken into 2 * 2:

$$2 * 2 * 2 * 3 = 24$$

When this process can be carried no further, the resulting factors are called *prime factors*. Each prime factor is a number that has no factors other than 1 and itself.

 An integer that has no integer factors other than 1 and itself is called a prime number.

Notice the prime factors of the first several positive integers in the following table:

Integer	Factors (other than 1 and itself)	
2	None	Prime
3	None	Prime
4	2 * 2	
5	None	Prime
6	2 * 3	
7	None	Prime
8	2 * 2 * 2	
9	3 * 3	
10	2 * 5	
11	None	Prime
12	2 * 2 * 3	
13	None	Prime
14	2 * 7	
15	3 * 5	
16	2 * 2 * 2 * 2 * 2	
17	None	Prime
18	2 * 3 * 3	
19	None	Prime
20	2 * 2 * 5	

Note: Except for 2, all of the prime numbers in the preceding table are odd. This is because even numbers (other than 2) have at least one factor of 2 and thus cannot be prime.

Primes are present throughout the number line. There is no highest prime number. However, primes are less frequently encountered as numbers become larger.

You will see in Section A-4 that finding the factors of an integer is important when adding fractions.

Properties of Numbers

Associative Law

When adding three numbers, it makes no difference which two are added first. This is called the associative law of addition.

$$(a + b) + c = a + (b + c)$$

Prime numbers are at the heart of encryption systems used to keep computer data private.

Given two large prime numbers, it is easy to multiply them to make a composite number. The reverse, however, is not always true. Finding the prime factors of large numbers is difficult. As the factors get very large, the process becomes extremely difficult. At some point the difficulty of factoring exceeds the capability of even the fastest computers. If a faster computer is ever built, making the factors larger will quickly defeat it.

This fact is used to advantage in public key encryption systems. The public key is a very large number that is used to convert any body of text into a private format. The encrypted text can only be undone by knowing the private key, which is composed of the prime factors of the public key.

Thus, by selecting prime factors that are sufficiently large, one can easily derive a public key with factors that cannot be discovered by anyone else. The public key is then given to anyone with whom you want to correspond, and they can use it to encrypt (scramble) their messages to you. Only persons who have the private key (the prime factors of the public key) can unscramble the message and read its content. Thus, the encrypted message is very secure.

This system fits especially well with the new form of electronic correspondence, e-mail. Other encryption systems have to deal with the problem of transmitting a secret key. This is especially difficult when your correspondent is a great distance away. The public key encryption system avoids this problem by allowing you to make public the key that you want others to use when sending secret messages to you.

E-mail is fast becoming the dominant mode of communication for government, business, and private communications alike. When these communications must be kept confidential, a system such as public key encryption is a necessity.

For example:

$$(3 + 4) + 5 = 7 + 5 = 12$$
$$3 + (4 + 5) = 3 + 9 = 12$$

When multiplying three numbers, it makes no difference which two are multiplied first. This is called the associative law of multiplication.

$$(a * b) * c = a * (b * c)$$

For example:

$$(3 * 4) * 5 = 12 * 5 = 60$$
$$3 * (4 * 5) = 3 * 20 = 60$$

Commutative Law

Two numbers can be added in either order. This is called the commutative law of addition.

$$a + b = b + a$$

For example:

$$3 + 4 = 7$$
$$4 + 3 = 7$$

Two numbers can be multiplied in either order. This is called the commutative law of multiplication.

$$a * b = b * a$$

For example:

$$3 * 4 = 12$$
$$4 * 3 = 12$$

Distributive Law

Multiplying the sum of several terms $(a + b + c \ldots)$ by some number (m) is the same as the sum of the individual products $am + bm + cm$. . . .

$$(a + b + c \ldots) m = am + bm + cm \ldots$$

This is called the distributive law of multiplication over addition.

For example:

Consider the following computation:	$(2 + 3 + 4) * 3$
Sum first, then multiply:	$= 9 * 3 = 27$
Multiply each term, then sum the products:	$= (2 * 3) + (3 * 3) + (4 * 3)$
	$= 6 + 9 + 12 = 27$
Both ways yield the same result:	27

The distributive law of addition over multiplication *does not hold:*

$$(3 * 2) + 4 \neq (3 + 4) * (2 + 4)$$

A-4 ■ Manipulating Fractions

Fractions have been previously defined in Section A-3. This section will discuss methods for manipulating fractions in calculations to simplify an expression or express a fractional value in an alternate form. Topics covered include:

- Multiplying fractions
- Finding the reciprocal of fractions
- Dividing fractions
- Reducing fractions to lowest terms
- Adding fractions
- Subtracting fractions

Multiplying Fractions

To multiply two fractions:

- Multiply both numerators to form a new numerator.
- Multiply both denominators to form a new denominator.

▶ Example:

$$\left(\frac{2}{3}\right) * \left(\frac{4}{5}\right) = \frac{(2*4)}{(3*5)} = \frac{8}{15}$$

Special Case. Multiplying the numerator and denominator of a fraction by the same number or expression is the same as multiplying the fraction by unity and leaves the value of the fraction unchanged.

$$\frac{3}{5} * \frac{2}{2} = \frac{6}{10} \qquad \text{or reduced to lowest terms} = \frac{3}{5}$$

The fractions $\frac{3}{5}$ and $\frac{6}{10}$ represent the same magnitude on the number line.

▶ Examples:

$$\left(\frac{1}{2}\right) * \left(\frac{3}{4}\right) = \frac{3}{8}$$

$$\left(\frac{2}{3}\right) * \left(\frac{6}{7}\right) = \frac{12}{21}$$

$$\left(\frac{1}{2}\right) * 2 = \frac{2}{2} = 1$$

$$\left(\frac{4}{5}\right) * \left(\frac{3}{10}\right) = \frac{12}{15}$$

Practice Problems

Multiply the following fractions.

A-4.1 $\left(\frac{5}{6}\right) * \left(\frac{3}{7}\right)$

A-4.2 $\left(\frac{2}{3}\right) * \left(\frac{4}{3}\right)$

A-4.3 $\left(\frac{4}{7}\right) * \left(\frac{2}{5}\right)$

A-4.4 $\left(\frac{5}{9}\right) * \left(\frac{3}{4}\right)$

A-4.5 $\left(\frac{7}{9}\right) * \left(\frac{2}{3}\right)$

Finding the Reciprocals of Fractions

The reciprocal of a number is that number divided into 1. To find the reciprocal of a fraction, exchange the denominator and numerator of the fraction.

> **Note:** The only exception to the rule of finding reciprocals of fractions is when the numerator is zero; then, no reciprocal exists. Reciprocals can be used to convert division of fractions into multiplication. See the section Dividing Fractions, which follows.

▶ Examples:

The reciprocal of $\frac{1}{2}$	is $\frac{2}{1}$	or	1
The reciprocal of $\frac{1}{3}$	is $\frac{3}{1}$	or	3
The reciprocal of $\frac{1}{5}$	is $\frac{5}{1}$	or	5
The reciprocal of $\frac{1}{10}$	is $\frac{10}{1}$	or	10
The reciprocal of $\frac{1}{100}$	is $\frac{100}{1}$	or	100

Practice Problems

Write the reciprocal of the following fractions. Do not simplify.

A-4.6 $\frac{3}{4}$

A-4.7 $\frac{6}{7}$

A-4.8 $\frac{1}{2}$

A-4.9 $\frac{3}{7}$

A-4.10 $\frac{2}{3}$

Dividing Fractions

The result of division (called the quotient) is found by dividing one quantity (called the dividend) by another quantity (called the divisor).

$$\text{quotient} = \frac{\text{dividend}}{\text{divisor}}$$

It is possible to convert any division problem into a multiplication problem by using reciprocals. This is especially useful when division involves fractions. To divide one fraction (the dividend) by another fraction (the divisor), simply multiply the dividend by the reciprocal of the divisor.

$$\text{quotient} = \text{dividend} * \text{reciprocal of divisor}$$

Recall that to find the reciprocal of a fraction, you simply flip the numerator and the denominator. Thus, a division problem with fractions can be converted into a multiplication problem, which is easier to carry out.

To divide $\frac{5}{3}$ by $\frac{4}{7}$, convert it to a multiplication problem:

The division is $= \dfrac{\frac{5}{3}}{\frac{4}{7}}$

The reciprocal of $\dfrac{4}{7}$ is: $\dfrac{7}{4}$

The equivalent multiplication is: $\left(\dfrac{5}{3}\right) * \left(\dfrac{7}{4}\right)$

Carry out the multiplication: $\dfrac{(5 * 7)}{(3 * 4)}$

The result is: $\dfrac{35}{12}$ or 2 and $\dfrac{11}{12}$

► More examples:

$$\frac{\left(\dfrac{3}{4}\right)}{\left(\dfrac{1}{2}\right)} = \left(\dfrac{3}{4}\right) * \left(\dfrac{2}{1}\right) = \frac{(3 * 2)}{(4 * 1)} = \frac{6}{4} = \frac{3}{2}$$

$$\frac{\left(\dfrac{2}{3}\right)}{\left(\dfrac{4}{5}\right)} = \left(\dfrac{2}{3}\right) * \left(\dfrac{5}{4}\right) = \frac{(2 * 5)}{(3 * 4)} = \frac{10}{12} = \frac{5}{6}$$

$$\frac{\left(\dfrac{1}{3}\right)}{\left(\dfrac{1}{5}\right)} = \left(\dfrac{1}{3}\right) * \left(\dfrac{5}{1}\right) = \frac{(1 * 5)}{(3 * 1)} = \frac{5}{3} = 1 \text{ and } \frac{2}{3}$$

$$\frac{\left(\dfrac{5}{7}\right)}{\left(\dfrac{3}{4}\right)} = \left(\dfrac{5}{7}\right) * \left(\dfrac{4}{3}\right) = \frac{(5 * 4)}{(7 * 3)} = \frac{20}{21} = \frac{20}{21}$$

$$\frac{\left(\dfrac{1}{4}\right)}{\left(\dfrac{1}{2}\right)} = \left(\dfrac{1}{4}\right) * \left(\dfrac{2}{1}\right) = \frac{(1 * 2)}{(4 * 1)} = \frac{2}{4} = \frac{1}{2}$$

Practice Problems

Divide the following fractions by multiplying the dividend by the reciprocal of the divisor.

A-4.11 Divide $\frac{3}{4}$ by $\frac{2}{5}$.

A-4.12 Divide $\frac{2}{5}$ by $\frac{6}{7}$.

A-4.13 Divide $\frac{4}{7}$ by $\frac{2}{7}$.

A-4.14 Divide $\frac{3}{21}$ by $\frac{7}{8}$.

A-4.15 Divide $\frac{1}{3}$ by $\frac{2}{3}$.

Reducing Fractions to Lowest Terms

The process of reducing a fraction to lowest terms involves canceling like factors that appear in both the numerator and the denominator of the fraction.

Prime Factors. The first step in the process of reducing a fraction to lowest terms is to write the prime factors of the numerator and the denominator.

Prime factors of a given number are prime numbers that when multiplied together result in the given number. Recall that prime numbers are integers that have no factors except 1 and themselves. Two factors of 24 are 3 * 8. The number 3 is a prime factor, but 8 is not, since 8 has factors 2 * 2 * 2. Thus, the prime factors of 24 are 2 * 2 * 2 * 3.

Dividing out Like Factors. Common factors that appear in both the numerator and denominator of a fraction can be divided out, thus simplifying the fraction. This process works because any number (except zero) divided by itself can be replaced with unity.

$$\frac{3}{6} = \frac{3}{2 * 3} = \frac{1}{2} * \frac{3}{3} = \frac{1}{2} * 1 = \frac{1}{2}$$

Lowest Terms Checklist

- Separate the denominator and numerator into prime factors.
- Divide out (cancel) like factors.
- Recombine the remaining (prime) factors.

▶ Examples:

$$\frac{2}{4} = \frac{2}{(2 * 2)} \qquad\qquad = \frac{1}{2}$$

$$\frac{9}{12} = \frac{(3 * 3)}{(3 * 2 * 2)} \qquad\qquad = \frac{3}{4}$$

$$\frac{8}{24} = \frac{(2 * 2 * 2)}{(2 * 2 * 2 * 3)} \qquad\qquad = \frac{1}{3}$$

$$\frac{37}{74} = \frac{37}{(2 * 37)} \qquad\qquad = \frac{1}{2}$$

Practice Problems
Reduce the following fractions to lowest terms.

A-4.16 $\frac{4}{8}$

A-4.17 $\frac{6}{24}$

A-4.18 $\frac{8}{64}$

A-4.19 $\frac{30}{72}$

A-4.20 $\frac{21}{30}$

Adding and Subtracting Fractions
Two fractions can be added or subtracted only if they have like denominators.

$$\left(\frac{1}{3}\right) + \left(\frac{1}{3}\right) = \frac{2}{3}$$

$$\left(\frac{3}{5}\right) - \left(\frac{2}{5}\right) = \frac{1}{5}$$

When the denominators are not alike, they can be made alike by multiplying both the denominator and the numerator of each fraction by an appropriate factor:

- For the first fraction, use a factor equal to the denominator of the second fraction.
- For the second fraction, use a factor equal to the denominator of the first fraction.

▶ Example:

$$\left(\frac{1}{2}\right) + \left(\frac{1}{3}\right) = \left(\frac{1}{2}\right) * \left(\frac{3}{3}\right) + \left(\frac{1}{3}\right) * \left(\frac{2}{2}\right)$$

$$= \left(\frac{3}{6}\right) + \left(\frac{2}{6}\right)$$

$$= \frac{5}{6}$$

The reason this adjustment works is that multiplying both the numerator and the denominator of a fraction by the same amount leaves the value of the fraction unchanged. Thus, the denominator of both adjusted fractions will be the product of both original denominators.

 Before you can add fractions with unlike denominators, you must make the denominators the same. To do this, multiply both the numerator and the denominator of each fraction by an amount equal to the denominator of the opposite fraction. Remember to reduce the result to lowest terms, if necessary.

▶ More examples:

$$\left(\frac{1}{12}\right) + \left(\frac{1}{3}\right) = \left(\frac{1}{12}\right) * \left(\frac{3}{3}\right) + \left(\frac{1}{3}\right) * \left(\frac{12}{12}\right)$$

$$= \left(\frac{3}{36}\right) + \left(\frac{12}{36}\right)$$

$$= \frac{15}{36} \quad \text{(not in lowest terms)}$$

$$= \frac{5}{12} \quad \text{(reduced to lowest terms)}$$

$$\left(\frac{1}{5}\right) + \left(\frac{3}{7}\right) = \left(\frac{1}{5}\right) * \left(\frac{7}{7}\right) + \left(\frac{3}{7}\right) * \left(\frac{5}{5}\right)$$

$$= \left(\frac{7}{35}\right) + \left(\frac{15}{35}\right)$$

$$= \frac{22}{35} \quad \text{(already in lowest terms)}$$

Practice Problems
Evaluate the following expressions and reduce the result to lowest terms.

A-4.21 $\frac{1}{2} + \frac{1}{5}$

A-4.22 $\frac{1}{7} - \frac{1}{8}$

A-4.23 $\frac{3}{5} - \frac{2}{9}$

A-4.24 $\frac{5}{8} + \frac{1}{14}$

A-4.25 $\frac{6}{11} - \frac{1}{6}$

A-5 ■ Decimal Numbers

Positional Notation

Whole Numbers

Whole numbers can be expressed as a string of one or more digits (0, 1, 2, 3, 4, 5, 6, 7, 8, or 9) using a positional notation. The position of the digit within the number determines the magnitude of the digit. The rightmost place of a whole number is called the units place and the digit in this place is multiplied by 1. The next place to the left is called the tens place and the digit there is multiplied by 10. Each place to the left counts 10 times more than the preceding place.

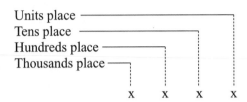

The magnitude of each position is a 10 raised to a power, starting with 10^0 in the units place. Thus, decimal numbers are also called base 10 numbers. In Chapter 3 we explore other number systems, that use bases other than ten.

To the Right of the Units Place

Decimal numbers that are not whole numbers require additional places to the right of the units place to define the portion of the number that is less than 1. Moving right, the tenths place comes first, followed by the hundredths place, the thousandths place, the ten thousandths place, and so on. Between the units place and the tenths place, a decimal point is inserted to indicate the beginning of the portion of the number that is less than 1. Each place to the right of the decimal point counts one-tenth of the preceding place.

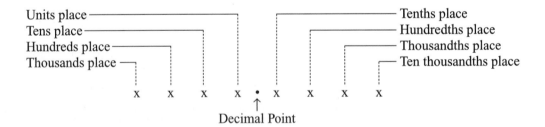

Significant Digits and Rounding

Significant Digits

Numbers that represent physical quantities may be exact or approximate. There are exactly 60 minutes in an hour or 1000 meters in a kilometer. But when time is measured with a clock or length is measured with a ruler, the result is approximate. A good stopwatch might be able to time the runner in a race to a tenth or even a hundredth of a second. A meter stick can measure a length to within a millimeter or so. Some readings can be made with more sophisticated measuring devices. An atomic clock may keep time to a millionth of a second, and a laser beam can be used to measure distance to within a millionth of a meter. But all human-made measuring devices have a limit to the accuracy of their measurements.

Significant digits provide a way to quantify the accuracy of a measurement. When we say that the length of a building is 151 feet, we imply that its length is closer to 151 feet than it is to 150 or 152 feet. When we say the length of a building is 151.3 feet, we imply that the length is closer to 151.3 feet than it is to 151.2 or 151.4 feet. In these examples 151 feet is said to have three significant digits, and 151.3 feet is said to have four significant digits.

However, if we are told that the weight of an automobile is 4000 pounds, we do not know the number of significant digits unless we are told.

- If the measurement implies than the weight is closer to 4000 than it is to either 3000 or 5000, then there is only **one** significant digit.

- If the measurement implies that the weight is closer to 4000 than it is to 4100 or 3900, then there are **two** significant digits.

- If the measurement implies that the weight is closer to 4000 than it is to 4010 or 3990, then there are **three** significant digits.

Sometimes a very accurate measurement happens to fall exactly on a multiple of one thousand. Other times the measurement has been rounded to the nearest thousand. Unless you are told, you don't know which is the case.

Rules for Determining Significant Digits

Cases in which digits in a decimal number are significant:

The digits 1–9 are not place holders and hence are always significant.

123.45 has five significant digits.   12.3 has three significant digits.

Zeros that fall between nonzero digits are significant.

105 has three significant digits.   2.305 has four significant digits.

If there are nonzero digits to the left of the decimal point, then all the zeros immediately to the right of the decimal point are significant.

2.05 has three significant digits.   17.003 has five significant digits.

Zeros on the right of the decimal part of a number are significant.

1.20 has three significant digits.   1.000 has four significant digits.

Cases in which digits in a decimal number are *not* significant:

Zeros to the left of the whole part of the number are not significant.

00345 has three significant digits.   056.2 has three significant digits.

If there is no whole part to the number, zeros immediately to the ~~left~~ right of the decimal point are not significant.

0.00034 has two significant digits.   0.01 has one significant digit.

Cases in which the number of significant digits in a decimal number is *unclear:*

If there are no digits to the right of the decimal point and the number ends in one or more zeros, then the number of significant digits is unclear.

6000 may have one, two, three, or four significant digits.

► More examples:

1	has one significant digit
21	has two significant digits
22.3	has three significant digits
2.345	has four significant digits

► Examples with trailing zeros:

1.0	has two significant digits
2.00	has three significant digits
10.00	has four significant digits

► Examples that are unclear:

100	has one, two, or three significant digits
2000	has one, two, three, or four significant digits

Practice Problems

Where possible, determine the number of significant digits for the following numbers. (Assume that all values are approximate.)

A-5.1 3	A-5.2 250000
A-5.3 36	A-5.4 0.12
A-5.5 306	A-5.6 0.12000
A-5.7 0000250.	A-5.8 0.1203
A-5.9 0.25000	A-5.10 0.00001

Rounding

It is often convenient to express a number to fewer significant digits. This is especially important when doing calculations that involve multiplication or division of values having different numbers of significant digits. This can be accomplished by rounding. There are three ways to round numbers:

1. Down

2. Up

3. To nearest place

Before the rules are explicitly laid out, here is an example:

Round 1.36 at the tenths position.

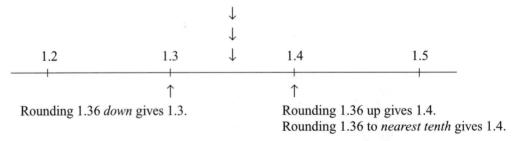

Rounding 1.36 *down* gives 1.3.

Rounding 1.36 up gives 1.4.
Rounding 1.36 to *nearest tenth* gives 1.4.

Rounding Down. Select the place (or digit) in the number at which you want to round. Drop (or zero-fill) all the digits that are to the right of the rounding place.

• Drop those that are to the right of the decimal point.

• Zero-fill those that are to the left of the decimal point.

Note: This is also called *truncating.*

► Examples of Rounding Down (rounding position is underscored):

Number	Rounding Place Place (RP)	Result
1.4̲0	Tenths	1.4
1.4̲1	Tenths	1.4
1.4̲9	Tenths	1.4
1̲0.00	Units	10
1̲0.1	Units	10
1̲0.9	Units	10
1̲0	Units	10
1̲0.0	Tens	10
1̲0.1	Tens	10
1̲1.5	Tens	10
1̲9.9	Tens	10
1̲23	Hundreds	100
1̲99	Hundreds	100

Practice Problems. Round the following values down to the places noted.

A-5.11	18 at tens place	A-5.12	27.17 to an integer
A-5.13	18.6 to an integer	A-5.14	39.965 at tenths place
A-5.15	999.999 at hundredths place	A-5.16	100.009 at hundredths place
A-5.17	656.1 at hundreds place	A-5.18	999.99 at tens place
A-5.19	127 at hundreds place	A-5.20	201.23 at tenths place

Rounding Up

If nonzero digits appear on the right of the rounding place:

- Add one to the digit in the rounding place. (Be sure to carry if necessary.)

Then, regardless of whether you added 1:

- Round down

► Examples of Rounding Up (rounding position is underscored):

Number	Rounding Place (RP)	Add 1 to RP (if needed)	Round Down at RP	Note
0.5̲9	Tenths	0.6̲9	0.6	1 added
0.5̲1	Tenths	0.6̲1	0.6	1 added
0.5̲0	Tenths	0.5̲0	0.5	1 not added
999.9̲9	Tenths	1000.0̲9	1000.00	1 added + carry
10̲0.000	Units	10̲0.000	100	1 not added
10̲0.001	Units	10̲1.001	101	1 added
2̲000	Thousands	2̲000	2000	1 not added

Practice Problems. Round the following values up at the places noted.

A-5.21 16.9 at units place
A-5.22 16.2 at tens place
A-5.23 16.872 at thousandths place
A-5.24 77.1 up at tens place

A-5.25 0.123 at tenths place

A-5.26 0.456 at units place

A-5.27 299 at hundreds place

A-5.28 500 at hundreds place

A-5.29 500.1 at hundreds place

Rounding to Nearest Place

Look at the check digit immediately to the right of the rounding place.

- Round up if check digit is 5 or greater.

- Round down otherwise.

Note: If there is no check digit, use zero.

▶ Examples of rounding to nearest place (rounding place is underlined)

Number	Rounding Place (RP)	Check Digit (right of RP)	Rounding Rule	Result
123.9999	Thousandths	9 (is 5 or greater)	Up	124 (carry)
123.4560	Thousandths	0 (not 5 or greater)	Down	123.456
123.4567	Thousandths	7 (is 5 or greater)	Up	123.457
123.901	Thousandths	0 (not 5 or greater)	Down	123.901
123.4510	Hundredths	1 (not 5 or greater)	Down	123.45
123.4590	Hundredths	9 (is 5 or greater)	Up	123.46
123.4160	Tenths	1 (not 5 or greater)	Down	123.4
123.4860	Tenths	8 (is 5 or greater)	Up	123.5
123.4560	Units	4 (not 5 or greater)	Down	123
123.7560	Units	7 (is 5 or greater)	Up	124
900	Hundreds	0 (not 5 or grater)	Down	900 (zero-fill)

Practice Problems. Round the following values to the nearest place noted.

A-5.30 7.65 at tenths place

A-5.31 235.16 at hundreds place

A-5.32 100 at tens place

A-5.33 19.278 at tens place

A-5.34 1.2 to an integer

A-5.35 99.999 at 100s place

A-5.36 32.999 at hundredths place

A-5.37 100 at hundreds place

A-5.38 16.213 at units place

A-5.39 250.164278 at thousandths place

Combining Significant Digits

The number of significant digits in a measured quantity is sometimes referred to as its accuracy.

The result of multiplication or division of measured quantities should be rounded to match the accuracy of the quantity in the calculation with the *least* number of significant digits. Values known exactly do not reduce the accuracy of the result.

The result of raising a base number to a positive integer power (e.g., square or cube) or taking a root of a base number (e.g., square root) should be rounded to match the accuracy of the base number.

▶ Examples with measured quantities:

$2.345 * 2.1 = 4.9245$	should be rounded to two significant digits:	4.9
$3.45 / 2.4 = 1.4375$	should be rounded to two significant digits:	1.4
$0.03 * 0.012 = 0.00036$	should be rounded to one significant digit:	0.0004
$\sqrt{2.04} * 0.1 = 0.142828$	should be rounded to one significant digit:	0.1
$(2.5)^2 * 1.034 = 6.4625$	should be rounded to two significant digits:	6.5

All the previous rounding rules apply to rounding at powers of ten (100, 10, 1, $\frac{1}{10}$, $\frac{1}{100}$, etc.). If you need to round at some point other than a power of ten, you must amend the rules.

Rounding a number x to the nearest 0.5 requires separating x into a whole number w and a decimal part d. Then use the following rules:

- $0 \le d < 0.25$ Round x down to w.
- $0.25 \le d < 0.75$ Round x to $w + 0.5$.
- $0.75 \le d < 0$ Round x to $w + 1$.

Thus, 2.1 will round to 2; 2.3 will round to 2.5; 2.7 will round to 2.5; and 2.8 will round to 3.0.

Rounding a number x to the nearest 25 requires separating x into a part p that is 100 and above, and a part q that is less than 100. Then use the following rules:

- $0 \le q < 12.5$ Round q to zero.
- $12.5 \le q < 37.5$ Round q to 25.
- $37.5 \le q < 62.5$ Round q to 50.
- $62.5 \le q < 87.5$ Round q to 75.
- $87.5 \le q < 0$ Round q to zero and add 100 to p.

Finally, add the new values of p and q to get the rounded value of x.

Thus, 112 will round to 100; 137 will round to 125; 138 will round to 150; 188 will round to 175; and 188 will round to 200.

Note: The range tests above break at multiples of half the rounding amount: 0.25 is half of 0.5 and 12.5 is half of 25.

▶ Examples with exact and measured quantities:

102 * 13 = 1326	should not be rounded if 102 and 13 are exact:	1326
Double 0.523 = 1.046	should be rounded to three (not two) significant digits:	1.05
902.01 * 3 = 2766.03	should be rounded to five significant digits:	2706.0

Note: Integers are considered to be the known exactly.

Practice Problems

Perform the following calculations. Unless the result is known exactly, round it to the appropriate number of significant digits. (Assume all integer values are exact.)

A-5.40 2.5 * 0.3456

A-5.41 12 * 0.345

A-5.42 50 * 0.1234

A-5.43 2.456 * 2.4

A-5.44 0.002 / 4.234

A-5.45 12.02045 / 2

A-5.46 10 * 20 * 2

A-5.47 6.23000 * 2.34

Percents, Decimals, and Fractions

A fraction can be expressed in the alternative forms of a decimal or a percent. In turn, these alternative forms can be expressed as a fraction.

Converting Fractions into Decimals

To convert a fraction into a decimal, carry out the division using a calculator. For example, the fraction $\frac{3}{4}$ can be evaluated on a calculator as 3 divided by 4. The result is 0.75. Decimal numbers converted from fractions have either

- a fixed number of digits (e.g., 2/5 = 0.4, 1/8 = .125) or
- a repeating block of digits (e.g., 1/3 = 0.333 . . . 2/3 = 0.6666 . . . , 1/7 = 0.142857142857 . . .)

Repeating decimals can be expressed in a shortened version by placing a bar over the repeating group of digits:

$$\frac{1}{7} = 0.\overline{142857}$$

▶ Examples:

$$\frac{1}{2} = 0.5$$

$$\frac{1}{3} = 0.33333 . . . = 0.\overline{3}$$

$$\frac{2}{5} = 0.4$$

$$\frac{4}{100} = 0.04$$

$$\frac{3}{7} = 0.428571 . . . = 0.\overline{428571}$$

Practice Problems

Convert the following fractions into decimals and round to three places.

A-5.48 $\frac{3}{5}$

A-5.49 $\frac{6}{13}$

A-5.50 $\frac{7}{8}$

A-5.51 $\frac{1}{9}$

A-5.52 $\frac{25}{30}$

A-5.53 $\frac{101}{1000}$

Converting Decimals into Fractions

To convert a decimal number (that is, a number less than 1) into an equivalent fraction:

- Let the **numerator** be the original number with the decimal point and leading zeros removed.
- Let the **denominator** be 1 followed by as many zeros as decimal places in the original number.

Where possible, reduce the resulting fraction to lowest terms.

► For example:

Decimal Number	Places	Numerator	Denominator	Fraction	Lowest Terms
0.5	1	5	10	$\frac{5}{10}$	$\frac{1}{2}$
0.25	2	25	100	$\frac{25}{100}$	$\frac{1}{4}$
0.125	3	125	1000	$\frac{125}{1000}$	$\frac{1}{8}$
0.015	3	15	1000	$\frac{15}{1000}$	$\frac{3}{200}$

Practice Problems

Convert the following decimals into fractions and, where possible, reduce to lowest terms.

A-5.54 0.1

A-5.55 0.3

A-5.56 0.999

A-5.57 0.35

A-5.58 0.375

A-5.59 0.045

Decimals Approximate to Fractions

The preceding procedure finds a fractional value that is *exactly equal* to a given decimal. Often you will prefer to know a simpler fraction that is approximately equal to a given decimal. For example, the decimal value 0.3333 is exactly equal to the fraction $\frac{3333}{10000}$. However, for most practical purposes it is far easier to say it is approximately $\frac{1}{3}$. Similarly, 0.6666 converts exactly to $\frac{6666}{10000}$, but is much more useful as the approximate fraction of $\frac{2}{3}$.

Use a calculator to examine fractions (less than 1) with denominators between 2 and 9. The decimal values equivalent to fractions with denominators of

2, 4, and 6	End after a few digits
3, 6, 9	Repeat a one-digit group indefinitely
7	Repeat a six-digit group indefinitely

The following table gives decimal numbers (expressed up to 10 places) for the fractions with denominators between 2 and 9. This table can be used to find an approximate fractional value for any decimal that is close to the value given in the table.

Decimal	Fraction	Decimal	Fraction
0.1111111111	$\frac{1}{9}$	0.5555555555	$\frac{5}{9}$
0.1428571428	$\frac{1}{7}$	0.5714285714	$\frac{4}{7}$
0.1666666666	$\frac{1}{6}$	0.6	$\frac{3}{5}$
0.2	$\frac{2}{5}$	0.625	$\frac{5}{8}$
0.2222222222	$\frac{2}{9}$	0.6666666666	$\frac{2}{3}$
0.25	$\frac{1}{4}$	0.7142857142	$\frac{4}{7}$
0.2857142857	$\frac{2}{7}$	0.75	$\frac{3}{4}$
0.3333333333	$\frac{1}{3}$	0.7777777777	$\frac{7}{9}$
0.375	$\frac{3}{8}$	0.8	$\frac{4}{5}$
0.4	$\frac{2}{5}$	0.8333333333	$\frac{5}{6}$
0.4285714285	$\frac{3}{7}$	0.8571428571	$\frac{6}{7}$
0.4444444444	$\frac{4}{9}$	0.875	$\frac{7}{8}$
0.5	$\frac{1}{2}$	0.8888888888	$\frac{8}{9}$

Note: Fractions with 9 as the denominator have decimal equivalents that are especially easy to identify. They all have a single repeating digit: $\frac{1}{9} = 0.1111$ . . . , $\frac{2}{9} = 0.2222$ . . . , $\frac{3}{9} = 0.3333$ . . . , (same as $\frac{1}{3}$), $\frac{4}{9} = 0.4444$ . . . , and so on to $\frac{8}{9} = 0.8888$ . . . . Fractions with 7 as the denominator have decimal equivalents that are the most difficult to recognize since they have a six-digit repeating group that continues indefinitely.

Use the preceding table to find an approximate fraction for the following decimal numbers.

Decimal Number	*Closest Decimal in Table*	*Fraction*
0.1662	0.1666666666	$\frac{1}{6}$
0.2521	0.25	$\frac{1}{4}$
0.3331	0.3333333333	$\frac{1}{3}$
0.374	0.375	$\frac{3}{8}$
0.4999	0.5	$\frac{1}{2}$
0.5715	0.5714285714	$\frac{4}{7}$
0.5998	0.6	$\frac{3}{5}$
0.7779	0.8	$\frac{7}{9}$

Practice Problems

Use the previous table to convert the following decimals into approximate fractions.

A-5.60 0.3334

A-5.61 0.6665

A-5.62 0.4447

A-5.63 0.5552

A-5.64 0.4286

A-5.65 0.1429

Converting Decimals into Percents

Percent means *parts per 100 parts*. Therefore, 100 percent represents some whole amount. A percentage less than 100 represents a quantity less than that whole amount. A percentage over 100 percent represents a quantity greater than that whole amount. Percents are often expressed using the symbol %, which means $\frac{1}{100}$. Thus, 100 percent is expressed as:

$$100\%$$

To convert any decimal number n to a percent, multiply it by 100 and attach the percent symbol:

$$n * 100\%$$

Since the percent symbol means $\frac{1}{100}$, the result is equivalent to the original number:

$$n * 100\% \quad = n * 100 * \frac{1}{100} \quad = n$$

Another way to accomplish multiplication by 100 is by moving the decimal point 2 places to the right. Thus, the conversion can be stated more simply as:

Move the decimal point 2 places to the right and append the percent symbol (%).

Practice Problems

Convert the following decimals into percents.

A-5.66 0.1

A-5.67 0.01

A-5.68 0.001

A-5.69 0.15

A-5.70 0.45

A-5.71 1.1

Converting Percents into Decimals

To convert a decimal into a percent, remove the percent sign and divide by 100.

▶ *Examples:*

1%	1 / 100	= 0.01
15%	15 / 100	= 0.15
50%	50 / 100	= 0.5
20%	20 / 100	= 0.2
100%	100 / 100	= 1.0
150%	150 / 100	= 1.5
200%	200 / 100	= 2.0

Note: You get the fractional form as an intermediate step.

Practice Problems

Convert the following percents into decimals.

A-5.72 27%

A-5.73 45%

A-5.74 16%

A-5.75 120%

A-5.76 30%

A-5.77 6%

Averages

When each item in a set of data represents some measurable quantity, an **average** can be thought of as the single value that best reflects that measurement for the set of data taken as a whole.

There are many different ways to compute an average, each bringing into focus some special quality of the data in the set. We will consider here the two ways that are encountered most often in everyday life, the arithmetic mean and the weighted average.

Arithmetic Mean

When you speak of an average value, most likely you are referring to the arithmetic mean. The arithmetic mean represents the ratio of the sum to the count of the items in a set of data:

$$\text{arithmetic mean} = \frac{\text{sum (individual data items)}}{\text{count (individual data items)}}$$

 The arithmetic mean applies only when all the data items have equal weight.

The average age of all the students in a class would be computed using the arithmetic mean. The age of each student contributes equally to the average. It clearly would be wrong to say that younger students don't count as much as older students, or that men don't count as much as women. While each student has a specific value for his or her age, each age has the same *weight* in the calculation.

However, if you try to use the arithmetic mean to compute the grade point average of a student, you run into a problem. All courses taken by the student may not represent the same number of credits. The grade for a one-credit course should not count as much as the grade for a five-credit course. At this point we need to consider the *weighted average*.

▶ *Example of Arithmetic Mean.* Compute the average daily sales for the week using the arithmetic mean.

Monday	$1000
Tuesday	$2500
Wednesday	$3000
Thursday	$1500
Friday	$2000

The sum of the daily sales data is $10,000. The count of data items is 5.

$$\text{Arithmetic mean} = \text{Sum of data / count of data}$$
$$= \$10,000 / 5$$
$$= \$2,000$$

The average daily sales amount for the week is $2,000.

Weighted Average

The weighted average must be applied in place of the arithmetic mean when data items in a set make different contributions to the average based on some weighting factor. For example, the grade point average (GPA) of a set of college courses having different numbers of credits is a weighted average with the credits being the weighting factor. The formula that applies is

$$\text{weighted average} = \frac{\text{sum of products (data value * weighting factor)}}{\text{sum (individual weighting factors)}}$$

Consider the following problem:

What is the GPA (grade point average) for a student who receives the following grades at the end of the first term:

Course	Grade	Credits
MATH 100	3.1	5
ART 100	4.0	2
PE 100	1	1
CIS 100	3.2	5
Total	14.3	13

The average computed with the arithmetic mean is (14.3)/4 = 3.575. But this average places the same weight on the one- and two-credit courses as on the five-credit courses and hence overstates the case.

A more representative average grade can be computed using the weighted average. (Use credits as the weighting factor.)

Grade	Credits	Grade * Credits
3.1	5	15.5
4.0	2	8.0
4.0	1	4.0
3.2	5	16.0
	13	43.5
	Sum of Credits	Sum of Products

$$\text{Grade point average: GPA} = \frac{43.5}{13} = 3.346$$

Units of a Weighted Average

The units of a weighted average will be the same as the units of the data items being averaged. The units of the weighting factor divide out and do not appear in the average. For example, the units of GPA are grade points, not credits. Since credits occur in both the numerator and the denominator of the division, they cancel out, leaving only the grade.

Beware When Averaging Rates

An arithmetic mean of quantities that are rates, such as miles per hour and miles per gallon, gives very misleading results. The weighted average is an improvement. But different weighting factors give different results. The worst weighting factor is one with units that match the units in the numerator of the rate. The best weighting factor is one with units matching the denominator of the rate. For example, hours for miles per hour or gallons for miles per gallon. In the end it depends on what data are available.

 When averaging rates, it is best to use a weighting factor with units matching the denominator of the rate. For example, if the rate is miles per hour, weight by hours.

Maximum and Minimum Averages

One useful piece of information when checking your calculations is that an average of a list of values cannot be larger than the maximum value in the list, nor can it be less than the minimum value in the list. This applies to both weighted averages and the arithmetic mean. So, when a computed average comes out larger than or smaller than any of the data items in the list, you know immediately that you have made a mistake.

▶ **Examples of Weighted Average**

1. The average age of students in several different classes has been recorded as follows:

Class	Class Size	Average Age
ENGL 101	29	27
MATH101	32	29
CIS101	34	32
PE101	40	23

Use a weighted average to compute the average age of students in the four classes. (Use class size as the weighting factor.)

Age	Class Size	Age * Class Size
27	29	783
29	32	928
32	34	1088
23	40	920
	135	3719
	Sum of Students	Sum of Products

$$\text{Average age} = \frac{3719}{135} = 27.5 \text{ years}$$

2. The following table shows the average hourly pay rate for the XYZ Company, by division:

Division	Number of Employees	Pay Rate
Production	300	$11.00
Sales	100	$14.00
Planning	50	$19.00

Use a weighted average to estimate the average pay rate for all hourly workers.

Pay Rate	Employees	Employees * Rate
11.00	300	3300.00
14.00	100	1400.00
19.00	50	950.00
	450	5650
	Sum of Employees	Sum of Products

$$\text{Pay rate} = \frac{5650}{450} = \$12.56 \text{ per hour}$$

Note: In all of the previous problems the sum of the quantity being averaged (i.e., grade, age, and pay rate) does not enter into the calculation.

Practice Problems

Introductory Problems. For each of the following problems compute the appropriate average (either arithmetic mean or weighted average). Limit answers to two decimal places, where necessary.

A-6.1 Compute the average of the following list of numbers: 2, 7, 3, 5.

A-6.2 Compute the average of the following list of numbers: 2, −3, 5, −7, 9.

A-6.3 Compute the average of the following list of numbers: 22, 33, 47, 16, 12.

A-6.4 Compute the average of the following list of numbers: 1, −1, 2, −2, 3, −3.

A-6.5 Compute the average of the following list of numbers: 3, −2, 4, −3, 5, −4.

A-6.6 What is the average number of credits for the following four courses?

Course	Credits
MATH 101	5
ENGLISH 105	5
PE 100	1
ART 100	2

A-6.7 Compute the grade point average (GPA) for a beginning student who has completed only the following courses:

Course	Grade	Credits
MATH 101	3.2	5
ENGLISH 105	3.4	5
PE 100	4.0	1
ART 100	4.0	2

A-6.8 Compute the grade point average (GPA) for a beginning student who has completed only the following courses:

Course	Grade	Credits
BUSINESS 101	3.0	5
BUSINESS 111	3.8	2
BUSINESS 123	4.0	1
BUSINESS 130	2.6	3

A-6.9 The ABC Company has a total of 600 employees working in three divisions. Compute the average age of the employees in the entire company, given the number of employees and the average age of each of its three divisions:

Division	Employees	Average Age
Sales	200	32
Manufacturing	300	27
Research	100	41

A-6.10 The PDQ Company reported the following daily sales for the periods indicated:

Period	Working Days	Daily Sales (Thousand $)
January	22	43
February	20	40
Part of March	5	54

Compute the average daily sales for the entire period of 47 working days.

A-6.11 What is the sum of the first 100 consecutive integers starting with 1? It is possible to work this problem quickly without having to do repetitive sums. A very simple formula exists that will compute the sum of consecutive integers from 1 to *n* for any value of *n*. Can you discover the formula by simple reasoning and a little arithmetic? (*Hint:* The solution of the general problem can be found using the definition for the arithmetic mean.)

A-6.12 A student has a 3.450 grade point average (GPA) after completing 90 credits of coursework. Compute the student's GPA after taking the following additional courses:

Course	Grade	Credits
MATH 201	3.2	5
ENGLISH 205	3.4	5
PE 200	4.0	1
ART 200	4.0	2

A-6.13 During the 11-month period from January through November the average monthly sales of Sally's Sandals was $18,500. Christmas shoppers helped make December the best sales month of the year, bringing in $24,200. Compute the average monthly sales for the entire year.

A-6.14 Carl's Clocks recorded average monthly sales during the first six months of the year of $36,125. Average monthly sales for the third quarter was $39,140 and for the fourth quarter was $42,670. Compute the average monthly sales for the entire year.

Problem Solutions

Section A-2

A-2.2 8	A-2.4 100
A-2.6 8	A-2.8 6.5
A-2.10 4	

Section A-4

A-4.2 $\frac{8}{9}$	A-4.4 $\frac{5}{12}$
A-4.6 $\frac{4}{3}$	A-4.8 $\frac{2}{1}$
A-4.10 $\frac{3}{2}$	A-4.12 $\frac{7}{15}$
A-4.14 $\frac{8}{49}$	A-4.16 $\frac{1}{2}$
A-4.18 $\frac{1}{8}$	A-4.20 $\frac{7}{10}$
A-4.22 $\frac{1}{56}$	A-4.24 $\frac{39}{56}$

Section A-5

A-5.2 2 to 6 (unclear)	A-5.4 2
A-5.6 5	A-5.8 4
A-5.10 1	A-5.12 27
A-5.14 39.9	A-5.16 100.00
A-5.18 990	A-5.20 201.2
A-5.22 20	A-5.24 80
A-5.26 1	A-5.28 500
A-5.30 7.7	A-5.32 100
A-5.34 1	A-5.36 33.00
A-5.38 16	A-5.40 0.86 (2 sig. digits)
A-5.42 6.170 (4 sig. digits)	A-5.44 0.0005 (1 sig. digit)
A-5.46 400 (exactly)	A-5.48 0.6
A-5.50 0.875	A-5.52 0.833
A-5.54 $\frac{1}{10}$	A-5.56 $\frac{999}{1000}$
A-5.58 $\frac{3}{8}$	A-5.60 $\frac{1}{3}$
A-5.62 $\frac{1}{2}$	A-5.64 $\frac{3}{7}$
A-5.66 10%	A-6.68 0.1%
A-5.70 45%	A-5.72 0.27
A-5.74 0.16	A-5.76 0.3

Section A-6

A-6.2 1.2	A-6.4 0
A-6.6 3.25	A-6.8 3.13
A-6.10 42.89 (1000$)	A-6.12 3.32
A-6.14 $38,515	

Solutions to Self-Test

1. a) 2   b) 36   c) 5
2. a) $\frac{1}{2}$ or 0.5   b) 2   c) 3   d) 4/8 $=\frac{1}{2}$ or 0.5
3. a) -5 b) 6   c) 0
4. a) 4   b) 1   c) 3/8
5. a) True   b) True   c) True
6. a) 2/15   b) 10/3 or = 3 1/3   c) 13/15
7. a) 1/3   b) 1/0 is undefined, since division by zero is not permitted
8. a) 1/3   b) 3/8
9. a) 2 * 2 * 3 * 5   b) 2 * 2 * 2 * 2 * 2 * 3
10. a) 75%   b) 0.25   c) 1/5
11. a) 13,000   b) 0.128

APPENDIX B

More Algebra

B-1 ■ Introduction

How to Use This Appendix

This appendix may be of interest to those students who want to extend the algebraic problem-solving tools introduced in Chapter 5.

Background

The background for this appendix is the same as that for Chapter 5.

Introductory Problem

Work this problem in teams of two:

A bottle and a cork cost $1.05 together. The bottle costs a dollar more than the cork. How much does each cost?

You may have seen this problem in Chapter 1 where you were asked to solve it by reasoning without algebra. Now, try your hand at translating the problem statement into two algebraic equations that employ the following letters to represent the unknown quantities:

Let b = cost of the bottle (in dollars).

Let c = cost of the cork (in dollars).

Work with the equations to come up with the correct value for the cost of the bottle and the cost of the cork. Carefully check your results to make sure they satisfy both conditions of the problem. Your instructor will provide the correct solution.

Solving Linear Equations Using Algebra

In Chapter 6 you saw how to solve two linear equations by plotting and finding the point at which they intersect. The same solution can be found using algebra.

Consider the following equations:

$$y = 2x + 1 \qquad \text{and} \qquad y = -x + 4$$

Notice that both equations are in the slope-intercept form with y, the dependent variable, on the left side. It is possible to form a new equation by combining the right sides of both equations. This is because both expressions are equal to the same quantity (y). Therefore, they must equal each other:

$$2x + 1 = -x + 4$$

The y term has been eliminated from the new equation, leaving an equation with one unknown, which can be solved for x:

$$2x + x = 4 - 1$$
$$3x = 3$$
$$x = 1$$

We have the value for x, but what is y? Simply substitute $x = 1$ into either of the original two equations. Substituting into the first equation:

$$y = 2x + 1$$
$$y = 2(1) + 1$$
$$y = 3$$

So $x = 1, y = 3$ is the solution to the set of two equations.

Because it is easy to make mistakes in these calculations, it is always wise to check the solution in the other original equation:

$$y = -x + 4$$
$$y = -(1) + 4$$
$$y = 3$$

This approach worked because both of the original equations were in slope-intercept form, with y alone on the left side. This may not always be the case; sometimes you must put them into this form.

Solving Simultaneous Linear Equations

Again we will limit ourselves to solving two linear equations with two unknowns. You have already seen that if the two equations represent intersecting lines, then the point of intersection marks the only point where the values of x and y will satisfy both equations *at the same time*. Thus, these equations are sometimes called *simultaneous* equations.

Mathematicians use many approaches to solve simultaneous equations. Here is the easiest way when you have *two linear equations with two unknowns:*

1. Slope-intercept form: Transpose both equations into the form: $y = mx + b$.

2. Eliminate y: Form new equation from right sides of original equations.

3. Solve for x: Solve new equation, which contains no y terms, for x.

4. Solve for y: Substitute x value into one original equation and solve
                                   for y.

5. Check: Substitute values of x and y into other original equation.

Two special cases:

1. If one original equation has no y terms, then solve this equation for x and use it at
   step 4.

2. If after step 1 the equations have the same slope, then there is no solution because
   the lines are parallel and therefore have no intercept.

 Remember that any letter can be used to represent a variable. Although x and y are
often used, any of the other letters could be used just as well.

Aside: Simultaneous Linear Equations with Many Unknowns

A single equation with one unknown can be solved for an explicit value of the
unknown. To solve explicitly for two unknowns requires two independent equations.
Independent linear equations will intersect on a graph. The point of intersection represents the values of the unknowns that solve both equations simultaneously. Dependent linear equations, on the other hand, graph as parallel lines, and therefore have no
simultaneous solution.

If you conclude that to solve for three unknowns requires three independent
equations, you would be correct. In fact, in order to find the explicit values that solve
any set of independent equations:

 The number of independent equations must match the number of unknowns.

▶ **Example Problems.** Use the five-step procedure to find the specific x and y that satisfies
the pairs of simultaneous equations, if one exists.

1. $2y = 4x - 2$

   $y - 2 = 3x$

   a) $y = 2x - 1$ Put into slope-intercept form.
      $y = 3x + 2$

   b) $2x - 1 = 3x + 2$ Eliminate y.

   c) $x = -3$ Solve for x.

   d) $2y = 4(-3) - 2$ Solve for y.
      $y = -7$

   e) $-7 - 2 = 3(-3)$ Check.
      $-9 = -9$

2. $3y = x - 6$
   $2y - 4 = 2x$

   a) $y = (1/3)x - (6/3)$ Put into slope-intercept form.
      $y = (1/3)x - 2$

$$y = (2/2)x + 4/2$$
$$y = x + 2$$

b) $(1/3)x - 2 = x + 2$ Combine both equations to eliminate y.

c) $(2/3)x = -4$ Solve for the single unknown, x.
$$x = 3\,(-4)/2$$
$$x = -6$$

d) $3y = (-6)-6$ Solve for y by substituting $x = -6$
$$y = -4$$
into second equation.

e) $3(-4) - 4 = -12 - 4$ Check by substituting $x = -6$ and $y = -4$
$$-8 = -8$$
into first equation.

3. $x = 4$
$$y - 2 = 2x$$

a) $x = 4$ Special case; use this value of x in step 4.
$$y = 2x + 2$$

b) Skip this step.

c) Skip this step.

d) $y = (2 * 4) + 2$ Solve for the single unknown, y.
$$y = 10$$

e) $12 - 2 = (2 * 4) + 2$ Check by substituting $x = 4$ in original equation.
$$10 = 10$$

4. $2y = 6x + 4$
$$-3x = 3 - y$$

a) $y = 3x + 2$ Slope-intercept form
$$y = 3x + 3$$
Notice that both equations have a slope of 3.

When slopes in both equations are the same, there is *no* solution. This is because the two lines represented by the equations are parallel to each other. Parallel lines never intersect and thus have no common point.

Practice Problems

Solve for x and y (or indicate that no solution is possible) using the five-step method.

B-2.1 $2y + 3 = 3y + 2x$ and $3x = 3y + 2$

B-2.2 $\frac{5}{2}x + \frac{3}{2}y = 2$ and $2x + 3y = 4$

B-2.3 $3x = y + 4$ and $3y = 6x + 2$

B-2.4 $100x + 100y = 100$ and $200x - 200y = 400$

B-2.5 $2x + 1 = 3x$ and $x + y = 5$

B-2.6 $3x + 2y = 4$ and $x + 2 = 5$

B-2.7 $1.2x + 3.5y = 2.1$ and $2.5x - 3.1y = 1.5$

B-2.8 $4y = 8$ and $2x + 2y = 6$

B-2.9 $-4x - 3y = 7$ and $x - 4y = 1$

B-2.10 $7y + 4 = 14x$ and $3y + 5 = 6x$

Equations that have unknowns raised to powers of 2 or greater are not linear and do not plot as straight lines. These equations cannot be solved with the techniques outlined so far. This book is not the place to cover this in detail. But it is worth mentioning the solution of second power equations (called quadratic equations) that have the following form:

$$ax^2 + bx + c = 0 \qquad (1)$$

The quadratic formula can be used to solve these equations for x:

$$x = \frac{-b \pm \sqrt{b^2 - 4ac}}{2a} \qquad (2)$$

There will be two solutions for x, one for $-b + \sqrt{b^2 - 4ac}$ and one for $-b - \sqrt{b^2 - 4ac}$. The pair of solutions will be either real (usual numbers) or imaginary (involving $\sqrt{-1}$).

Although you may never need to use the quadratic equation, there is a special case that is much easier to solve that you are more likely to encounter. The special case of the quadratic equation is when $a = 1$, $b = 0$, and $c = -k$. Then the equation reduces to

$$x^2 = k \qquad \text{(where } k \text{ is any positive number)} \qquad (3)$$

Solutions can be found by taking the square root of both sides of the equation. Because x^2 is positive for all values of x, including negative values, there will be two solutions to this equation:

$$x = \sqrt{k} \qquad \text{and} \qquad x = -\sqrt{k}$$

Or simply

$$x = \pm \sqrt{k} \qquad (4)$$

Can you show equation (4) to be true by substituting $a = 1$, $b = 0$, and $c = -k$ into equation (2)?

Equations such as $x^2 = k$ are encountered in Chapter 6, Graphing, where the Pythagorean theorem is introduced to compute the sides of a right triangle.

Time-Rate-Distance Problems

The Formula

All time-rate-distance problems, even the difficult ones, involve an easy-to-remember formula that relates time, rate, and distance:

$$\text{distance} = \text{rate} * \text{time} \qquad \text{or} \qquad d = rt$$

If you travel at a rate of 60 miles per hour for two hours, the distance covered is 120 miles.

$$d\,(\text{mi}) = 60\,(\text{mi/}\cancel{\text{hr}}) * 2\,(\cancel{\text{hr}}) = 120\,(\text{mi})$$

Notice that the hour unit cancels leaving miles on both sides of the equation. If time is given in minutes, you will need to convert it into hours in the calculation. If you travel at 60 miles per hour for 30 minutes, what is the distance covered?

$$d\,(\text{mi}) = 60\,(\text{mi/}\cancel{\text{hr}}) * 30\,(\cancel{\text{min}}) * 1/60\,(\cancel{\text{hr}}/\cancel{\text{min}}) = 30\,(\text{mi})$$

 Review Chapter 3 if you are not clear about balancing units.

Equations for the Two-Object Problem

More challenging problems involve two objects moving in relation to each other, or perhaps the same object moving under two different conditions. In either case, two equations can be written:

$$d_1 = r_1 * t_1$$
$$d_2 = r_2 * t_2$$

These two equations have a total of six unknowns, so at this stage no exact solution is possible. You have to translate the information in the problem statement into values that can be substituted for terms in the two equations, eliminating all but two terms. Then, with two equations and two unknowns, the problem can be solved for the desired quantity.

Most problems require you to find one of these six unknowns as the final result. Occasionally, the unknown term in the equations becomes an intermediate result that leads through another step to the final result. An example of a problem that requires an additional step is one in which you are asked to find the time of day that two travelers meet. The previous equations above will only give you the travel time in hours. As a final step you need to add the travel time to the starting time to get the meeting time expressed as a time of day.

Solving the Two-Object Problem

The following problem will be solved as an example:

> Bill leaves Seattle traveling south on Interstate 5 at 50 miles per hour. One hour later, Jim leaves Seattle traveling the same route at 70 miles per hour. How long must Jim travel before passing Bill?

After you identify the input and the output of the problem, you must identify which of the two objects will be object$_1$ and object$_2$ in the following equations. Then you must draw a diagram and develop the approach to the solution. In the example let Bill be object$_1$ and Jim be object$_2$.

Here is a four-step method for the approach:

1. **Basic equations** (the two time-rate-distance equations with six general terms):

$$d_1 = r_1 * t_1$$
$$d_2 = r_2 * t_2$$

2. **Values to substitute** (derived from information in problem statement):

First, assign the variable x to the term you are trying to find as the final result. (Sometimes corresponding terms in both equations will be equal to x.) In the example, let $t_2 = x$. Now find Jim's travel time.

Look for any other variables that can be expressed in terms of x. In the example let $t_1 = x + 1$, since Bill's time is one hour more.

Look for terms that have values given in the problem statement. In the example $r_1 = 50$ (Bill's speed) and $r_2 = 70$ (Jim's speed).

Of the remaining terms, look for ways to express one in terms of another.

Usually, both times or both rates or both distances will be equal. In the example $d_1 = d_2$ since both cover the same distance.

3. **Substitute values:**

In the example:

$$d_1 = 50(x + 1)$$
$$d_1 = 70x$$

Because there are two equations, there must be only two unknown terms at this point. If you have more than two unknowns, return to step 2 and look for other relationships.

4. **Solve for x:**

First, solve both equations for the non-x unknown in terms of x. In some cases this will already be done. In others, you will have to move terms from one side of the equation to the other. In the example:

$$d_1 = 50(x + 1)$$
$$d_1 = 70x$$

Both equations are already solved for d_1 in terms of x.

Next, combine the equations to eliminate the non-x unknown and solve explicitly for x:

$$70x = 50(x + 1)$$

This equation has only one unknown and can be solved explicitly for x:

$$70x = 50(x + 1)$$
$$70x = 50x + 50$$
$$70x - 50x = 50$$
$$(70 - 50)x = 50$$
$$20x = 50$$
$$x = 50/20$$
$$x = 2.5 \text{ hours}$$

Either this is the output you desire, or with additional steps, it will lead you to that output.

Check the Solution. With all the number manipulations, it is quite possible that an error exists. Never fail to check the solution.

Check the solution by substituting the value of x, along with the other substitutions from step 2, back into the original equations in step 1 to confirm some requirement of the problem.

$$d_1 = 50(2.5 + 1) = 175 \text{ mi}$$

$$d_2 = 70 * 2.5 = 175 \text{ mi}$$

In this case, both distances are the same. Thus, the solution checks.

 Be sure to check that the units of both equations balance according to the rules of unit analysis given in Chapter 3.

Example Problems

Example 1: Here is the same problem solved using the IPO method in Chapter 1.

Problem Statement	Bill leaves Seattle traveling south on Interstate 5 at 50 miles per hour. One hour later, Jim leaves Seattle traveling the same route at 70 miles per hour. How long will it take Jim to overtake Bill?	
Output	The time in hours that Jim travels before overtaking Bill	
Input	Bill's speed is 50 mph and Jim's speed is 70 mph. Jim starts one hour after Bill.	
Process	Notation	Let x = Jim's travel time in hours. Let Bill be object$_1$ and Jim be object$_2$.
	Additional Information	Bill's distance is the same as Jim's distance. The time-rate-distance equation is: $d = rt$.
	Diagram	———————————————Bill at 50 mph——————————— Seattle ————— >>>> —————— >>>> ————X ——————————Jim (one hour later) at 70 mph———— X marks the spot where Jim overtakes Bill.
	Approach	Use the four-step approach given earlier: 1. Basic equations 2. Determine values to substitute 3. Substitute values 4. Solve for x
Solution	1. Basic equations: $\quad d_1 = r_1 * t_1 \quad$ and $\quad d_2 = r_2 * t_2$ 2. Values to substitute: $\quad$ Let $t_2 = x$ $\qquad\qquad\qquad$ Jim's time is the unknown to be found. $\quad$ Let $t_1 = x + 1$ $\qquad\qquad$ Bill's time is one hour more. $\qquad r_1 = 50$ and $r_2 = 70$ $\qquad$ Both speeds are given. $\qquad d_1 = d_2$ $\qquad\qquad\qquad$ Both cover the same distance. 3. Substitute values: $\quad d_1 = 50(x + 1) \quad$ and $\quad d_1 = 70x$ 4. Solve for x: $\qquad 70x = 50 (x + 1)$ $\qquad$ Combine to eliminate d_1 $\qquad 70x = 50x + 50$ $\qquad\quad$ $(70 - 50)x = 50$ $\qquad 20x = 50$ $\qquad\qquad\quad$ $x = 50 / 20$ $\qquad\quad x = 2.5$ hours $\qquad\quad$ Jim travels 2.5 hours to overtake Bill.	
Check	Substitute x and other values back into general equations: $\qquad d_1 = 50(2.5 + 1) = 175$ mi $\qquad$ and $\qquad d_2 = 70 * 2.5 = 175$ mi Both distances are the same as required.	

Example 2

Problem Statement	Two trains approach each other on adjacent tracks. Train 1 is traveling south at 30 miles per hour and train 2 is traveling north at 60 miles per hour. How long after they are 10 miles apart will they pass each other?	
Output	How long until the two trains pass each other	
Input	Initial distance apart is 10 miles. Train 1 is traveling at 30 mph. Train 2 is traveling at 60 mph. Train 1 is approaching train 2	
Process	Notation	Let x = time until they pass in hours.
	Additional Information	Train 1 time is the same as train 2 time. The time-rate-distance equation is $d = rt$.
	Diagram	$\overline{\qquad\qquad\qquad\text{10 miles}\qquad\qquad\qquad}$ Train 1 ____>>>____X____<<<<____Train 2 X marks the spot where they meet.

	Approach	Use the four-step approach given earlier:
		    1. General equations
		    2. Determine values to substitute
		    3. Substitute values
		    4. Solve for x

Solution	1. General equations:
	
	$$d_1 = r_1 * t_1 \quad \text{and} \quad d_2 = r_2 * t_2$$
	
	2. Values to substitute:
	
	    $x = t_1 = t_2$ Both trains take the same time to pass.
	    $r_1 = 30$ and $r_2 = 60$ Speeds are given in mph.
	    $10 = d_1 + d_2$ Both distances add to 10 miles.
	
	3. Substitute values:
	
	    $d_1 = 30x \quad \text{and} \quad d_2 = 60x \quad \text{and} \quad 10 = d_1 + d_2$
	
	4. Solve for x:
	
	    $10 = 30x + 60x$ Substitute $d_1 = 30$ and $d_2 = 60$ into $10 = d_1 + d_2$.
	
	    $10 = 90x$
	
	    $x = 10 / 90 = 1/9$ hour
	
	    $x = 1 / 9$ (hr) $* 60$ (min/hr) $= 6.666 \ldots$ (min)
	
	Thus, time for trains to pass is 6 and 2/3 minutes.	

Check	Substitute x and other values back into general equations:
	
	    $d_1 = 30 / 9 = 3\ 1/3$ miles $d_2 = 60 / 9 = 6\ 2/3$ miles
	
	    $d_1 + d_2 = 3\ 1/3 + 6\ 2/3 = 10$ miles as required	

Example 3

Problem Statement	Two planes leave Seattle at the same time. Plane 1 flies east and plane 2 flies west. Plane 1 flies at 400 miles per hour. How fast does plane 2 need to fly for both planes to be 900 miles apart after an hour?	
Output	Speed of plane 2 in mph	
Input	Planes fly in opposite directions. Travel time for both is one hour. Plane 1 travels at 400 mph. Distance apart at the end is 900 miles.	
Process	Notation	Let x = speed of plane 2 in mph.
	Additional Information	The time-rate-distance equation is $d = rt$.
	Diagram	────────── 900 miles ────────── Plane 1 ___ <<<<___ Seattle ___ >>>> ___ Plane 2 ─────────── 1 hour ───────────

	Approach	Use the four-step approach given earlier:
		  1. General equations
		  2. Determine values to substitute
		  3. Substitute values
		  4. Solve for x

Solution	1. General equations: $$d_1 = r_1 * t_1 \quad \text{and} \quad d_2 = r_2 * t_2$$ 2. Values to substitute: $\quad$ Let $x = r_2$ $\qquad$ Speed of plane 2 $\qquad r_1 = 400$ $\qquad$ Given speed of plane 1 in mph $\qquad d_1 = 900 - d_2$ $\quad$ Both distances add to 900 miles $\qquad t_1 = t_2 = 1$ $\qquad$ Both travel for one hour 3. Substitute values: $$d_1 = 400 * 1 \quad \text{and} \quad d_2 = x * 1 \quad \text{and} \quad 900 = d_1 + d_2$$ 4. Solve for x: $\qquad 900 = 400 + x \qquad$ Substitute $d_1 = 400$ and $d_2 = x$ into $900 = d_1 + d_2$. $\qquad x = 500$ mph Thus, rate of plane 2 is 500 mph.	
Check	Substitute x and other values back into general equations: $$d_1 = 400 * 1 \quad \text{and} \quad d_2 = 500 * 1$$ $$d_1 + d_2 = 400 + 500 = 900 \text{ miles, as required}$$	

Practice Problems

Solve the following time-rate-distance problems.

B-3.1 Bill travels 200 miles in four hours. What is his average speed in miles per hour?

B-3.2 Bill leaves Chicago at 4:00 P.M. traveling east on Interstate 90 toward Toledo, 200 miles away. Jim leaves Toledo at the same time, traveling west toward Chicago on the same road. If Bill travels at 50 miles per hour and Jim travels at 60 miles per hour, at what time will Bill pass Jim?

B-3.3 Bill leaves Chicago at 4:00 P.M. traveling east on Interstate 90 toward Toledo, 200 miles away. Jim leaves Toledo at 5:00 P.M., traveling west toward Chicago on the same road. If Bill travels at 50 miles per hour and Jim travels at 60 miles per hour, at what time will Bill pass Jim?

B-3.4 Bill leaves Chicago at 4:00 P.M. traveling east on Interstate 90 toward Toledo, 200 miles away. Jim leaves Toledo at the same time, traveling west toward Chicago on the same road. They pass each other at 6:00 P.M. If Bill travels at 55 miles per hour, how fast is Jim traveling?

B-3.5 Bill leaves Chicago at 4:00 P.M. traveling east on Interstate 90 at 70 miles per hour toward Toledo. Also at 4:00 P.M., Jim leaves Toledo, 200 miles east of Chicago, and heads west for Chicago at 50 miles per hour. At 5:00 P.M. they both stop one hour for dinner and then continue at their previous speeds. At what time will Bill pass Jim?

B-3.6 Bill leaves Chicago at 4:00 P.M. traveling east at 70 miles per hour on Interstate 90. At 6:00 P.M. Jim leaves Toledo, 200 miles east of Chicago, also traveling east on the same road at 50 miles per hour. At what time will Bill pass Jim?

B-3.7 Bill leaves Chicago at 4:00 P.M. traveling east at 70 miles per hour on Interstate 90. At 6:00 P.M. Jim leaves Toledo, 200 miles east of Chicago, also traveling east on the same road at 50 miles per hour. How far from Toledo will Bill pass Jim?

More Difficult Problems

B-3.8 Two trains are heading toward each other on the same track, both traveling at 1 mile per minute. A super bee is riding on the headlight of one of the trains. When the trains are exactly two miles apart, the bee flies off to the other train. Upon reaching the headlight of the other train, the bee immediately reverses direction and flies back to the first train. The bee flies back and forth as the two trains continue on their collision course. If the bee maintains an average speed of 2 miles per minute, how far does the bee travel before it is squished?

B-3.9 While traveling on a section of Montana highway, where there is no speed limit and very little traffic, you drive up a 1-mile hill at 30 miles per hour. How fast must you come down the 1-mile on the other side of the hill to average 45 miles per hour for the entire 2-mile stretch of highway? Can you work the problem so the average speed for the 2-mile stretch is 60 miles per hour?

B-3.10 A boy walks across a railroad bridge. After he has crossed $\frac{3}{8}$ of the bridge, he looks back to see a train approaching the bridge from behind. Being a mathematical whiz, he quickly sizes up the situation. If he runs back toward the train, he will reach the beginning of the bridge and be able to jump clear just as the train starts across the bridge. If he runs forward, he will reach the far end of the bridge and be able to jump clear just as the train reaches that end. If the boy runs at 10 miles per hour, how fast is the train going?

Linear Interpolation

Proportions can be used to estimate data values between given data points using a technique called linear interpolation.

Suppose you have a table of data. The data can be most anything, as long as they meet the following criteria:

- The data come in corresponding pairs recorded at different points.
- The data have meaningful values between the recorded points.

Example Problem

The cost of 1 megabyte of computer memory, recorded every two years, over an eight-year period is as follows. (This data is made up and not actual.)

Year	*Cost*
1988	$700
1990	$300
1992	$150
1994	$85
1996	$40

 Linear interpolation works with the ratio of differences between the data points, not the data values themselves.

To estimate the cost in 1995, we need to examine the data on either side of 1995: 1996 and 1994. Take the difference between the data values for these years.

The difference in years is $1996 - 1994 = 2$.

The difference in cost is $40 - 85 = -45$.

Now what are the differences between one of these years and the year we wish to estimate? Either the year 1994 or 1996 will do. Using 1994:

The difference between years is $1995 - 1994 = 1$.

The difference between cost is $x - 85$ where x is the cost in 1995.

Linear interpolation assumes that the ratio (years to cost) of differences between the two recorded data points (1996 and 1994) *are in the same proportion* as the ratio (years to cost) of differences between the 1996 and a new data point for 1995. Thus

$$(x - 85) / (1995 - 1994) = (40 - 85) / (1996 - 1994)$$

$$x = 85 + (40 - 85) * (1995 - 1994) / (1996 - 1994)$$

$$x = 85 + (1)(-45) / (2)$$

$$x = 85 - (45/2) = 62.50$$

Solving for x gives the cost of memory in 1995 as $62.50.

Remember, this is just an estimate since actual data for 1995 is not in the data set.

Things to Keep in Mind

- Linear interpolation is a means of estimating missing data. It usually will not give you the exact value, only an estimate.

- Linear interpolation works best when the data changes are fairly regular.

- Linear interpolation works best when the estimate is made between two data points that are as close together as possible.

- When using linear interpolation it is assumed that the data between any two points will trace a straight line if plotted. This is where the term *linear* comes from. The closer the recorded data is to being linear (following a straight line when plotted), the better the estimation.

- The direction you use to calculate the differences must be the same for all the differences. Notice that in the example, both differences were computed between the later year and the earlier year. The opposite direction is also OK; just be consistent.

Linear Interpolation Formula

Suppose you have a set of *recorded* data for two related quantities, x and y. Select two adjacent points within this set and call them point 1 and point 2. Then the x-y pairs of data recorded at these two points can be referred to as (x_1, y_1) and (x_2, y_2). This data can be shown as:

$$x_1 \qquad y_1$$
$$x_2 \qquad y_2$$

Suppose you want to *estimate* the value of x_p at some point p between points 1 and 2, where you know the value of y_p. Adding these new points, the data set becomes:

$$x_1 \qquad y_1$$
$$x_p \qquad y_p$$
$$x_2 \qquad y_2$$

Linear interpolation is a method that can be used to make this estimate. The change in the value y between points 1 and p represents a certain proportion of the total change recorded between points 1 and 2. Linear interpolation works under the assumption that this proportion for values of y is the same for values of x. Brackets show these differences in the following data set. The differences are then written in a proportional equation.

$$x_2 - x_1 \left\{ \begin{array}{c} x_p - x_1 \left\{ \begin{array}{cc} x_1 & y_1 \\ x_p & y_p \end{array} \right\} y_p - y_1 \\ \\ x_2 \qquad y_2 \end{array} \right\} y_2 - y_1 \qquad\qquad \frac{x_p - x_1}{x_2 - x_1} = \frac{y_p - y_1}{y_2 - y_1}$$

Rearranging terms yields a formula for the estimated value of x_p for a given y_p:

$$x_p = x_1 + (x_2 - x_1)\frac{(y_p - y_1)}{(y_2 - y_1)} \tag{5}$$

Similarly, given the value for x_p, a value for y_p can be calculated by

$$y_p = y_1 + (y_2 - y_1)\frac{(x_p - x_1)}{(x_2 - x_1)} \tag{6}$$

When x and y represent physical quantities, their associated units must be consistent so the units on both sides of the equation balance. Notice that in equation (5) the y terms on the right side cancel leaving only the x terms on the right to balance the x term on the left. Similarly, in equation (6), the x terms cancel, leaving the desired y term on both sides.

What the Formula Means

Examining equation (5) yields a step-by-step procedure than produces the same result as the formula.

Given a value of y at point p between points 1 and 2, estimate the corresponding value of x:

1. Compute the difference of y between point 2 and point 1.

$$y_2 - y_1$$

2. Compute the difference of x between point 2 and point 1.

$$x_2 - x_1$$

3. Compute the rate that x changes per unit of y. (This rate is like the slope of a line, thus the term *linear*.)

$$(x_2 - x_1) / (y_2 - y_1)$$

4. Compute the change of y between point p and point 1.

$$y_p - y_1$$

5. Estimate the change of x between point p and point 1.

$$(y_p - y_1) * (x_2 - x_1) / (y_2 - y_1)$$

6. Add the estimated change in x to the value of x at point 1.

$$x_1 + (y_p - y_1) * (x_2 - x_1) / (y_2 - y_1)$$

7. Rearrange the terms to give the same equation as equation (5).

$$x_p = x_1 + (x_2 - x_1) * (y_p - y_1) / (y_2 - y_1)$$

This step-by-step procedure presents a verification of equation (5). This procedure can be used to check the results found by using the equation. A similar procedure exists for equation (6).

Remember: All the x values must be expressed in the same units. All the y values must be expressed in the same units. However, the units of x will be different from the units of y.

▶ Example Problems

While a database is being processed, the computer prints the total number of records processed so far at 100-second intervals:

Total Seconds	Total Records Processed
100	527
200	2640
300	4721
400	6240
500	8529

1. Use linear interpolation to estimate the time in seconds needed to process 5000 records.

> 5000 records occurs between 300 and 400 seconds.
> Let x represent seconds and y represent minutes.

Then

$$x_1 = 300 \qquad y_1 = 4721$$
$$x_p = ? \qquad y_p = 5000$$
$$x_2 = 400 \qquad y_2 = 6240$$

Use equation (5):

$$x_p = x_1 + (x_2 - x_1) \frac{(y_p - y_1)}{(y_2 - y_1)}$$

$$x_p = 300 + (400 - 300) * (5000 - 4721) / (6240 - 4721)$$

$$x_p = 300 + (100 * 279 / 1519)$$

$$x_p = 318.4 \text{ (seconds)}$$

2. Use linear interpolation to estimate the number of records processed at two minutes.

Two minutes is 120 seconds, which occurs between 100 and 200 seconds. Let x represent seconds and y represent minutes.

Then:

$$x_1 = 100 \qquad y_1 = 527$$
$$x_p = 120 \qquad y_p = ?$$
$$x_2 = 200 \qquad y_2 = 2640$$

Use equation (6):

$$y_p = y_1 + (y_2 - y_1) \frac{(x_p - x_1)}{(x_2 - x_1)}$$

$$y_p = 527 + (2640 - 527) * (120 - 100) / (200 - 100)$$

$$y_p = 527 + (2113 * 20 / 100)$$

$$y_p = 949.6 \text{ (records)}$$

Practice Problems

Introductory Problems. The XYZ Company makes widgets. Here is a table of production data recorded at several times throughout the day:

Time	Total Units
2:05 P.M.	9874
3:10 P.M.	11240
4:07 P.M.	14371
5:01 P.M.	16110

Note:

- Differences in time must be converted from hours:minutes to minutes.

- The calculation should be done in minutes.

- Convert the final result back into hours:minutes.

B-3.11 Use linear interpolation to estimate total units at 2:30 P.M.

B-3.12 Use linear interpolation to estimate total units at 3:00 P.M.

B-3.13 Use linear interpolation to estimate total units at 4:00 P.M.

B-3.14 Use linear interpolation to estimate total units at 5:00 P.M.

B-3.15 Use linear interpolation to estimate the time when total production reached 10,000 units.

B-3.16 Use linear interpolation to estimate the time when total production reached 12,000 units.

B-3.17 Use linear interpolation to estimate the time when total production reached 14,000 units.

B-3.18 Use linear interpolation to estimate the time when total production reached 16,000 units.

More Difficult Problems. Linear interpolation is used to estimate data points *between* two recorded points. Compute a ratio equation similar to linear interpolation to estimate data points *just outside* the range of recorded data. This process is called linear extrapolation. Use this ratio equation to solve the following problems.

B-3.19 Use linear extrapolation to estimate total units completed at 1:00 P.M.

B-3.20 Use linear extrapolation to estimate total units completed at 6:00 P.M.

B-3.21 Use linear extrapolation to estimate the time when 9,000 units were completed.

B-3.22 Use linear extrapolation to estimate the time when 17,000 units were completed.

Problem Solutions

B-2.2 $x = 0, y = \frac{4}{3}$

B-2.4 $x = \frac{3}{2}$ and $y = -\frac{1}{2}$

B-2.6 $x = 3$ and $y = -\frac{5}{2}$

B-2.8 $x = 1$ and $y = 2$

B-2.10 no solution exists

B-3.2 5:49 P.M.

B-3.4 45 mph

B-3.6 9:00 P.M.

B-3.8 2 miles

B-3.10 40 mph

B-3.12 11029.8 units

B-3.14 16077.8 units

B-3.16 3:24 P.M.

B-3.18 4:58 P.M.

B-3.20 18010 units

B-3.22 5:29 P.M.

APPENDIX C

Geometry

C-1 ■ Introduction

How to Use This Appendix

In Chapter 5 you learned how to use algebra to translate words into equations, solve problems with these equations, and then translate the mathematical result back into words. In this appendix you will learn an additional step in the problem-solving process: how to use geometry to translate words into diagrams and then to use formulas associated with these diagrams to translate the diagrams into equations.

This appendix is not a course in geometry, but rather an introduction to some basic geometric principles. The concepts of length, area, and volume are covered. A few of the most commonly encountered plane and solid figures are presented along with formulas that relate their length, area, and volume. The Pythagorean theorem, which gives the relationship between the sides of a right triangle, is also covered.

This material is intended to give you the background necessary to apply geometry to problem solving. You already know the importance of drawing a diagram as the first step in solving a problem. This material provides the next step in that process by giving you additional tools to analyze and make calculations based on that diagram.

Background

The word *geometry* is from the Greek meaning "earth measure." Ancient civilizations knew a great deal about geometry and used it for such things as land surveying, navigation, astronomy, carpentry, and drawing. Euclid organized all that was known about the field in his text, *Elements,* in about the year 325 B.C. The concepts from this text have survived with little change over the past 2300 years.

Midway through the 20th century the study of Euclid's geometry was a staple of the high school education. Many a student was first introduced to the concepts of critical thinking and logical reasoning by taking such a course. Sadly, this important experience is not nearly as common in today's schools.

Geometry for Problem Solving

One of the features of Euclid's geometry is that it is a subject stripped of the complexities of tedious calculation encountered with many other branches of mathematics, leaving the concepts of formal reasoning clearly exposed and unencumbered. Starting from a handful of first principles, and using rigorous step-by-step reasoning, a series of propositions are proved. As each is proved, it becomes a building block available for the construction of more complex proofs. As such, Euclid's geometry is a splendid introduction to problem solving.

For those of you who have not had the benefit of this learning experience, this book is not the appropriate place to allow you to do so. There will be no geometric proofs here. Only some basic geometric tools, useful in problem-solving definitions, will be covered. But, if you are ever in a position to influence the education of a future generation, consider recommending a course in Euclidean geometry.

Introductory Problem

1. Use graph paper, a compass, a ruler, and some paper tape to estimate the value of π (the ratio of circumference to diameter of a circle) in two ways (using length then area).
   Estimate π using length:

   - Use a compass to draw the largest circle that will fit on a piece of notebook paper.

   - Measure the distance across the circle along a line through the center (diameter).

   - Measure the distance around the circle (circumference) by fitting a length of paper tape around the path of the circle. Mark where the tape overlaps. Lay the tape flat and measure the distance to the mark. This measures the circumference.

   - Estimate π as the ratio of the circumference to the diameter you measured.

   Estimate π using area:

   - Use a compass to draw the largest circle that will fit on a piece of graph paper.

   - Measure the radius of the circle by counting grid lines on the graph paper.

   - Estimate the area of the circle by counting the squares on the graph paper.

   - Note that the actual area of a circle is found by the formula $a = \pi r^2$

   - Substitute the measured values of area and radius into the formula.

   - Compute an estimate of π.

   Compare your estimates of the value of π given in Section C-4.

2. Find a relationship between the sides of a right triangle:

   - Draw a right triangle on a piece of graph paper.

   - Construct three squares, one on each side of the triangle.

   - Estimate the area of each square by counting the small squares on the graph paper.

   - Describe a relationship between the area of the largest square and the areas of the other two squares.

Point

A point indicates only a position. It has no length, no width, and no thickness. As such it has no dimensions. It can be represented by a dot. But, the dot is only a device that enables us to visualize what would otherwise be invisible.

Curve

A curve is formed by moving a point along a path according to a specified set of rules. Curves include both lines and curved lines. The path can be represented in various ways, for example, by a mark of a pencil on a piece of paper or by pixels on a computer screen. But again, these are only devices that allow us to see the invisible.

Line

A line is the curve formed by moving a point in a single direction. In its most fundamental form it has no beginning or end, extending from infinity in one direction to infinity in the opposite direction. It has only one dimension, length.

Curved line

A curved line is formed by moving a point along a path in a continuously changing direction. The path may be in a two-dimensional plane (e.g., parabola or circle) or in three-dimensional space (e.g., helix).

Straight line

A straight line is the part of an infinite line that occurs between two points. Sometimes it is called a line segment. One of the basic principals of Euclidean geometry is that *a straight line is the shortest distance between two points.*

Angle

An angle is formed between two straight lines meeting at a point. Angles may be measured in degrees or radians, as defined later.

Degree

Degrees are one way to measure angles. One degree is 1/360th part of a full circle. Degrees can be expressed using the ° symbol. For example, an angle of 90° forms a quarter circle.

Pi

The Greek letter π is used to represent the ratio of the circumference of a circle to its diameter.

The circumference of a circle is the length of the curved line that marks a path equidistant from the center. The diameter of a circle is the distance across its widest part that passes through the center. The value of π can never be calculated exactly but is approximately 3.141592654.

Radian

Another way to measure an angle is by the amount of circular arc it forms. One radian (about 57.3 degrees) corresponds to the angle at the center of a circle that captures a portion of the circumference equal to the radius. A full circle (360°) is 2π radians. A half circle (180°) is π radians. A quarter circle (90°) is $\pi/2$ radians.

Straight angle

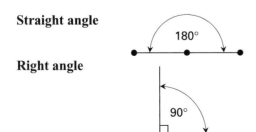

A straight angle is an angle that opens to form a straight line. It measures a half circle or 180 degrees or π radians.

Right angle

A right angle measures a quarter circle or 90 degrees or $\pi/2$ radians. Placing a small square in the corner of an angle indicates that it is a right angle. Lines at right angles to each other are said to be perpendicular.

Surface

A surface has length and width, but no thickness. As such it is two dimensional. Moving a curve through space forms a surface. Representations of a surface might be the top of a table, one side of a sheet of paper, or the outside of a basketball. But again, these are only representations of a surface, not the surface itself.

Plane

A plane is a surface formed by moving a line through space in a constant direction, other than the direction of the line. Like the line, a plane is infinite in extent. Any three points define a plane.

Parallel lines

Parallel lines are straight lines that lie entirely in the same plane and do not intersect however far they are extended.

Perpendicular

90°

When two lines intersect to form a right angle, they are said to be perpendicular to each other.

Plane figures

Plane figures are formed by connecting straight or curved lines in a plane so they enclose a given area. They have length and width, but no thickness. Plane figures have a surface area, but no volume. Examples are triangles, squares, rectangles, and circles.

Solid figures

Solid figures have length, width, and thickness. As such they are three-dimensional figures. They have both a surface area and a volume.

Right solids

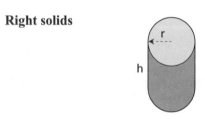

Right solid figures are formed by moving a plane figure in a direction perpendicular to its plane. In this way, a cube is formed from a square, a rectangular solid is formed from a rectangle, a prism is formed from a triangle, and a cylinder is formed from a circle. The sides of a right solid figure are perpendicular to its base and top.

Rotated solids

Solid figures may also be formed by rotating a plane figure about a line (an axis). Examples: rotating a circle forms a sphere; rotating a triangle forms a cone; rotating a rectangle forms a cylinder.

C-3 ■ Triangles

Perhaps the most important figure in geometry is the triangle. It comes up all the time in problem solving.

Definition: Three straight lines form the boundaries of a figure called a triangle. Each side has an opposite angle formed by the remaining two sides.

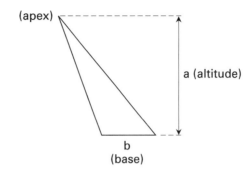

Components:
Vertex:	The point where two adjacent sides meet
Included angle:	The angle associated with a vertex
Base:	Any side can be called the base
Apex:	The vertex opposite the base
Altitude:	The perpendicular distance from the base to a line parallel to the base that passes through the apex

Relationships: Perimeter = sum of the lengths of the three sides
Sum of three angles = 180°
Sum of two shorter sides is greater than the longest side
Area = $\frac{1}{2}$ * base * altitude

Notice that the area of any triangle, regardless of its shape, can be found by $\frac{1}{2}$ * base * altitude. Because all of the following triangles have the same base (b) and the same altitude (a), they also have the same area.

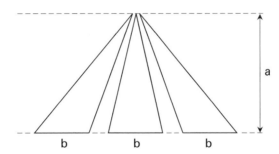

For each triangle:
area = 1/2 ab.

All three areas are equal.

Right Triangles

When two sides of a triangle are perpendicular to each other (form an angle of 90 degrees), the triangle is said to be a right triangle.

Definitions: Hypotenuse: The longest side, always opposite the right angle (c)
Base: One of the sides opposite the hypotenuse (b)
Altitude: The side perpendicular to the base (a)

Relationships: The sum of two angles opposite the right angle is 90 degrees.
Area = $\frac{1}{2}$ ab
Sides are related by the Pythagorean theorem (see the following section)

Pythagorean Theorem

The sides of any right triangle have a fixed relationship to one another. This relationship is expressed in the *Pythagorean theorem:*

 The square of the hypotenuse equals the sum of the squares of the other sides.
$$c^2 = a^2 + b^2$$

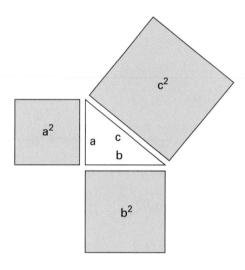

Given any two sides of a right triangle, the remaining side can be calculated:

$$c = \sqrt{a^2 + b^2} \qquad b = \sqrt{c^2 - a^2} \qquad a = \sqrt{c^2 - b^2}$$

Geometric Confirmation of the Pythagorean Theorem

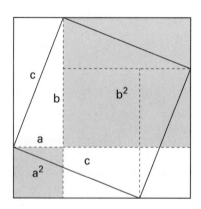

The whole area is made up of

- A small shaded square, a^2

- A large shaded square, b^2

- Two white rectangles, each ab

 Thus, the whole area is equal to $a^2 + 2ab + b^2$.

Looking at it another way, the whole area is also made up of

- The rotated square, c^2
- Four corner triangles, each $\frac{1}{2}ab$

 Thus, the whole area is also equal to $c^2 + 2ab$.

Equating these two ways to look at the whole:

$$c^2 + 2ab = a^2 + 2ab + b^2$$

Subtracting the term 2ab from both sides of the equation gives the Pythagorean theorem:

$$c^2 = a^2 + b^2$$

▶ Example Problems

1. Find the hypotenuse of a right triangle with the two smaller sides 3 feet and 4 feet.

$$a = 3 \text{ ft} \qquad \text{and} \qquad b = 4 \text{ ft}$$
$$c^2 = a^2 + b^2$$
$$c^2 = (3 * 3) + (4 * 4) \qquad = 9 + 16 = 25$$
$$\sqrt{c^2} = \sqrt{25}$$
$$c = 5 \text{ ft}$$

Mathematically, the square root of 25 has two solutions, $+5$ and -5. However, in this case a negative length makes no physical sense and therefore it is discarded.

    This is an example of a right triangle in which all sides are whole numbers. Most right triangles do not have this property. The 3-4-5 right triangle is the smallest right triangle that does.

2. Find the hypotenuse of a right triangle with the two shorter sides each 1 foot long.

$$a = 1 \text{ ft} \qquad \text{and} \qquad b = 1 \text{ ft}$$
$$c^2 = a^2 + b^2$$
$$c^2 = (1 * 1) + (1 * 1) = 2$$
$$\sqrt{c^2} = \sqrt{2}$$
$$c = 1.41 \text{ ft} \qquad \text{(rounded to the nearest hundredth foot)}$$

This is an example of a right triangle with a hypotenuse that has a length that is neither a whole number nor a number that can be expressed as a ratio of whole numbers. Thus, the length of the hypotenuse is not a *rational* number. It is instead a new kind of number called irrational. The square root of 2 is an irrational number.

3. Find the altitude of a right triangle with a hypotenuse of 10 feet and a base of 5 feet.

$$c = 10 \text{ ft} \qquad \text{and} \qquad b = 5 \text{ ft}$$
$$c^2 = a^2 + b^2$$

$$(10 * 10) = a^2 + (5 * 5)$$

$$a^2 = 100 - 25 \ \ = 75$$

$$\sqrt{a^2} = \sqrt{75} \ \ = \sqrt{5 * 5 * 3}$$

$$a = 5\sqrt{3}$$

$$a = 5 * 1.732 \ldots$$

$$a = 8.66 \text{ ft} \qquad \text{(rounded to the nearest hundredth foot)}$$

4. Find the side of a square with a diagonal of 2 feet.

$$c = 2 \text{ ft} \qquad \text{and} \qquad a = b \qquad \text{(since a and b are the equal sides of a square)}$$

$$c^2 = a^2 + b^2$$

$$2^2 = a^2 + a^2 \qquad \text{(substitute } a^2 \text{ for } b^2\text{)}$$

$$2^2 = 2a^2$$

$$a^2 = (4 / 2) = 2$$

$$\sqrt{a^2} = \sqrt{2} \qquad \text{(taking the square root of both sides)}$$

$$a = 1.41 \text{ ft} \qquad \text{(rounded to the nearest hundredth foot)}$$

Similar Triangles

Similar triangles have the same shape but not necessarily the same size. For any two triangles to be similar, corresponding angles must be equal. Then corresponding sides will have

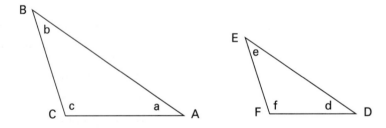

proportional lengths.

Triangles ABC and DEF are similar if corresponding angles are equal:

$$a = d \qquad b = e \qquad c = f$$

Corresponding sides of similar triangles are proportional to each other. This means that the ratio of any two sides of triangle ABC is equal to the ratio of the corresponding sides in triangle DEF. There are six possible ratios in each triangle:

$$
\begin{array}{ll}
AB / AC = DE / DF & BC / AC = EF / DF \\
AB / BC = DE / EF & BC / AB = EF / DE \\
AC / AB = DF / DE & AC / BC = DF / EF
\end{array}
$$

Thus, if the lengths of any two sides are known in one triangle and the length of only one of the corresponding sides is known in the similar triangles, the remaining side can be calculated using the previous ratios.

Similar Right Triangles

For two right triangles to be similar, corresponding angles opposite the right angle must be equal. Consider the two right triangles formed by dividing triangle ABC with line DE parallel to line AC. Triangles ABC and DBE are similar because corresponding angles are equal:

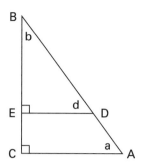

- Both triangles contain angle b.

- Both triangles contain a right angle (90°).

- Angles a and d are equal because three angles in a triangle add to 180°.

Therefore: d = 180° − 90° − b
a = 180° − 90° − b

The following ratios will be equal between the similar right triangles ABC and DBE:

base / hypotenuse

base / altitude

altitude / hypotenuse

▶ **Example Problem.** Suppose the right triangle ABC has a base of 3 feet, an altitude of 4 feet, and a hypotenuse of 5 feet. If side DE is 2.5 feet, find the length of side BE:

$$AC / BC = DE / BE$$

$$3 / 4 = 2.5 / BE$$

$$BE = 2.5 * (4 / 3)$$

$$BE = 3.33 \text{ ft}$$

Aside: Trigonometry

Trigonometric functions are defined as a ratio of sides of a right triangle.

If the three sides of the right triangle are hypotenuse, adjacent, and opposite, then the functions for the angle x, included between the hypotenuse and the adjacent side, are

Sine	sin x = opposite side / hypotenuse
Cosine	cos x = adjacent side / hypotenuse
Tangent	tan x = opposite side / adjacent side

Given a few parts (sides, angles) of a triangle, trigonometry can be used to find the others. You can use your scientific calculator to determine the sine, cosine, and tangent for any angle.

C-4 ■ Circles and Ellipses

Circles

A circle is another geometric figure that is commonly encountered in problem solving.

Definition: A circle is a figure bounded by all the points that are equidistant from its center.

Components: Center: The point that is equidistant from all points on the boundary

   Circumference: The distance around the boundary, the perimeter
   Radius: The distance from the center of the edge (r)
   Chord: A straight line that intersects the circle at two points
   Diameter: A chord passing through the center, twice the radius (d)

Relationships: Diameter $= 2r$
   Circumference $= \pi d = 2\pi r$ (π is defined in the following section)
   Area $= \pi r^2 = \frac{1}{4}\pi d^2$

The Value of Pi

The Greek letter pi, written π, designates the ratio of the length of the circumference of a circle to its diameter:

$$\pi = c / d$$

This ratio is the same for circles of any size. The value of π can never be expressed exactly as the ratio of two whole numbers. Thus, if the radius is a whole number, the circumference cannot be a whole number. Or, if the circumference is a whole number, the radius cannot be a whole number.

Mathematicians call π a transcendental number. It can only be expressed as a never-ending and never-repeating decimal number. There are, however, fractions that approximate the value of π. The fraction 22/7 is correct to two decimal places. The fraction $\frac{355}{113}$ is even closer, being correct to six decimal places. Most scientific calculators will display the value of π rounded to 10 decimal places:

$$\pi = 3.141592654 \ldots$$

Ellipses

The ellipse is a "stretched out" circle.

Definition: An ellipse is a figure with two central points called foci, f_1 and f_2, having the following property:

Any point p on the boundary of the figure is positioned such that the sum distances from the point to each focus remains constant.

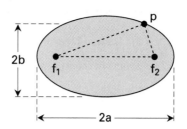

An ellipse is formed by all the points where

$$pf_1 + pf_2 = 2a$$

The terms are defined as follows:

- pf_1 is the distance from any point p to focus f_1.

- pf_2 is the distance from any point p to focus f_2.

- 2a, the constant, is the length of the line that passes through the two foci and extends to the opposite edges of the ellipse.

Components: Foci: Two points, f_1 and f_2, that lie on the major axis such that $pf_1 + pf_2$ remains constant for any point p on the ellipse.

Major axis: The widest diameter that passes through f_1 and f_2 (2a)

Minor axis: The widest diameter perpendicular to the major axis (2b)

Semi-major axis: Half the major axis (a)

Semi-minor axis: Half the major axis (b)

Relationships: $2a = pf_1 + pf_2$
Area $= \pi ab$

How to Draw an Ellipse

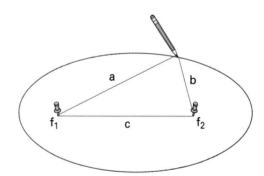

Materials you will need:

A piece of paper

A pencil

Length of string

Two push pins

A flat board

Follow these steps:

1. Lay the paper on the board.

2. Push the pins into the paper about three inches apart.

3. Tie the string into a loop about 8 inches long.

4. Put the loop around the pins.

5. Put the pencil in the loop and stretch the loop tight.

6. Keeping the string taught, draw a curved path around the pins.

The curve you draw will be an ellipse. Try to draw other ellipses by adjusting the distance between the pins or the size of the loop.

Notice that:

- The positions of the pins represent the two foci of the ellipse, f_1 and f_2.

- For a given loop size, $a + b + c$ is constant and the sum of the distance from each focus to the path of the ellipse is also a constant. This is a fundamental geometric requirement for any ellipse.

- Keeping the pins in the same position:

  As the loop gets larger, the ellipse becomes rounder (more like a circle).

  As the loop gets smaller, the ellipse becomes narrower (more like a line).

- Keeping the loop the same length:

  As the pins get closer together, the ellipse becomes rounder (more like a circle).

  As the pins get farther apart, the ellipse becomes narrower (more like a line).

 A circle is a special case of the ellipse in which the two foci merge to become the center.

C-5 ■ Other Plane Figures

Plane figures, as their name implies, lie entirely in a single plane. Straight lines or curves are used to form a boundary that encloses the area of the figure. The triangle, the circle, and the ellipse have already been discussed. Here, the properties of a few other plane figures will be summarized.

Perimeter
The length of the boundary of a figure is called the perimeter. For the circle, this perimeter is called the circumference. The perimeter of a figure is measured in units of length.

Area
The area of a surface is measured in units of length squared. As such, the calculation of an area involves multiplying two lengths together. The surface area of a geometric figure represents the number of square units (and parts of these units) that can be "tiled" to exactly cover the surface bounded by the edges of the figure.

Measuring Area by Tiling
Count the unit squares needed to cover the following rectangle and compare this to the area computed by height times width. The following tiles are needed to cover this rectangle:

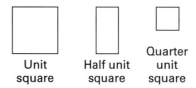

| Unit square | Half unit square | Quarter unit square |

Count the tiles needed to cover the rectangle $6\frac{1}{2}$ units wide and $4\frac{1}{2}$ units high:

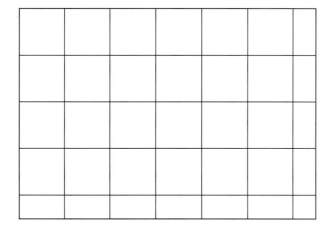

Count the tiles used:

		Area
Full squares	24	24
Half squares	10	5
Quarter squares	1	$\frac{1}{4}$

Total square units $= 24 + 5 + \frac{1}{4}$

Area of tiles used $= 29.25$ squre units

Compute area as height times width and compare:

$$\text{area} = \text{width} * \text{height} = 4.5 \text{ units} * 6.5 \text{ units} = 29.25 \text{ square units}.$$

The answer is the same.

Polygons

Polygons are figures bounded by straight lines. The term means "many sides." A few of the most commonly encountered polygons will be shown first. Then the special case in which all sides of a polygon are of equal length, a regular polygon, will be covered.

Rectangle

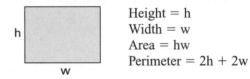

Height = h
Width = w
Area = hw
Perimeter = 2h + 2w

Square

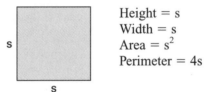

Height = s
Width = s
Area = s^2
Perimeter = 4s

Triangle

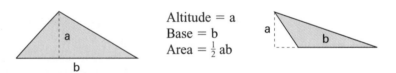

Altitude = a
Base = b
Area = $\frac{1}{2}$ ab

Trapezoid

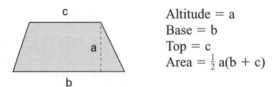

Altitude = a
Base = b
Top = c
Area = $\frac{1}{2}$ a(b + c)

Parallelogram

Altitude = a
Base = b
Area = ab

Regular Polygons

Regular polygons are plane figures with equal sides, equal central angles, and equal interior angles where

The altitude is the perpendicular distance from center to side.

The central angle is the angle between the center point and the ends of any side.

The interior angle is the angle between adjacent sides.

Regular polygons can have any number of sides greater than three. The first few are presented here:

Equilateral Triangle

Number of sides: 3
Sides all $= s$
Central angle $= \frac{1}{3} * 360° = 120°$
Interior angle $= \frac{1}{3} * 180° = 60°$
Altitude $= (\frac{1}{2} s) / \sqrt{3} = 0.28867s$
Perimeter $= 3s$
Area $= \frac{1}{2}$ altitude * perimeter $= (\frac{3}{4} s^2) / \sqrt{3} = 0.43301 \ s^2$

Square

Number of sides $= 4$
Sides all $= s$
Central angle $= \frac{1}{4} * 360° = 90°$
Interior angle $= \frac{2}{4} * 180° = 90°$
Altitude $= \frac{1}{2} s$
Perimeter $= 4s$
Area $= \frac{1}{2}$ altitude * perimeter $= s^2$

Pentagon

Number of equal sides $= 5$
Sides all $= s$
Central angle $= \frac{1}{5} * 360° = 72°$
Interior angle $= \frac{3}{5} * 180° = 108°$
Altitude $= 0.68819 * s$
Perimeter $= 5s$
Area $= \frac{1}{2}$ altitude * perimeter $= 1.72048 \ s^2$

Hexagon

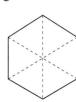

Number of sides $= 6$
Sides all $= s$
Central angle $= \frac{1}{6} * 360° = 60°$
Interior angle $= \frac{4}{6} * 180° = 120°$
Altitude $= 0.86602 * s$
Perimeter $= 6s$
Area $= \frac{1}{2}$ altitude * perimeter $= 2.59808 \ s^2$

General Regular Polygon

Number of sides: $= n$
Sides all $= s$
Central angle $= 360° / n$
Interior angle $= 180° -$ central angle $= 180° * (n - 2) / n$
Altitude $= \frac{1}{2} s / \tan (360° / 2n)$
Perimeter $= n * s$
Area $= \frac{1}{2} *$ altitude * perimeter

C-6 ■ Solid Figures

Solid figures lie in three-dimensional space. Some solid figures are formed by moving (or rotating) a plane figure in space to sweep out a volume. Other solid figures have boundaries formed by several plane figures, each in a different plane.

Volumes

Volumes of solid figures are measured in units of length cubed. As such, the calculation of a volume involves multiplying three lengths together. The volume of a solid figure represents the number of cubic units (and parts of these units) that can be assembled as building blocks to exactly fill the space inside the figure.

 Three-dimensional solid figures have volumes measured in length units cubed.

Using Unit Cubes to Compute Volumes

Like tiling area with unit squares, units volumes can be calculated by counting the number of unit cubes that will fit inside. Consider the right rectangular solid (box) that has the following dimensions: $3\frac{1}{2}$ units by $5\frac{1}{2}$ units by 4 units. How many unit cubes will fit inside?

Look at the unit cubes as being like building blocks assembled in layers. We will need full cubes, half cubes, and quarter cubes to assemble each layer.

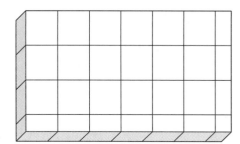

The first layer of blocks is made up of

15 full cubes	15 cubic units
8 half cubes	4 cubic units
1 quarter cube	$\frac{1}{4}$ cubic units

The volume of the first layer is $19\frac{1}{4}$ cubic units.

Since each layer is one unit high, exactly four layers are needed to fill the 4-unit-high box. Four layers of 19.25 units each is 77 cubic units in all.

Computing the volume by multiplying length, width, and height is $3\frac{1}{2} * 5\frac{1}{2} * 4 = 77$ cubic units.

 The volume of a box that is found by counting the number of unit blocks needed to fill it is exactly the same as the volume found by multiplying the height, width, and depth of the box.

Surface Area of Solids

The surface area of a solid figure is similar to the area of a plane figure, except that the surface of a solid does not lie entirely in the same plane. Rather, it exists in three-dimensional space. Some solid figures, for example, the cube, have surface areas that are the sum of several plane figures. In the case of the cube, the surface is made up of the six squares that form the sides. Other solid figures, for example, the sphere, have surface areas that do not have parts but are "wrapped" continuously around the figure in three-dimensional space.

 The surface area of a solid figure is measured in units of length squared.

Right Solid Figures

Right solid figures have a base that is a plane figure. The base figure is raised to a perpendicular height above the plane to form the solid. Right solid figures have the following properties:

- The top and bottom surfaces of the solid are the same.

- The sides are at right angles to the base.

- The volume is equal to the area of the base times the perpendicular height.

- The surface area is the sum of twice the area of the base plus the area of the sides.

- The area of the sides is equal to the perimeter of the base times the perpendicular height.

Rectangular Solid

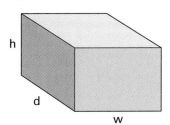

Base is a rectangle

Base has: Width = w
Depth = d
Perimeter = 2w + 2d
Area = wd

Solid has: Height = h
Volume = wdh
Surface area
= 2wd + h(2w + 2d)

Cube

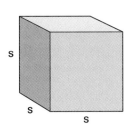

Base is a square

Base has: All sides = s
Area = s^2
Perimeter = 4s

Solid has   Height = s
Volume = s^3
Surface area = $6s^2$

Equilateral Prism

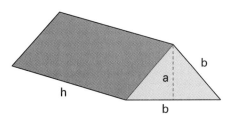

Base is an equilateral triangle

Base has: All sides = b
Altitude = a
Area = $\frac{1}{2}$ ab

Solid has: Height = h
Volume = $\frac{1}{2}$ abh
Surface area = ab + 3bh

Cylinder

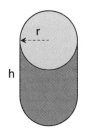

Base is a circle

Base has: Radius = r
Circumference = $2\pi r$
Area = πr^2

Solid has: Height = h
Volume = $\pi r^2 h$
Surface area = $2\pi r^2 + 2\pi rh$

Other Solid Figures

Rotating a plane figure about an axis can form other solid figures.

Right Cone

Formed by rotating right triangle about its altitude, (h)

Triangle has: Base = r
                    Altitude = h

Circle has: Radius = r
                Area = $\pi r2$

Solid has: Height = h
               Volume = $\frac{1}{3}\pi r^2 h$
               Surface area
                   $= \pi r^2 + \pi r \sqrt{r^2 + h^2}$

Sphere

Formed by rotating circle about a diameter, (d)

Circle has: Diameter = d
                Radius = $r = \frac{1}{2}d$

Solid has: Maximum height = 2r
               Volume = $\frac{4}{3}\pi r^3$
               Surface area = $4\pi r^2$

C-7 ■ Problem Solving with Geometry

In Chapter 5, A Little Algebra, you saw how to translate a problem statement directly into equations. Geometry gives you some additional problem-solving tools. A variety of geometric figures are now at your disposal to draw diagrams to help explain a problem graphically. The preceding sections introduced many of these figures and presented geometric properties for each. This information can be used to develop additional equations for problem solving that complement those already available from algebra.

Diagrams

A picture is worth a thousand words. Geometric figures such as lines, triangles, rectangles, and circles are available to construct diagrams to explain a problem in a graphic fashion. Learn to make use of the geometric figures introduced in the earlier part of this appendix to construct these diagrams.

Analyzing Diagrams

Each geometric figure has properties such as length, area, and volume with associated formulas that are available to write equations to help you solve a problem. Once the diagram is built, look for relationships within the diagram that will yield equations. Recall that in algebra, one independent equation is needed for each unknown in the problem. These "geometric" equations may be the key that allows you to solve for the last remaining unknown of the problem.

The following four-step process will prove useful in attacking problems with geometry:

1. Draw the diagram by combining one or more geometric figures.

2. Use the formulas available for each figure to write equations.

3. Substitute values of input into the formulas.

4. Solve the equations for the desired output.

► Example Problem

There are many situations in which you are given the horizontal and vertical distances between two points and need to calculate the diagonal distance. Consider this example:

Starting from the intersection of Fifth Avenue and Main Street, you walk four blocks north and three blocks west. How far are you from your starting point?

Along the city streets you are seven blocks away. The distance "as the bird flies" is shorter. Because the city streets are at right angles to one another, we can calculate the shortest distance between two points using the Pythagorean theorem.

Assume that all blocks are the same length and that the distance "as the bird flies" is measured in the same unit.

Conceptual Diagram

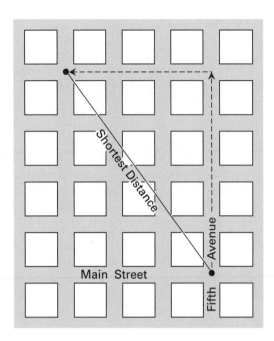

Main Street

1. Draw a simplified diagram:

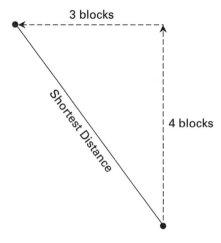

3 blocks

4 blocks

Shortest Distance

Let d = diagonal distance (in blocks).

2. Write an equation:

$$d^2 = a^2 + b^2$$

3. Substitute values:

$$d^2 = 4^2 + 3^2$$

4. Solve for the desired output:

$$d = \sqrt{16 + 9} = \sqrt{25}$$

$$d = 5 \text{ blocks}$$

▶ **Another Example**

What is the distance between point p_1 (2, 3) and point p_2 (5, 7) on a rectangular coordinate system in which each unit of distance represents 1 foot?

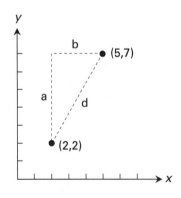

1. Draw a diagram.
   Let d = diagonal distance from p_1 to p_2.
   Construct right triangle at given points:
   Let a = $y_2 - y_1 = 7 - 2 = 5$.
   Let b = $x_2 - x_1 = 5 - 2 = 3$.

2. Write an equation:

$$d^2 = a^2 + b^2$$

3. Substitute values:

$$d^2 = 5^2 + 3^2$$

$$\sqrt{d^2} = \sqrt{25 + 9} = \sqrt{34}$$

4. Solve for the desired output.

$$d = 5.83 \text{ ft.}$$

Three-Dimensional Example

What is the diagonal distance across a cube? The diagonal of a cube is the distance between the two corners that are farthest apart. Assume the length of a side is one unit.

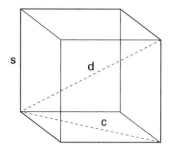

1. Draw a diagram.
   Let s = length of the side of the cube.
   Let c = diagonal of one face.
   Let d = longest diagonal across the cube.

2. Write equations:

$$c = \sqrt{s^2 + s^2}$$
$$d = \sqrt{s^2 + c^2}$$

3. Substitute values:

Length of side = 1 (unit cube)

$$c = \sqrt{1^2 + 1^2}$$
$$c = \sqrt{2}$$
$$d = \sqrt{1^2 + \sqrt{2}^2}$$

4. Solve for desired output:

$$d = \sqrt{1 + 2}$$
$$d = \sqrt{3}$$

Adding Figures

Many times a diagram for a problem cannot be made with a single geometric figure. By adding several simple geometric figures together, a more complex figure can be constructed.

▶ *Example Problems*

1. What is the area of the floor plan for this irregularly shaped room? Lengths are in feet.

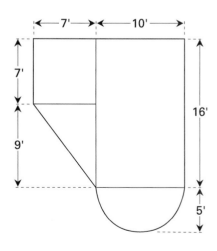

   a) Draw a diagram. (Room area is the sum of partial areas.)
   b) Write formulas:

   Area of square $= s^2$
   Area of rectangle $= ab$
   Area of triangle $= \frac{1}{2} ab$
   Area of semicircle $= \frac{1}{2} \pi r^2$

   c) Substitute values:

   Area of square $= 7 * 7 = 49$ ft^2
   Area of rectangle $=$
       $16 * 10 = 160$ ft^2
   Area of triangle $=$
       $\frac{1}{2} * 7 * 9 = 31.5$ ft^2
   Area of semicircle $=$
       $\frac{1}{2} * 3.1416 * 5 * 5$
       39.3 ft^2

d) Solve for output:

Room area = area of square
+ area of rectangles
+ area of triangle
+ area of semicircle

Room area = 49 + 160 + 39.5 + 39.3

Room area = 287.8 ft^2

2. Estimate the volume of the building. Lengths are in feet.

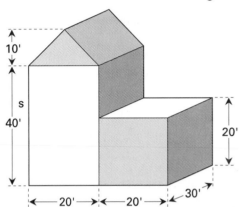

Parts of the building are:

Volume of prism = $\frac{1}{2}$ abd
= $\frac{1}{2}$ 10 * 20 * 30 = 3000 ft^3

Volume of large box = wdh
= 20 * 30 * 40 = 24,000 ft^3

Volume of small box = wdh
= 20 * 30 * 20 = 12,000 ft^3

Building volume = sum of parts
= 3,000 + 24,000 + 12,000 ft^3

Building volume = 39,000 ft^3

3. How many feet of metal rods are needed to construct a crab trap in the shape of a cube with sides three feet long?

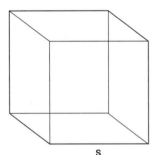

The "perimeter" of a cube is the length of all edges.
Let s = side of the cube.
Let p = perimeter of square = 4s.
Let p = perimeter of cube which is made up of:
 A square base with 4 edges = 4s
 A square top with 4 edges = 4s
 Additional edges connect 4 corners = 4s

p = 4s + 4s + 4s = 12s
Since s = 3 ft
p = 12 * 3 ft
p = 36 ft

4. What is the area of the following shape?

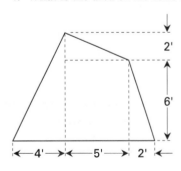

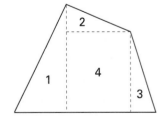

Divide the shape into triangles and rectangles, each with a known base and altitude:

Triangle 1:   A1 = $\frac{1}{2}$ (4 * 8)   = 16 ft^2

Triangle 2:   A2 = $\frac{1}{2}$ (2 * 5)   = 5 ft^2

Triangle 3:   A3 = $\frac{1}{2}$ (2 * 6)   = 6 ft^2

Rectangle 4:   A4 = 5 * 6   = 30 ft^2

Total area = 16 + 5 + 6 + 30 = 57 ft^2

Subtracting Figures

Subtracting figures to arrive at more complex shapes also proves to be an important problem-solving tool.

▶ *Example Problems*

1. What is the area of a one-inch-wide ring of crust added to a 10-inch pizza to make it a 12-inch pizza? Express the area of the ring as a percent of the 10-inch pizza.

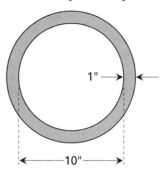

Area of 10-in. pizza $= \pi(r_1)^2 = \pi(5)^2$
                                     $= 78.5 \text{ in}^2$

Area of 12-in. pizza $= \pi(r_2)^2 = \pi(6^2)$
                                      $= 113.1 \text{ in}^2$

Area of ring $= 113.1 - 78.5 = 34.6 \text{ in}^2$

Area of ring as a percent of 10-in. pizza:
      34.6 is what percent of 78.5?
      $34.6 = (x / 100) * 78.5$
      $x = 100 * 34.6 / 78.5$
      $x = 44\%$

Area of ring is 44% of 10-in. pizza.

2. Using the same diagram to represent a cross section of a spherical shell 1-inch thick that is added to a 10-inch-diameter sphere, express the volume of the shell as a percent of the 10-inch sphere.

Radius of 10-in. sphere is 5-in.
Radius of 12-in. sphere is 6-in.
Volume of a sphere is $(4/3)\pi r^3$.

Volume of 10-in. sphere is $(4/3)\pi(5)^3$ $= (4/3)\pi(125)$ $= 523.6 \text{ in.}^3$
Volume of 12-in. sphere is $(4/3)\pi(6)^3$ $= (4/3)\pi(216)$ $= 904.8 \text{ in.}^3$

Volume of shell is the difference between two volumes $= 904.8 - 523.6 = 381.2$.

Volume of shell as a percent of the smaller volume:
      381.2 is what percent of 523.6?
      $381.2 = (x / 100) * 523.6$
      $x = 100 * 381.2 / 523.6$
      $x = 72.8\%$

Area of shell is 72.8% of smaller volume.

Notice that increasing the diameter 20% increases

- The area of a circle by a factor of 1.2^2 or 44% (from example 1)
- The volume of a sphere by a factor of 1.2^3 or 72.8%

Practice Problems

Solve the following problems using the *informal* IPO method. Just draw a diagram and use the appropriate formula presented earlier in the appendix. Compute the answer in units that correspond to the units given. For example, if length is given in feet to compute area, the result should be in square feet. Results that are not whole numbers should be rounded to two decimal places.

Distance Problems

C-7.1   Compute the perimeter of a square 2 inches on each side.

C-7.2   Compute the perimeter of a rectangle with a 3-inch base and a 4-inch altitude.

C-7.3   Compute the circumference of a circle with a 2-foot diameter.

C-7.4   Compute the circumference of a circle with a 2-foot radius.

C-7.5 Compute the diameter of a circle with a circumference of 12 inches.

C-7.6 Compute the length of the hypotenuse of a right triangle with a base of 2 feet and an altitude of 3 feet.

C-7.7 Compute the altitude of a right triangle with a 12-foot base and a 13-foot hypotenuse.

C-7.8 Compute the diagonal distance between opposite corners of a square 1 foot on a side.

C-7.9 Compute the diagonal distance between opposite corners of a 2-foot by 1-foot rectangle.

C-7.10 Compute the altitude of a triangle in which all three sides are 2 feet in length.

C-7.11 Compute the sum of the lengths of all the edges in a rectangular solid 5 by 4 by 3 inches.

C-7.12 Compute the height of a right cylinder with a 1-foot radius and a volume of 10 cubic feet.

C-7.13 Compute the perpendicular distance from the center to the side of a regular polygon with four sides, each 8 inches in length.

Area Problems

C-7.14 Compute the area of a square with sides of 2 yards each.

C-7.15 Compute the area of a rectangle with a 5-inch height and a 6-inch width.

C-7.16 Compute the area of a triangle with a 2-inch base and a 4-inch altitude.

C-7.17 Compute the area of half a 3-inch by 6-inch rectangle cut along its diagonal.

C-7.18 Compute the area of a circle with a diameter of 3 feet.

C-7.19 Compute the area of a circle with a radius of 2 feet.

C.7.20 Compute the area of a circle with a circumference of 10 feet.

C-7.21 Compute the surface area of a cube 3 centimeters on each side.

C-7.22 Compute the surface area of a rectangular solid 2 by 4 by 6 meters.

C-7.23 Compute the surface area of a sphere with a 1-foot radius.

C-7.24 Compute the surface area of a cone with a base 2 feet in diameter and a height of 6 feet.

C-7.25 Compute the area of a square with a diagonal of 3 feet.

C-7.26 Compute the surface area of a cube with the diagonal of a face 2 feet in length.

C-7.27 Compute the surface area of a cube in which the longest diagonal (from a top corner to the opposite bottom corner) is 3 feet.

C-7.28 Compute the area of an equilateral triangle each side of which is 1 foot in length.

Volume Problems

C-7.29 Compute the volume of a cube 2 inches on each side.

C-7.30 Compute the volume of a cube with the diagonal of a side 2 meters in length.

C-7.31 Compute the volume of a right rectangular solid 6 by 12 by 3 feet.

C-7.32 Compute the volume of a sphere 2 feet in diameter.

C-7.33 Compute the volume of a right cylinder with a 2-foot diameter base and a 6-foot height.

C-7.34 Compute the volume of a right solid with a base area of 6 square inches and a height of 4 inches.

C-7.35 Compute the volume formed by rotating a right triangle with sides 3 feet, 4 feet, and 5 feet. Rotate the triangle about the 4-foot side.

More Difficult Problems

C-7.36 Compute the length of the longest diagonal (from a top corner to the opposite bottom corner) in a rectangular box 3 feet by 2 feet by 1 foot.

C-7.37 Compute the volume of water needed to fill a circular pipe with walls $\frac{1}{8}$-inch thick, outside diameter of 1 inch, and length of 12 inches.

C-7.38 Compute the volume of plywood sheets needed to make a box in the shape of a cube with edges 8 inches long. Assume the thickness of the plywood is exactly $\frac{1}{4}$ inch.

C-7.39 Compute the internal volume of a cube with outside edges of 8 inches and walls $\frac{1}{4}$-inch thick.

C-7.40 How many cubic yards of concrete should you order to make a foundation of a building? The concrete is to be poured into a trench with straight sides 3 feet deep and a flat bottom 3 feet wide. The outside of the trench forms a square 30 feet on a side, when viewed from the top.

C-7.41 Billy invites four friends over for a birthday party. His cake is made in the shape of a 10-inch by 10-inch square 1 inch high, with frosting on the top and sides. How can the cake be cut with just five slices of the knife so that each of the five persons gets a piece with the same amount of cake and the same amount of frosting?

C-7.42 Each month, in its elliptical orbit around the earth, the moon ranges from a maximum of about 248,000 miles to a minimum of about 217,000 miles above the surface of the earth. The farther the moon is from the earth, the smaller it appears in the sky. The closer it is, the larger it appears. It is a wonderful accident that somewhere within this range is the precise distance at which the moon appears exactly the same size as the sun. Every few years when the earth, moon, and sun are aligned in a straight line and the moon is at the correct distance, the moon's shadow sweeps across the face of the earth. This is the spectacle of a total eclipse of the sun.

Given the following information, use proportional right triangles to determine the distance of the moon from the surface of the earth so the disk of the moon will appear in the sky to be the same size as the disk of the sun.

Radius of sun:	435,000 miles
Radius of moon:	1,080 miles
Distance from earth to sun:	93,000,000 miles

Problem Solutions

C-7.2   14 in.
C-7.4   12.57 ft
C-7.6   3.60 ft
C-7.8   1.41 ft
C-7.10 1.73 ft
C-7.12 3.18 ft
C-7.14 4 yd^2
C-7.16 4 $in.^2$
C-7.18 7.07 ft^2
C-7.20 7.96 ft^2
C-7.22 88 m^2
C-7.24 25.12 ft^2
C-7.26 12 ft^2
C-7.28 0.22 ft^2
C-7.30 2.83 m^3
C-7.32 4.19 ft^3
C-7.34 24 $in.^3$
C-7.36 3.74 ft.
C-7.38 90.13 $in.^3$
C-7.40 36 yd^3
C-7.42 230,896 mi

Glossary

Words in italics are defined elsewhere in the glossary. The chapter where a term is introduced is given in parentheses at the end of each definition.

Absolute Value The value of a number without regard to its sign. Thus, -100 has the same absolute value as $+100$. This can be written as $|-100| = |+100| = 100$. (Ap A)

Accumulating An *assignment* technique used in computer programs, usually inside a *loop,* to compute the sum of a set of values. Each time the accumulating statement is executed, the value stored in an accumulating variable is increased by an amount stored in another *variable.* The amount added changes each time the statement is executed. Accumulating differs from *incrementing* in that the amount added or subtracted is a *variable* rather than a *constant.* (Ch 7)

Algebraic Expression An expression formed by combining algebraic *variables, coefficients,* and *constants* with *arithmetic operators,* though not necessarily all at once. Two examples are $3x - 1$ and $x^2 + y^2$. (Ch 5)

Algorithm A set of instructions that tells how to accomplish some *process.* The instructions must be unambiguous for the situation in which they are used. If the instructions are telling a cook how to bake a cake, the algorithm is called a recipe. If the instructions are telling a computer how to produce an annual sales report, the algorithm is called a *computer program.* (Ch 7, Ch 9)

Altitude The *straight line* that measures the height of a geometric figure such as a *polygon, cone,* or *cylinder.* For a *triangle,* it is the *perpendicular* distance from the *vertex* to the line containing the *base.* For a *right triangle,* it is the side *perpendicular* to the base. (Ap C)

AND Operator A logical operator that works between two *conditional expressions, a* and *b.* The true

(T) -false (F) outcome of the expression (a AND b) depends on the logical outcome of the two conditional expressions: Thus, T AND T = T, T AND F = F, F AND T = F, and F AND F = F. (Ch 8)

Angle A measure of the *circular arc* enclosed between two *straight lines* meeting at a *point.* Angles may be measured in *degrees* or *radians.* (Ap C)

Area A measurement of the size of a *surface* having *units* of length squared. This measurement applies to the surface of a two-dimensional or *plane figure* such as a *rectangle* or *circle.* It also applies to the two-dimensional outer surface of a three-dimensional figure such as a *cube* or *sphere.* In this case it is called *surface area.* (Ap C)

Arithmetic Expression An expression used in a *computer program* that combines *constant* values and values stored in *variables* with *arithmetic operations* such as addition, subtraction, multiplication, division, and exponentiation. The expression is evaluated using an *order of precedence* to give a specific numeric value. The simplest arithmetic expression is a single *variable* or *constant.* (Ch 7)

Arithmetic Mean The *average* of a set of numbers derived by dividing their sum by their count. (Ap A)

Arithmetic Operator An operator that performs arithmetic. Examples are addition $(+)$, subtraction $(-)$, multiplication $(*)$, division $(/)$, exponentiation (x^2), or roots $(\sqrt{x})$. (Ap A)

Assignment A technique used in *computer programs* to store data in *variables.* The data may come from outside the program as *input,* from other *variables,* or from calculations. (Ch 7)

Associative Law A property of numbers that says: When adding several numbers, it makes no difference which two are added first. When multiplying several numbers, it makes no difference which two are multiplied first. (Ap A)

Average A statistical measure of a set of data that gives the value of a typical member of the set. There are various ways to calculate this value. This book includes two: the *arithmetic mean* and the *weighted average*. (Ap A)

Bar Chart A *graphing* technique in which a *data series* is displayed as a set of bars, each bar representing one data item. The length of each bar is proportional to the value of the data item it represents. (Ch 6)

Base (arithmetic) A number to which an exponent is applied. The number raised to a power. (Ap A)

Base (geometry) The *straight line* (usually horizontal) on which a *polygon* stands. Also, for *solid figures* with at least one flat side, the *area* (usually horizontal) on which the figure stands. (Ap C)

Base (number system) The number that is raised to successive powers to define the magnitude of each position within a *positional number system.* For *decimal numbers* the base is 10; for *binary numbers* the base is 2; for *hexadecimal numbers* the base is 16. (Ch 3)

Binary Number A number generated by the positional number system using the base 2. Binary numbers are used exclusively in the internal operation of computers. This is because computers are built from transistors, electronic devices having just two states, on and off. (Ch 3)

Bit A binary digit (zero or 1). Digits in binary numbers are called bits. In computer memory, eight bits make up a *byte.* (Ch 4)

Boolean Algebra An algebra of logic developed by 19th-century mathematician George Boole. (Ch 8)

Boolean Variable *Variables* used in *programming languages* that contain a logical outcome of true or false. Some languages have a dedicated type of variable for this purpose. Other languages allow a regular variable to store a zero for false and some other value (usually-1) for true. These variables can be used in *conditional expressions.* (Ch 8)

Byte A collection of eight *bits.* A measure of computer memory. (Ch 4)

Circle A *plane figure* bounded by the path of all *points* equidistant from some central *point.* The bounding edge is called the *circumference.* Any

straight line between the center and the *circumference* is called a *radius.* Any *straight line* from edge to edge, passing through the center, is called a *diameter.* The value formed by the *ratio* of the *circumference* and the *diameter,* denoted by the symbol π (called *pi*) is constant for all circles. (Ap C)

Circular Arc The part of the *circumference* of a *circle* contained between two *radii.* (Ap C)

Coefficient A term that multiplies an algebraic *variable.* Usually a coefficient is a *constant* value or a letter that represents a *constant* value. (Ch 5)

Commutative Law A property of numbers that says: Two numbers can be added or multiplied in either order. (Ap A)

Complex Number A number formed by a *real* part and an *imaginary* part. (Ch 2)

Computer An electronic device that accepts data as *input,* processes that data into information, and delivers the information as *output.* The *input, process,* and *output* are under the control of a set of instructions called a *computer program.* Both the data and the program reside in a storage area called memory. Programs and their data can be easily changed so the computer can be applied to a variety of problems. This concept of the stored program makes the computer the most versatile problem-solving machine yet devised. (Ch 7)

Computer Graphing Coordinates A special case of a two-dimensional *rectangular coordinate system* used to plot points on a computer screen. Points along the *x*- and *y*-axis are measured in *pixels.* No negative or fractional pixels are allowed. The *origin* is at the upper left corner of the screen. The *y*-axis is positive down, ranging from zero to some maximum value. The *x*-axis is positive to the right, ranging from zero to some maximum value. Together, the maximum *x*- and *y*-values represent the resolution on the screen. (Ch 6)

Computer Language Another name for *programming language.* (Ch 7)

Computer Program A set of instructions, written in a *programming language,* that controls the operation of a computer. Programs fall into two principal types: systems programs and application programs. Systems programs, also called the operating system, control the most primitive aspects of the computer hardware and allow application programs and their data to be loaded into memory and run. Application programs are run under the control of the operating system to apply the computer to specific tasks such as spreadsheets, word processors, games, or airline reservation systems. (Ch 7)

Conditional Expression   An expression that can be evaluated as either true or false to direct processing in *IF-THEN-ELSE* and *LOOP* statements of *computer programs.* The expression may contain *arithmetic expressions, variables, relational operators,* and *logical operators* and is evaluated using an *order of precedence* to give a single unambiguous result. (Ch 8)

Cone   A geometric *solid figure* formed by rotating a *triangle* about its *altitude.* The *base* of a cone is a *circle.* The sides of a cone taper down to a *point.* If the rotated *triangle* is a *right triangle,* the figure formed is a right cone (the one most familiar) in which the *altitude* is *perpendicular* to the *base.* (Ap C)

Constant (algebra)   A term in an algebraic *equation* that does not depend on any of the other variables in the equation. Usually it is a specific value or a letter that represents that value. (Ch 5)

Constant (computer)   A value used in *arithmetic expressions* that never changes. (Ch 7)

Conversion Factor   A quantity derived from an *equivalence expression* that can be inserted into a calculation to change one *unit* to another related *unit.* Conversion factors have *units* expressed as a *ratio:* unit *a* / unit *b.* The conversion factor represents the amount of unit *a* that is equivalent to exactly one of unit *b.* (Ch 4)

Cosine   A relationship between the sides of a *right triangle* that depends on the *angle* (x) formed between the *hypotenuse* and the adjacent side. The cosine of the angle (x), abbreviated cos(x), is defined by the ratio: adjacent / hypotenuse. The cosine is one of the *trigonometric functions.* (Ap C)

Cross Multiplication   A technique used to solve *ratio equations* in which the numerator on each side of the equation multiplies the denominator on the opposite side. The outcome is an equivalent equation without ratios. Thus, $(a\ /\ b) = (c\ /\ d)$ can be cross multiplied to give $(a * d) = (b * c)$. (Ch 5)

Counting   An *assignment* technique used in *computer programs,* usually inside a loop, to *increment* a variable. Each time the counting statement is executed, the value stored in the variable is increased by 1. When the counting process is stopped, the variable contains a count of the number of times the loop was executed. (Ch 7)

Cube (arithmetic)   A number multiplied by itself and by itself again. The third power of a number. A number with the exponent 3. (Ch 2)

Cube (geometry)   A geometric *right solid* figure with a base that is *square.* A cube has 6 equal square sides and 12 equal edges. (Ap C)

Curve   The path formed by moving a *point* along a path according to a specified set of rules. The curve is a general concept that includes *straight lines* and *curved lines.* The path can be represented in various ways, for example, by a mark of a pencil on paper or by *pixels* on a computer screen. But these are only devices that allow us to see the invisible. (Ap C)

Curved Line   Formed by moving a point along a path in a continuously changing direction. The path may be in a two-dimensional *plane* (e.g., parabola or *circle*) or in three-dimensional space (e.g., a helix, a figure resembling a spring or a corkscrew). (Ap C)

Cylinder   A geometric *right solid* figure that has a *circle* as a *base.* (Ap C)

Data Series   A collection of related observations of some quantity. All the data in a series must be measured in a consistent manner and with the same physical *units.* When observations are made at equal time intervals, it is called a time series. (Ch 6)

Decimal Numbers   The *positional number system* with the base 10 that is in common use. (Ch 3, Ap A)

Degree (geometry)   A unit of angular measurement that divides the *circular arc* of a full *circle* into 360 equal parts or degrees, represented by the symbol °. Thus, no arc is 0°, a quarter *circle* (right angle) is 90°, a half *circle (straight angle)* is 180°, and a full *circle* is 360°. (Ap C)

Degree (temperature)   A unit of measurement for temperature. The size of the unit depends upon the scale. The Celsius scale, measured in °C, has 100 degrees between the freezing point of water (0°C) and boiling point of water (100°C). The Fahrenheit scale, measured in °F, has 180 degrees between the freezing point of water (32°F) and the boiling point of water (212°F). (Ch 4)

Dependent Variable   The *variable* in an *algebraic equation* that is defined in terms of the other, *independent variables.* It is usually denoted by the letter *y* and occurs alone on the left side on the equation. (Ch 5)

Desk Checking   A method of manually checking the correctness of an *algorithm* (usually in pseudocode before it is converted into a *programming language*). The checking involves carefully keeping track of the data assigned to each *variable* as the steps of the *algorithm* are processed. (Ch 7)

Diagonal   The *straight line* between opposite *vertices* of a geometric figure. On a *square* or *rectangle* it connects opposite corners. On a *cube* or *rectangular solid* it connects opposite corners of opposite faces. (Ap C)

Distributive Law   A property of numbers that says: The sum of several terms ($a + b + c$ . . .) multiplied by some number (m) is the same as the sum of the products ($a * m + b * m + c * m$ . . .). (Ap A)

Ellipse   A geometric *plane figure* that often looks like a stretched-out *circle*. It has two central points called foci, f_1 and f_2, such that any point p on the boundary is positioned so the sum of the distance from f_1 to p to f_2 remains constant. Depending on the placement of the foci, the ellipse will take on different shapes. The *circle* is a special case in which the two foci coincide. (Ap C)

Equation   Two *algebraic expressions* that represent the same quantity. An equal sign separates the two expressions and signifies that they are in balance. One or both of the expressions contain algebraic *variables* representing unknown quantities. An equation may sometimes be thought of as a mathematical sentence that is true only for certain values of the *variables*. (Ch 5)

Equilateral Triangle   A *regular polygon* with three sides. A *triangle* with equal sides. (Ap C)

Equivalence Expression   An English language statement that relates the same physical measurement in two different *units*. Within this book, equivalence expressions are used to avoid equations with *units* that do not balance. Expressed mathematically as a *conversion factor.* (Ch 4)

Exponents   A way to represent *powers* by using superscript notation. The second *power* of b is represented by b^2. The third *power* of b is represented by b^3. Negative exponents are used to represent *reciprocals* (e.g., the *reciprocal* of b is represented by b^{-1}), and fractional exponents are used to represent roots (e.g., the *square root* of b is represented by $b^{1/2}$). (Ch 2)

Factor   A value that multiplies another value to form a composite result. Conversely, a value that divides evenly into a composite value. Factors may also be *algebraic expressions.* (Ap A)

Flag   A *computer programming* technique that uses a *Boolean variable* to communicate an event recorded in one part of a program to another part of the program. It functions like the red flag on many rural postboxes, hence the name. The postbox flag is raised to indicate that there is mail to be picked up. Later the postman lowers the flag to indicate the mail has been picked up. (Ch 8)

Flowchart   An alternative to using *pseudocode* as a *computer program* development tool, flowcharts provide a means to diagram an algorithm. Using just a few standard symbols, a step-by-step process can

be displayed in a graphic means that is easy to understand. (Ch 9)

Fraction   One *integer* (numerator) divided by another *integer* (denominator). The division can be indicated by either a horizontal fraction bar or a slash. The numerator represents a number of parts of a whole (represented by the denominator). (Ap A)

Function (arithmetic)   A predetermined operation that takes a given input (operand) and returns a result. Examples are the *radical* (which returns the positive *square root* of a number), the *absolute value* (which returns the positive value associated with a signed number), and the *trigonometric functions* (which evaluate a trigonometric ratio for a given angle). (Ch 2, Ap A, Ap C)

Function (computer)   A predetermined operation that takes a given input (operand) and returns a result. Many functions are built into a computer language. Others may be custom designed by the programmer. Arithmetic functions do such things as calculate a *square root,* determine an absolute value, or *truncate* a decimal value into an integer. String functions perform various operations on strings of text, for example, returning a few characters from a longer string of text. (Ch 7)

Gigabyte   Approximately one billion *bytes.* More precisely, 2^{30} (or 1,073,741,824) *bytes.* Also used to represent one million *kilobytes* or one thousand *megabytes.* Abbreviated as GB. (Ch 4)

Graph   A way to display data in graphic form to show trends and relationships among different data values. Often uses a coordinate system, for example a *rectangle coordinate system,* to locate a specific value at a point. Also called *charts; pie charts, bar charts, line charts,* and *x-y charts* are examples. (Ch 6)

Hexadecimal Number   A number generated by the *positional number system* using the base 16. Hexadecimal numbers are easily converted to and from *binary numbers,* but are more compact and easier to read. Therefore, they are commonly used by *computers* to display such things as memory addresses. (Ch 3)

Hexagon   A *polygon* with six sides. If all sides are equal, it is called a regular hexagon. (Ap C)

Hypotenuse   The longest side of a *right triangle.* The side opposite the *right angle.* The hypotenuse is related to the other two sides of the right triangle by the *Pythagorean theorem.* Also used in the computation of *trigonometric functions.* (Ap C)

IF-THEN-ELSE   A *programming language* statement that will process other statements depending on a *conditional expression* being true or false. (Ch 7)

Imaginary Number   The part of a *complex number* that is associated with the quantity *i,* such that $i^2 = -1$. Then *i* represents the square root of -1. Imaginary numbers are not counted among the real numbers and are not discussed further in this book. (Ch 2)

Incrementing   An *assignment* technique used in *computer programs,* usually inside a *loop,* to add (or subtract) some fixed amount to a variable. A special case of *counting,* incrementing allows the program to count by 2s or 10s or any value, even negative values. (Ch 7)

Independent Variable   The *variable* (or variables) in an *equation* that does not depend on another *variable* for its value. With two variables, the independent *variable* is usually denoted by *x* and the *dependent variable* is denoted by *y.* (Ch 5)

Initialization   An *assignment* technique used in *computer programs* to assign some initial data to a *variable* before that *variable* is used in another statement. (Ch 7)

Input (computer)   A *programming language* statement used to assign data from outside the program to variables inside the program. Also can be used as a noun to represent the data coming into the program. Also can be used as a verb to represent the action of bringing in the data. (Ch 7)

Input (problem solving)   Information given to solve a problem. (Ch 1)

Integer   A positive or negative whole number. (Ap A)

IPO Chart   A traditional tool used in the design of *computer programs* to distill the complexity of the program into a simple chart. The chart outlines the essential components of the program in terms of *input, process,* and *output.* (Ch 9)

IPO Method   A problem-solving methodology developed for this book that is especially appropriate for information technology. The method helps identify the essential steps needed to solve a problem in terms of *input, process,* and *output.* The method then applies these steps to determine a solution. Finally, the solution is checked for correctness. (Ch 1)

IPO Worksheet   A worksheet to be filled in when using the *IPO method* to plan the solution of a problem. The process of filling in the worksheet helps answer the essential questions needed to arrive at a correct solution. (Ch 1)

Irrational Number   An infinite nonrepeating decimal number that cannot be formed as the ratio of two integers. Counted among the *real numbers,* examples are $\sqrt{2}$ and π. (Ap A)

Kilobyte   Approximately 1000 *bytes.* More precisely, 2^{10} (or 1024) *bytes.* Abbreviated as KB. (Ch 4)

Line   A special case of a curve, formed by moving a point in a single direction. In its most fundamental form it has no beginning or end, extending from infinity in one direction to infinity in the opposite direction. It has only one dimension, length. The part of a line between any two points is called a *straight line.* (Ap C)

Line Chart   An alternate way to display a data series suitable for a *bar chart.* However, a dot is placed in the center of the top of each bar, the dots are connected by lines, and the bars are removed. (Ch 6)

Linear Algebraic Equation   An equation in which all *variables* are to the first power. A linear equation with two *variables* plots as a *straight line* in a *rectangular coordinate system.* (Ch 6, Ap B)

Linear Extrapolation   A technique for estimating an additional value for a table or *graph,* at a point just outside two other known values. It is assumed that the data between two known points lie on a straight line when *graphed* and that the line can be extended to locate some new point. (Ap A)

Linear Interpolation   A technique for estimating an additional value for a table or *graph* at a point between two known values. It is assumed that the data between the known points lie on a straight line when *graphed* and that the new point will also lie on that line. (Ap A)

Logical Condition   Another name for a *conditional expression.* (Ch 7, Ch 8)

Logical Operator   Translates the logical input of one or more *conditional expressions,* according to specific rules, to produce a logical output. Examples are AND, OR, and NOT. Each operator has an associated *truth table* that specifies the various logical outputs for all combinations of the logical inputs. (Ch 8)

LOOP   A *programming language* statement that repeats some processing, depending on a *conditional expression* being true or false. (Ch 7)

Megabyte   Approximately one million *bytes.* More precisely, 2^{20} (or 1,048,576) *bytes.* Also used to represent 1000 *kilobytes* (or 1,024,000 *bytes*). Abbreviated as MB. (Ch 4)

Modular Design   The decomposition of tasks into subtasks during *top-down development* is not haphazard. The division of the entire task into smaller processing modules is done along functional lines, with like functions grouped together into the same module. (Ch 9)

NOT Operator   A logical operator that negates a conditional expression. If the logical outcome of an expression *a* is true, then "NOT(a)" is false. If the logical outcome of the expression *a* is false, then "NOT(a)" is true. (Ch 8)

Number Line   A way of visualizing the relationships among all the real numbers by locating them in proper order along a horizontal line. The line stretches from negative infinity on the left to positive infinity on the right. Since real numbers are ordered, any number (whether rational or irrational) will correspond to only one point on the line. Conversely, each point corresponds to only one number. A part of a number line is used as an axis on a graph. (Ap A)

Object-Oriented Design   A *computer program* design methodology that views the program as a set of interacting objects having properties quite different from modules. Objects combine both data and programming instructions (called methods) in a way that allows them to be reused in many programs. One object can be built from several other objects. The methods of the related objects are shared in a process called inheritance, so that the parent object assumes the combined characteristics of its component objects. (Ch 9)

OR Operator   A logical operator that works between two conditional expressions, *a* and *b*. The true (T) - false (F) outcome of the expression (a OR b) depends on the logical outcome of the two conditional expressions: Thus, T OR T = T, T OR F = T, F OR T = T, and F OR F = F. (Ch 8)

Order of Precedence (arithmetic)   Within an *arithmetic expression,* calculations proceed in a predetermined order. The various parts of the expression are evaluated in the following order: (1) parentheses, innermost first; (2) exponentiation; (3) multiplication or division; (4) addition or subtraction. If the expression includes several parts of the same type, they are evaluated left to right. Using this order of precedence, complex calculations will always yield a single result. Parentheses may be used to force a different order of calculation. (Ch 7, Ap A)

Order of Precedence (logic)   Conditional expressions are evaluated in the following order: (1) expressions within parentheses, innermost first; (2) expressions with NOT operators; (3) expressions with AND operators; (4) expressions with OR operators. If the expression includes several parts of the same type, they are evaluated left to right. Parentheses may be used to force a different order of calculation. (Ch 8)

Origin   The point on a *rectangular coordinate system* at which the *x*-axis and the *y*-axis intersect at the values $x = 0$ and $y = 0$. (Ch 6)

Output (computer)   The processing results that are delivered to a device outside the program. Also can be used as a noun to represent the data coming out of a program, or as a verb to represent the action of sending out the data. (Ch 7)

Output (problem solving)   The result you are trying to find. The solution of a problem. (Ch 1)

Parallel   Two *lines* in the same *plane* that never intersect are said to be parallel. Two *planes* that never intersect are also said to be parallel. The distance between parallel lines and planes is the same at all points (Ap C)

Pentagon   A *polygon* with five sides. If all sides are equal, it is called a regular pentagon. (Ap C)

Percent   A notation used to express fractions and decimal numbers with a special symbol (%) that means *parts per hundred.* Thus, $\frac{35}{100}$ can be expressed as 35 parts per hundred or 35%. Also, $\frac{1}{2}$ can be expressed as 50%, and 0.25 can be expressed as 25%. Percents have no units. (Ch 5)

Percent Change   A change in a value from some initial amount to some final amount, expressed as a *percent* of the initial value. (Ch 5)

Perfect Square   Numbers formed by multiplying an integer by itself. Examples include 25 (formed by 5 * 5), 49 (formed by 7 * 7), and 100 (formed by 10 * 10). (Ch 2)

Perimeter   The length of the boundary of a *plane figure.* For a rectangle it is twice the length plus twice the width. The perimeter of a circle, called the *circumference,* is 2π. (Ap C)

Perpendicular   Two lines are said to be perpendicular if they are at *right angles* to each other. A line is perpendicular to a plane if it is at *right angles* to all possible lines that lie in that plane. Two planes are perpendicular if a line perpendicular to one plane lies entirely in the other plane. (Ap C)

Pi   The Greek letter π used to designate the ratio of the length of the *circumference* of a circle to its diameter: $\pi = c / d$. This ratio is the same for circles of any size. The value of π, which can never be expressed exactly as the ratio of two whole numbers, is approximately 3.1415926. (Ap C)

Pie Chart   A *graphing* technique that divides a *circle* (the whole pie) into several sectors (pieces of pie).

The whole pie represents the sum of all the items in a data series, and each piece of pie represents one data item. The size of each piece, determined by the angle of its wedge, is proportional to the share of the whole for that data value. (Ch 6)

Pixel   Short for picture element. A pixel is the smallest dot that can be plotted on a computer screen for a given screen resolution. The higher the resolution, the smaller the pixel size. (Ch 6)

Plane   A surface formed by moving a line through space in a constant direction, other than the direction of the line. In its most general form, a plane is infinite in extent. Often, however, any flat surface is called a plane. Unless they are *parallel,* two planes intersect in a single line. Any three points define a single plane. (Ap C)

Plane Figure   Two-dimensional geometric figures that lie entirely in a single *plane. Straight lines* or *curved lines* are used to form a boundary that encloses the area of the figure. The *polygon* (e.g., *triangle, square, pentagon,* etc.) and the *circle* are examples. (Ap C)

Polygon   A geometric *plane figure* bounded by three or more straight lines as sides. Some examples are triangle (three sides), square or rectangle (four sides), pentagon (five sides), and hexagon (six sides). If all the sides have equal length, it is called a *regular polygon.* (Ap C)

Positional Number System   A number system composed of a string of digits, in which the position of each digit is weighted by a power of some *base* number. Adjacent digits have weighting factors that vary by a factor of the base, b: (. . . b^3, b^2, b^1, b^0, b^{-1}, b^{-2} . . .). The number of symbols available to represent any digit is equal to the base. The decimal number system uses the base 10 and has 10 symbols. The binary number system uses the base 2 and has two symbols. The hexadecimal number system uses the base 16 and has 16 symbols. (Ch 3)

Power   The product of equal base numbers. The degree of a power is the number of times the base number is multiplied. The second power of the number b is b squared. The third power of the number b is b cubed. This concept can be generalized to allow negative and fractional powers. Also see *exponents.* (Ch 2)

Prefix   Terms appended to the beginning of some *unit* of measurement that scale the *unit* by some factor. Decimal prefixes multiply the *unit* by powers of 10. Binary prefixes multiply the *unit* by powers of 2. Positive *powers* make the unit larger. Negative *powers* make the unit smaller. (Ch 4)

Prime Number   An *integer* that has no integer *factors* other than 1 and itself. (Ap A)

Priming Read   A technique used in computer programming to *input* (read) the first data item immediately before entering a processing *loop.* That way the next data can be read at the bottom of the *loop* after processing the first. Eventually, after all the data has been processed, an end-of-data mark (or sentinel value) is read. The *loop* condition detects this special value and exits the *loop* in time to avoid a processing error. (Ch 7)

Process (computer)   The steps specified in a *computer program* that convert data brought into the program (input) into results to be sent to some device outside the program (output). (Ch 7)

Process (problem solving)   The steps needed to convert what you are given (*input*) into what you are trying to find (*output*). (Ch 1)

Programming Language   Used to express an *algorithm* in a language that a computer can process. A computer can only run programs written in a special code consisting of zeros and 1s called machine language. While this binary code presents no difficulty for computers to process, it is very difficult for humans to understand. Therefore, other higher-level *programming languages* are available for human programmers. Before these higher-level languages can be run on a computer, their English-like statements must be translated into machine language. A special program called a compiler is available to do this translation. Examples of higher-level programming languages are BASIC, COBAL, C, C++, Java, and Visual Basic. (Ch 7)

Proportion   Ratio used to show a relationship between one quantity and another. It can be used to compare a part of something with the whole. (Ch 5)

Pseudocode   A tool used to design *algorithms* that will ultimately become *computer programs.* Pseudocode avoids the strict rules of grammar required by *computer programs,* in which every comma, semicolon, and period must be precisely in place. On the other hand, its similarity to a *programming language* makes it easy to translate once the details of the *algorithm* have been described. (Ch 9)

Pythagorean Theorem   A relationship between the lengths of the three sides of a *right triangle* that says that the *square* of the *hypotenuse* is equal to the sum of the squares of the other two sides. Useful in finding the distance between two *points* on a *graph.* (Ch 6, Ap C)

Radian A unit of angular measurement. One radian (about 57°) corresponds to the *angle* at the center of a *circle* that captures a portion of the circumference equal in length to the radius. A full *circle* (360°) is 2π radians. A half *circle* (180°) is π radians. A quarter *circle* (90°) is $\pi / 2$ radians. (Ap C)

Radical A symbol used to represent the positive *square root* of a number. For example, the positive *square root* of b is represented by $\sqrt{b}$. (Ch 2)

Rate A quantity associated with a *ratio* of *units* such as miles per hour or feet per second. (Ch 4)

Ratio A measure of the relative size of two quantities (in similar *units*) expressed as a proportion. Ratios, like *fractions,* involve dividing one number by another. *Ratios,* however, are not restricted to dividing *integers.* (Ap A)

Ratio Equation An *algebraic equation* with a *ratio* on either side of the equal sign. It is solved by using *cross multiplication.* (Ch 5)

Rational Number A number that can be formed as the *ratio* of two integers. Rational numbers include the *integers* (which can be formed by some whole number divided by 1). (Ap A)

Real Numbers All *rational* numbers (which include *integers*) and *irrational* numbers. (Ap A)

Reciprocal The reciprocal of a number is 1 divided by that number. All numbers have a reciprocal, except zero. For example, the reciprocal of 2 is $\frac{1}{2}$. Reciprocals of an expression with an exponent can be found by multiplying the exponent by -1. For example, the reciprocal of 2^1 is 2^{-1} and the reciprocal of 2^{-2} is 2^2. (Ch 2, Ap A)

Rectangle A geometric *plane figure* bounded by four sides. Opposite sides are equal and adjacent sides are at *right angles* to each other. (Ap C)

Rectangular Coordinate System A system used for *graphing* data that has two (or three) scales, or axes, at right angles to one another. Each axis is a part of a *number line,* which shows the ordered location of values along its length. For two dimensions, there is an x-axis (horizontal) and a y-axis (vertical) centered in some rectangular space. Data to be plotted come in x, y pairs, and each data pair locates a specific point within the rectangular space. The point where $x = 0$ and $y = 0$ is called the origin. (Ch 6)

Rectangular Solid A geometric *right solid* figure that has a *rectangle* as a base. Shaped like a box, it has a volume determined by the product of its length, width, and height. The *cube* is a special case in which all sides are equal. (Ap C)

Regular Polygon A geometric *plane figure* bounded by three or more *straight line* sides. All the sides are equal. All the interior *angles* between the sides are also equal. (Ap C)

Relational Operator An operator used in *conditional statements* that states the relationship between positions on a *number line.* The following six relationships are possible: (1) greater than ($>$), (2) equal to ($=$), (3) less than ($<$), (4) greater than or equal to ($> =$), (5) less than or equal to ($< =$), (6) not equal to ($< >$). Then, for example, a statement such as $x < 5$ implies that x can assume any value to the left of the number 5 on the *number line.* (Ch 8)

Right Angle An *angle* equal to a quarter *circle* or 90 degrees. Lines (and planes) that are *perpendicular* to each other are at right angles. (Ap C)

Right Solid A geometric *solid figure* having a *plane figure* as a *base* and sides at *right angles* to the *base.* Examples are the *cube, rectangular solid,* and *cylinder.* (Ap C)

Right Triangles A triangle in which one of the interior angles is a right angle. The longest side, the one opposite the right angle, is called the hypotenuse. The other two sides are called the base (horizontal) and the altitude (vertical). The base and the altitude are interchangeable. (Ap C)

Rounding The process of reducing the number of *significant digits* of a number to bring it into line with an appropriate accuracy. Three rules of rounding are covered in this book: (1) Rounding Down, (2) Rounding Up, and (3) Rounding to Nearest Place. (Ap A)

Rounding Place The position (or power of 10) to which a number is to be rounded. (Ap A)

Rounding Down A type of rounding in which all the digits to the right of the rounding place are dropped (or zero-filled if they are also left of the decimal place). For example, 3.789 rounded down at the tenths place becomes 3.7. This type of rounding is also called *truncating.* (Ap A)

Rounding to Nearest Place A type of rounding that depends upon the digit immediately right of the *rounding place:* If this digit is 5 or greater, then follow the rules for *rounding up.* Otherwise, follow the rules for *rounding down.* For example, rounding 2345 to the nearest thousand would change it to 2000. Rounding 2.1256 to the nearest hundredth would change it to 2.13. (Ap A)

Rounding Up A type of rounding that adds 1 to the digit in the rounding place if there are nonzero digits to the right. Then, regardless of whether you added 1, round down. For example: 4.321 rounded up at the tenths place becomes 4.4. However, 3200 rounded up at the hundreds place stays 3200, since there are no digits (other than zeros) to round. (Ap A)

Scientific Notation   A way to express values that makes it easier to do calculations with very large and very small magnitudes. The notation has two parts: a decimal value between 1 and 10 (but not equal to 10) and an appropriate power of 10. (Ch 2)

Segmented Address   A memory addressing scheme used to allow up to 1 *megabyte* of addressable memory for the early 16-*bit* processors built by Intel. Modern Intel processors can address memory well above 1 *megabyte*. But, for reasons of backward compatibility, they are still able to use segmented addressing when needed. (Ch 4)

Sentinel Value   A special data value placed at the end of a data list to indicate that there is no more data to be processed. The sentinel value must be different in some fundamental way from all of the other data items. Programs must be designed so that the sentinel value is not processed with the other data. One way to do this is by using a *priming read.* (Ch 7)

Significant Digits   Significant digits provide a way to quantify the accuracy of a measurement. When we say that the length of a building is 151 feet, we imply that its length is closer to 151 feet than it is to 150 or 152 feet. When we say that the length of a building is 151.3 feet, we imply that the length is closer to 151.3 feet than it is to 151.2 or 151.4 feet. In these examples 151 feet is said to have three significant digits, and 151.3 feet is said to have four significant digits. Also see *rounding.* (Ap A)

Simultaneous Equations   Two or more *algebraic equations* that together have a unique solution for the *variables* they contain. The equations must be independent of one another; that is, no equation can be a simple multiple of another. For linear (first power) equations having two *variables,* a solution exists only when each equation has a different slope when plotted as a line on a *graph.* That way the two lines will intersect at the point corresponding to the solution. (Ap B)

Sine   A relationship between the sides of a *right triangle* that depends on the *angle* (x) formed between the *hypotenuse* and the adjacent side, the third side being opposite the angle. The sine of the angle (x), abbreviated sin(x), is defined by the ratio: opposite / hypotenuse. The sine is one of the *trigonometric functions.* (Ap C)

Slope   A measure of the direction that a *straight line* takes when it is *graphed.* Slope can be calculated by dividing the rise (change of *y*-values) by the run (change of *x*-values). Knowing the slope and the *y-intercept* of the line allows you to write its equation in the *slope-intercept form.* The slope is not defined for vertical lines, as the run is zero. (Ch 6)

Slope-Intercept Form   A convenient way to write a *linear algebraic equation* with two variables (equations that plot as a straight line on a *graph*). When the equation is placed in this special form, its *slope* and the *y-intercept* are explicitly shown. In general terms the form is $y = mx + b$, where *m* represents the *slope* and *b* represents the *y-intercept.* This form works for all straight-line equations, except vertical lines. (Ch 6)

Solid Figure   A geometric figure that occupies three-dimensional space. Some solid figures are formed by moving (or rotating) a plane figure in space to sweep out a *volume.* Other solid figures have boundaries formed by several *plane figures,* each in a different *plane.* (Ap C)

Sphere   A geometric *solid figure* formed by rotating a *circle* about one of is diameters. All the points on the surface of a sphere are equidistant from the central point that lies inside. (Ap C)

Square (arithmetic)   A quantity multiplied by itself. The second *power* of a number. A quantity with the *exponent* 2. (Ch 2)

Square (geometry)   A *regular polygon* (equal sides and equal interior *angles*) with four sides. A special case of the *rectangle,* in which all sides are equal. (Ap C)

Square Root   The square root of a number is that value which when multiplied by itself results in the original number. Square roots come in pairs having the same *absolute value* but different signs. The positive square root of a number (*x*) is often indicated using a *radical* sign as $\sqrt{x}$. Also, the square root of a number can be indicated with an *exponent* of one half ($x^{\frac{1}{2}}$). (Ch 2)

Stepwise Refinement   A program development approach that starts with a simple and limited solution to a problem. Then, through a series of refinement steps, the program evolves into a fully functioning version. This step-by-step approach usually arrives at a much better result than could have been achieved in a single step. (Ch 7)

Straight Angle   An *angle* that opens to form a *straight line.* It measures a half circle (180 degrees or π radians). (Ap C)

Straight Line Part of an infinite line that occurs between two *points*. Sometimes it is called a line segment. One of the basic principles of Euclidean geometry is that a straight line is the shortest distance between two *points*. (Ap C)

Structure Theorem The structure theorem states that any *algorithm,* and hence any *computer program,* can be written using only three basic ways to assemble instructions into what are known as control structures: (1) sequence (one step after another); (2) selection (also called *IF-THEN-ELSE*), and (3) repetition (*looping*). Each of these structures may contain substructures composed of any of the other structures. (Ch 9)

Structured English A way to describe an *algorithm* using English language phrases arranged in hierarchical structure, much like an outline. A refined type of structured English, called *pseudocode,* is often used to plan the design of computer programs. (Ch 7, 9)

Surface Area The *area* of the "outside skin" of a geometric *solid figure.* Unlike the *area* of *plane figures,* surface area does not lie in a single *plane* but wraps around the *solid figure* in three dimensions. Although it occupies three-dimensional space, its area is measured in the same units as a plane *area,* length *squared.* (Ap C)

Tangent (geometry) A *straight line* that touches a *curved line,* such as a *circle,* at a single point. (Ap C)

Tangent (trigonometry) A relationship between the sides of a *right triangle* that depends on the *angle* (x) formed between the *hypotenuse* and the adjacent side, the third side being opposite the angle. The tangent of the angle (x), abbreviated tan(x), is defined by the ratio: opposite / adjacent. The tangent is one of the *trigonometric functions.* (Ap C)

Top-Down Development A *computer program* design technique that takes a big task (the overall problem) and breaks it down into a number of smaller tasks. These smaller tasks are called modules. If necessary, any of these smaller tasks can be broken down into still smaller tasks. In this way, the programmer starts with the major (top-level) tasks, then proceeds to the intermediate (midlevel) tasks next, and the complete detailed (low-level) tasks last. Also see *modular design.* (Ch 9)

Triangle A geometric *plane figure* formed by three *straight lines* (sides) that intersect at three points or *vertices.* The sum of the three interior *angles* (between the sides) is half a *circle* or 180 *degrees.* (Ap C)

Trigonometric Function Defined for a given *angle* as a *ratio* of lengths for two sides of a *right triangle.*

Three trigonometric functions are defined in this book: *sine, cosine,* and *tangent.* (Ap C)

Truncating Another name for *rounding down.* For example, truncating 3.1279 to the hundredths place gives 3.12. (Ap A)

Truth Table A table showing the outcomes of a logical operation, given various logical inputs. For example, the truth table for (a AND b) would display the different logical outcomes for the four possible combinations of *a* and *b* being assigned either true (T) or false (F) values: T AND T = T, T AND F = F, F AND T = F, F AND F = F. (Ch 8)

Unit A previously defined standard to which a measured physical quantity is compared. (Ch 4)

Unknown Another name for an *algebraic variable.* (Ch 5)

Variable (algebra) *Unknown* quantities in algebraic expressions and equations, having values yet to be determined. Unlike constant numbers, which never change, variables can take on several values depending on the situation. Variables are represented by letters such as *x, y,* and *z.* (Ch 5)

Variable (computer) A named location in a computer's memory used in a *computer program* to store a result of one operation until it is needed by another operation. As the name implies, the data stored in a variable can be retrieved, changed, and stored again as many times as needed for the successful operation of the program. However, at any instant, only one value may be stored in a given variable. (Ch 7)

Vertex The point farthest from the base of a geometric figure (e.g., *triangle* or *cone*). Also, the intersection of two sides of a *plane figure* (e.g., the corner of a *square*). Also, the *point* shared by three or more sides of a *solid figure* (e.g., the corner of a *cube*). (Ap C)

Volume A measurement of the space inside a three-dimensional figure having units of length *cubed.* (Ap C)

Weighted Average An *averaging* method used in place of the arithmetic mean when data items in a set make different contributions based on some weighting factor, for example, the grade point average (GPA) of a set of college courses, each possibly having a different number of credits. The GPA is an average of grades in which credits are the weighting factor. The calculation involves dividing one sum by another. The first is the sum of the products of each data item multiplied by its corresponding weighting factor. The second is the sum of the individual weighting factors. (Ap A)

x-y Chart   A *graphing* technique that uses a *rectangular coordinate system* to display two corresponding *data series* (*x* and *y*). Corresponding data from the two series form (*x*-*y*) pairs. Each data pair defines one point on the coordinate system. When all the data points are connected, they show trends and relationships within the data. (Ch 6)

y-Intercept   The value of *y* when $x = 0$ for a *straight line* plotted on a *rectangular coordinate system*. Knowing this and the *slope* of the line allows you to write the *equation* of the line in *slope-intercept form*. (Ch 6)

Index

Information Technology, *(Contd.)*
  square root function, 55
  subroutines, 29
  systems development lifecycle, 34
  time equals money, 117
Initialization. *See* Assignment tasks
Input, 411. *See also* IPO worksheet sections;
    Programming language statements
  of algorithms, 295
  computer program, 231
  flowchart symbols, 300–301
  IPO chart, 13, 296–298
  IPO method, 12
  IPO worksheet, 14
Integers. *See* Numbers
Interior angle, 395
Interpolation, linear, 394–398
Introductory problems. *See* Problems,
    introductory
Inverse operations, 52–54, 56, 69, 70. *See also*
    Reciprocal
IPO chart, 13, 291, 296, 299, 303–307, 322, 411
IPO method, 12–34, 35–43, 44–45, 411
IPO worksheet, 14–34, 35–40, 44, 411
IPO worksheet sections, 14
  additional information, 24–25
  approach, 28–29
  check, 32–33
  diagram, 26–27
  input, 20–21
  notation, 22–23
  output, 18–19
  problem statement, 16–17
  process, 22–29
  solution, 30–31
Irrational numbers. *See* Numbers
IT. *See* Information technology

JAVA programming language, 227, 281–282

Kilo (unit prefix)
  binary, 127–128
  decimal, 127
Kilobyte (KB), 107–109, 113, 127–130,
    132–134, 149, 152, 158, 160, 411

Larson, Gary, 117
Least significant bit (LSB), 91, 92
Least significant digit (LSD), 86, 96, 97
Less is more, 297
LET statement, 233–234
Line
  chart, 196–217, 41
  curved, 383, 411
  straight, 383, 411
  straight line equation, 203–207, 217
Linear equations, 411. *See also* Equations
  beyond, 367
  general form of, 207
  graphing of, 200–210
  simultaneous, 394–397

slope-intercept form of, 203–210, 217
  solving two, 209–210
Linear extrapolation, 397, 411
Linear interpolation, 394–398, 411
Logic, direct comparison of, 284
Logical expression. *See* Conditional expression
Logical operator, 411
  AND operator, 267–268, 407
  flipping, 274–275
  NOT operator, 266–267, 236–240, 412
  OR operator, 268–269, 412
  order of precedence for, 269
  removing NOTs, 236–240
LOOP statement, 236–240, 261, 302, 411

Machine language, 227
Macro language, 224
Maximum value. *See also* Minimum value
  finding in data list, 258
  scale for graph, 196
Mega (unit prefix)
  binary, 127
  decimal, 128
Megabyte, 107–109, 127–128, 130, 132–135,
    158–160, 411
Memory addressing. *See* Computer memory
Metric system of measurement, 117, 124, 127
Micro (unit prefix), 127
Milli (unit prefix), 127
Minimum value. *See also* Maximum value
  finding in data list, 258
  scale for graph, 192–193, 199
  sorting by finding lowest card, 294
Modular design, 411
  calling modules, 315
  defined, 292, 310–311, 322, 411
  flowchart symbols for, 318–319
  structure chart for, 315–316, 320
Most significant bit (MSB), 91, 92
Most significant digit (MSD), 86, 96, 97
Multiplication
  of algebraic expressions, 167–169
  of arithmetic expressions, 229
  associative law for, 339, 408
  computer notation, 329
  converting number systems by repeated, 111
  cross, 170
  notation for, 25, 165, 329
  order of precedence for, 229, 257, 330, 412
  significant digits for, 350
  of units, 117, 119, 136–137

Names vs. content. *See* Computer variables
Naming conventions
  for Boolean variables, 285
  for variables, 228–229
Nano (unit prefix), 127
NOT operator. *See* Logical operators; Truth tables
Notation. *See also* IPO worksheet sections
  algebra, 135
  arithmetic, 329–331

Square root, *(Contd.)*
  of negative numbers, 54
  pairs of, 51
  of perfect squares, 51
Squaring a sum, 49
Step-by-step methods
  converting binary to decimal, 98–99
  converting decimal to binary, 91
  converting decimal to hex, 96
  converting hex to decimal, 94
  checking a solution, 33
  ellipse, how to draw, 391–392
  evaluating conditional expressions, 270
  IPO method, 12
  IPO worksheet, 14
  linear interpolation, 376–377
  percent change, 183
  Polya's four phases, 4–5
  prefixes, changes, 130
  program development, 314
  removing NOTs, 275
  rounding, 348–350
  scientific notation, 75
  simultaneous equations, 364–365
  solving simple equations, 169–170
  stepwise refinement, 35
  time-rate-distance problems, 369–370
  two's complement, 104
  unit analysis problems, 136–138
Stepwise refinement, 35, 250, 258, 415
Story problems, selected. *See* Problems, story
Straight angle, 383, 415
Straight line, 383, 416
String handcuffs, 2–3
Structure charts, 292, 315–316, 320
Structure theorem, 311–313, 416
Structured English, 291, 299, 322, 416
Structured programming. *See also* Algorithm
    design tools; Structure theorem
  beyond, 312–313
  defined, 292, 310
Subscript notation, 88, 94
Substituting values in equations, 170, 181
Subtraction
  in 2's compliment notation, 105–107
  converting to addition, 105
  of figures, 403
  of fractions, 344–345
  of numbers, 334–335
  order of precedence for, 229, 257
Surface, 384
Surface area, 396–397, 404, 416
Systems development life cycle, 34
System international (SI), 117

Tables
  computer vs. math notation, 329
  English-to-Algebra Dictionary, 177
  fractions, decimal approximate to, 353
  relational operators, summary of, 264
  truth. *See* Truth table

unit abbreviations, 123
unit conversion factors, 124
unit prefixes, 127, 128
Tangent, 416
  geometry, 416
  trigonometry, 389, 416
Temperature, 120, 123, 124, 174–175
Tera (unit prefix)
  binary, 127
  decimal, 128
Testing ranges, 279–282
Time-rate-distance, 40, 43, 368–374
Top-down development, 292, 310, 322, 416
Translating words into equations. *See* English-to-
    algebra dictionary
Transposing equations, 167
Trapezoid, 394
Triangle, 211–213, 385–389, 394–395, 404–405,
    416
Trigonometric functions, 389, 416
Truncating, 348, 416
Truth tables, 416
  AND operator, 367
  NOT operator, 366
  OR operator, 368
TV. *See* High-definition TV
Two's complement notation
  adding, 105
  overflow, 106, 107
  subtracting, 105

Units of measurement, 416. *See also* Prefixes
    used with units
  abbreviations for, 123, 124
  addition of, 119, 160
  balancing in equation, 118, 119
  beware when adding, 156
  composite, 117, 118, 160
  conversion factors for, 118, 122–124, 160,
    409
  conversion of, 122, 126, 136–159, 160
  derived, 119, 120
  dividing out, 137, 160
  in equation, 118, 119
  equivalence expressions, 122, 125, 132, 136,
    139, 160, 410
  hidden, 120
  IPO worksheet for, 139–140
  metric measures, 117, 124, 127
  missing factors, 157
  multiplication of, 119
  in nonsense equations, 156
  problem input, 21
  problem output, 19
  rates, 118, 160
  simple, 117, 160
  weight vs. mass, 120, 121
Unity, 335
Unknown. *See* Algebraic variable
UNTIL. *See* LOOP statement